International Series in Heat and Mass Transfer

Editors: Arthur E. Bergles and Franz Mayinger

Grigull and Sandner, *Heat Conduction*
(Translated by J. Kestin)

Stephan, *Heat Transfer in Condensation and Boiling*
(Translated by C.V. Green)

Karl Stephan

Heat Transfer in Condensation and Boiling

Translated by
Curtis V. Green
Rensselaer Polytechnic Institute
Troy, New York

With 169 Figures

Springer-Verlag
Berlin Heidelberg GmbH

Professor Dr. Karl Stephan

Institut für Technische Thermodynamik
und Thermische Verfahrenstechnik
Universität Stuttgart
Pfaffenwaldring 9
W-7000 Stuttgart 80, FRG

Translated from the first German Edition "Wärmeübergang beim Kondensieren und beim Sieden" (Springer-Verlag, 1988) by Curtis V. Green, Rensselaer Polytechnic Institute, Troy, New York, USA.

ISBN 978-3-642-52459-2

Library of Congress Cataloging-in-Publication Data
Stephan, K. (Karl), 1930– [Wärmeübergang beim Kondensieren und beim Sieden. English] Heat transfer in condensation and boiling / Karl Stephan. p. cm. – (International series in heat and mass transfer) Translation of: Wärmeübergang beim Kondensieren und beim Sieden. Includes bibliographical references and indexes.
ISBN 978-3-642-52459-2 ISBN 978-3-642-52457-8 (eBook)
DOI 10.1007/978-3-642-52457-8
1. Heat – Transmission. 2. Condensation. 3. Ebullition. I. Title. II. Series.
TJ260.S774513 1992 621.402'2 – dc20 91-45689

Originally published by Springer-Verlag Berlin Heidelberg New York in 1992

Typesetting: Thomson Press (India) Ltd., New Delhi, India.

61/3020-5 4 3 2 1 0 – Printed on acid-free paper

Preface to the English-Language Edition

I welcome the opportunity to have my book translated, because of the great emphasis on two-phase flow and heat transfer in the English-speaking world, as related to research, university education, and industrial practice.

The 1988 Springer-Verlag edition of "Wärmeübergang beim Kondensieren und beim Sieden" has been enlarged to include additional material on falling film evaporation (Chapter 12) and pressure drop in two-phase flow (Chapter 13). Minor errors in the original text have also been corrected.

I would like to express my sincere appreciation to Professor Green, Associate Professor of German at Rensselaer, for his excellent translation and cooperation. My thanks go also to Professor Bergles for his close attention to technical and linguistic details. He carefully read the typescript and made many comments and suggestions that helped to improve the manuscript.

I hope that the English edition will meet with a favorable reception and contribute to better understanding and to progress in the field of heat transfer in condensation and boiling.

February 1992 K. Stephan

Preface to the German-Language Edition

This book is a continuation of the series "Heat and Mass Transfer" edited by U. Grigull, in which three volumes have already been published. Its aim is to acquaint students and practicing engineers with heat transfer during condensation and boiling, and is intended primarily for students and engineers in mechanical, chemical, electrical, and industrial processing engineering.

They are frequently confronted with the task of dealing with problems of heat and mass transfer in which the processes of condensation and boiling are dominant. The pertinent literature has increased greatly in recent years, but it is often accessible only in a form that makes it difficult for the user to determine the areas of applicability and the limits of validity of the theoretical and experimental findings. It, therefore, seems appropriate to make a systematic presentation of the material for the user. This has required the author, on the one hand, to select those areas to be treated and, on the other hand, to exclude several subareas, particularly those which are still too empirical in nature.

The book is divided into two parts: heat transfer during condensation and heat transfer during boiling, both of which have the same structure.

The part on heat transfer during condensation begins with a survey of the different types of condensation and their basic mechanisms. It treats first the film and drop condensation of stagnant vapors and then the film and drop condensation of flowing vapors, including metal vapors.

When vapors of multi-component mixtures are condensed, in most condensers a condensate of miscible or immiscible liquids is formed. A relatively large amount of space is devoted to this area so important for technology. It turns out that, in spite of the progress made recently, there are still many problems to be solved. Also, the many possibilities for the augmentation of heat transfer during condensation have in no way been exhausted.

In order to understand heat transfer during boiling, it is necessary to understand the fundamentals of vapor production and vapor bubble formation. Consequently, these are treated in the chapter on heat transfer during boiling. The further discussions treat heat transfer during free convection boiling, falling film evaporation, and forced convection boiling. Multi-component mixtures of miscible and immiscible liquids are treated extensively, with regard to heat transfer during boiling in free flow and forced flow. In conclusion, the different possibilities for the augmentation of heat transfer during boiling are discussed.

In the presentation, the physical fundamentals are always in the foreground. The limits and ranges of validity of physically based models are discussed. It seems just as important, however, to treat in sufficient detail the technical applications and to prepare for the user the critical tools necessary for heat transfer calculations. To this end, experimental results, application formulas, and calculation models for complicated problems are discussed.

I wish to express my thanks to H. Auracher and J. Mitrović for their helpful suggestions and advice during the writing of the manuscript. I wish also to thank D. Butz for reading and correcting textual errors in the manuscript, and also to thank the Springer-Verlag for its pleasant cooperation and careful preparation of this book.

Stuttgart, October 1987 K. Stephan

Table of Contents

List of Symbols

a) Latin letters

Symbol	Name	SI-unit (as an example)
a, a_0, a_1	constants	–
a	thermal diffusivity, $a = \lambda/(\varrho c_p)$	m^2/s
a_{ij}	coefficients	–
A	area, tube cross section	m^2
A, A'	factor	–
$\boldsymbol{A}$	matrix with coefficients a_{ij}	–
$\boldsymbol{A}_{ad}$	adjoint matrix of $\mathbf{A}$	–
$\lvert \boldsymbol{A} \rvert$	determinant of matrix $\mathbf{A}$	–
A_0	total surface area of a finned tube	m^2
A_0	total surface area of a condenser	m^2
A_G	surface area of smooth tube	m^2
A_R	surface area of fins	m^2
A_T	surface area of liquid drops	m^2
b	width	m
b, b_1	constants	–
b	Laplace constant, $b = [2\sigma/g(\varrho_L - \varrho_G)]^{1/2}$	m
$b = \sqrt{\lambda \varrho c_p}$	thermal penetration coefficient	$Ws^{1/2}/Km^2$
b_{ij}	coefficients	–
b_{ST}	thermal penetration coefficient of rust-free steel	$Ws^{1/2}/Km^2$
B, B'	factor	–
$\boldsymbol{B}$	matrix with coefficients b_{ij}	–
c	concentration	$mole/m^3$
c	specific heat	J/kgK
c, c_0, c_1, c_2	constants, coefficients	–
c_v	specific heat at constant volume	J/kgK
c_p	specific heat at constant pressure	J/kgK
C	factor	–
d	diameter	m
d	radius of curvature at the tip of a vapor bubble	m

Symbol	Name	SI-unit (as an example)
d_A	break-off diameter of a vapor bubble	m
d_e	equivalent diameter, $d_e = 4A/U_b$	m
$\boldsymbol{d}_i$	gradient, $\boldsymbol{d}_i = \text{grad } y_i$	1/m
$\boldsymbol{d}_i^*$	abbreviation form: $\boldsymbol{d}_i^* = \boldsymbol{d}_i c\delta$	mole/m^3
d_K	diameter of drop dome, Eq. (13.169)	m
D	diffusion coefficient	m^2/s
D	outer diameter	m
$\mathfrak{D}_{ik}$	binary diffusion coefficient	m^2/s
D_{ik}	multi-component diffusion coefficient	m^2/s
E	suction factor	–
f	bubble frequency	1/s
f	condensation coefficient, correction factor	–
f	mean amplification factor	–
f	degree of irregularity, $f = q_{max}/\bar{q}$	–
f, F	symbol for function	–
F	force	N
F	enhancement factor in two-phase flow	–
F_1, F_2, F_3	factors in two-phase flow	–
F_A	buoyancy force	N
F_R	resulting force	N
F_T	inertia force	N
g	gravitational constant; normal acceleration due to gravity $g_n = 9.80665\,\text{m/s}^2$	m/s^2
G	friction parameter for turbulent film condensation, $G = (\Gamma \sigma_w)/(\varrho_L \sigma)$	m^2/s
h	specific enthalpy	J/kg
h	height	m
Δh	specific enthalpy difference	J/kg
Δh_v	specific enthalpy of vaporization	J/kg
Δh_v^*	enthalpy difference, Eq. (2.24)	J/kg
Δh_{vu}^*	enthalpy difference, Eq. (2.26)	J/kg
h'	specific enthalpy of saturated liquid	J/kg
h''	specific enthalpy of saturated vapor	J/kg
H	fin height	m
H_i	partial molar enthalpy of component i	J/mole
$\Delta \bar{H}_v$	molar enthalpy of evaporation	J/mole
$\Delta \bar{H}_u$	molar enthalpy of transfer	J/mole
$\boldsymbol{I}$	unit matrix	–
$\boldsymbol{j}$	diffusional flux	mole/m^{2}s
j	diffusional mass flux	kg/m^{2}s

Symbol	Name	SI-unit (as an example)
j_G^*	dimensionless vapor velocity, $j_G^* = x^*M/A[gd\varrho_G(\varrho_L - \varrho_G)]^{1/2}$	–
k	constant	–
k	overall heat transfer coefficient	W/m^2K
k'	heat transfer coefficient between wall and coolant	W/m^2K
$\boldsymbol{k}_b$	mass transfer matrix, Eq. (6.97)	–
K	constant	–
K	curvature, $K = 1/r$	1/m
K	number of components	–
K_1, K_2, K_3	factors in two-phase flow	–
l, L	length	m
m	exponent	–
$\dot{m}$	mass flux	kg/m^2s
M	molar mass	kg/mole
$\dot{M}$	mass flow rate	kg/s
$\dot{M}_{FL}$	mass flow rate, entire cross section flooded by condensate	kg/s
n	exponent	–
n	number of tubes	–
n_1, n_2, n_3	exponents	–
n_i	quantity of drops of class i	–
$\dot{n}$	molar flux	$mole/m^2s$
N	total of all drops	–
$\dot{N}$	molar flow rate	mole/s
p	pressure, $1 Pa = 1 N/m^2$	N/m^2
p_0	partial pressure of inert gas	N/m^2
p_1	partial pressure of substance 1	N/m^2
p_{cr}	critical pressure	N/m^2
p^*	reduced pressure, $p^* = p/p_{cr}$	–
Δp	pressure difference	N/m^2
q	heat flux	W/m^2
q_B	heat flux in nucleate boiling	W/m^2
q_{cr}	critical heat flux	W/m^2
q_G	heat flux of condensing vapor with inert gas	W/m^2
q_K	convective heat flux	W/m^2
q_{max}	peak heat flux in nucleate boiling	W/m^2
q_{min}	minimum heat flux in nucleate boiling	W/m^2
$\bar{q}$	average heat flux	W/m^2
q_0	heat flux with no resistance for mass transfer in vapor	W/m^2

Symbol	Name	SI-unit (as an example)
Q	total heat transfer	W
r	radius of curvature of condensate film	m
r_c	critical bubble radius	m
$(r_0)_c$	smallest pore radius necessary for bubble formation	m
r_1, r_2	main curvature radii	m
r_s	thermal resistance of fouling deposits	m^2K/W
R	radius	m
R	universal gas constant, $R = 8.31443$ J/mole K	J/mole K
R_P	depth of surface smoothness DIN 4762 (DIN is the German Industry Standard)	m
R_w	thermal resistance	m^2K/W
R_z	peak to valley height of a rough surface	m
s	arc length	m
s	slip factor, $s = w_G/w_L$	–
s	exponent	–
s	wall thickness	m
S	correction factor, two-phase flow	–
t	time	s
t_1	interior fin intervall	m
T	absolute temperature	K
T^+	absolute temperature at $y^+ = \delta^+$	K
U	circumference	m
U_w	wetted perimeter	m
U_h	heated perimeter	m
v	specific volume	m^3/kg
v_G	specific volume of gas	m^3/kg
v_L	specific volume of liquid	m^3/kg
v'	specific volume of saturated liquid	m^3/kg
v''	specific volume of saturated vapor	m^3/kg
V	volume	m^3
V_A	break-off volume of a vapor bubble	m^3
V_i	partial molar volume of component i	m^3/mole
$\dot{V}$	volumetric flow	m^3/s
$\dot{V}_s$	volume flux of water	dm^3/m^2s
$\bar{V}$	molar volume	m^3/mole
V'_1, V'_2	partial molar volumes of the components 1, 2 in the liquid phase	m^3/mole
V''_1, V''_2	partial molar volumes of the components 1, 2 in the vapor phase	m^3/mole
w	velocity	m/s

Symbol	Name	SI-unit (as an example)
w_τ	friction velocity, $w_\tau = \sqrt{\tau_w/\varrho_L}$	m/s
$\bar{w}$	average velocity	m/s
$\mathbf{w}$	average molar velocity, Eq. (6.72)	m/s
$\mathbf{w}_i$	average velocity of component i, Eq. (6.71)	m/s
w'	momentary velocity in a turbulent flow	m/s
w^+	relative velocity, $w^+ = w_x/w_\tau$	–
x	coordinate, liquid mole fraction	–
x	quality	–
x	factor, Eq. (11.49)	–
x_i	mass flux of drops of class i	$1/m^2s$
x^*	quality in two-phase flow, $x^* = \dot{M}_G/\dot{M}$	–
x^*_{real}	actual quality	–
x^*_{th}	thermodynamic quality	–
X	mass flux of drops = number of drops/m^2s	$1/m^2s$
X_{tt}	Martinelli-parameter, defined by Eq. (4.78a)	–
y	coordinate, vapor mole fraction	–
y	factor, Eq. (11.49)	–
y^+	relative wall distance, $y^+ = yw_\tau/\nu$	–
Y	parameter	–
Y	factor, Eq. (5.3)	–
z	coordinate	–
Z	parameter, Eq. (2.46a)	–

b) Greek letters

α	heat transfer coefficient	W/m^2K
α_B	heat transfer coefficient in nucleate boiling	W/m^2K
α_C	convective heat transfer coefficient	W/m^2K
α_{2Ph}	heat transfer coefficient in two-phase flow	W/m^2K
α_f	heat transfer coefficient for film boiling	W/m^2K
α_G	heat transfer coefficient in a condensing vapor flow	W/m^2K
α_G	heat transfer coefficient on a smooth tube	W/m^2K
α_{hor}	heat transfer coefficient, film condensation on horizontal wall	W/m^2K
α_I	heat transfer coefficient at the interface	W/m^2K
α_K	heat transfer coefficient for free convection	W/m^2K
α_l	heat transfer coefficient, laminar flow	W/m^2K
α_{Nu}	heat transfer coefficient according to the Nusselt film condensation theory	W/m^2K
α_r	heat transfer coefficient, thermal radiation	W/m^2K

Symbol	Name	SI-unit (as an example)
α_R	heat transfer coefficient on a finned tube of copper	W/m^2K
α_S	heat transfer coefficient, stratified flow	W/m^2K
α_t	heat transfer coefficient, turbulent flow	W/m^2K
α_0	heat transfer coefficient in stagnant vapor	W/m^2K
$\bar{\alpha}$	average heat transfer coefficient	W/m^2K
$\bar{\alpha}_{tot}$	entire heat transfer coefficient	W/m^2K
$\bar{\alpha}_\gamma$	heat transfer coefficient, film condensation on wall inclined to the horizontal by the angle γ	W/m^2K
$\bar{\alpha}_\varphi$	heat transfer coefficient averaged over the circumferential angle φ	W/m^2K
$\hat{\alpha}$	ideal heat transfer coefficient in boiling of mixtures, $\hat{\alpha} = \sum_k x_k \alpha_k$	W/m^2K
β	mass transfer coefficient	m/s
β_{ij}	mass transfer coefficient of the binary mixture ij	m/s
β_0	contact angle between bubble and wall, wetting angle	–
β_G	mass transfer coefficient in a condensing vapor flow	m/s
γ	angle of inclination against the horizontal	–
Γ	condensate mass flow rate per unit of length	kg/s m
Γ	parameter in a two-phase flow	–
δ	film thickness	m
δ_{ij}	Kronecker-symbol	–
δ^+	relative film thickness, $\delta^+ = \delta w_t/\nu$	–
ε	volumetric vapor content, or void fraction	–
ε	turbulent viscosity	m^2/s
ε_c	parameter, two-phase flow	–
ε_D	turbulent diffusion coefficient	m^2/s
ε_D^+	relative turbulent diffusion coefficient, $\varepsilon_D^+ = \varepsilon_D/\nu + 1$	–
ε_L	emissivity of a liquid surface	–
ε_q	turbulent thermal diffusivity	m^2/s
ε_w	emissivity of a wall	–
ε^+	reduced turbulent viscosity, $\varepsilon^+ = \varepsilon/\nu + 1$	–
ε^*	flowing volumetric quality, $\varepsilon^* = \dot{V}_G/\dot{V}$	–
ζ	friction factor	–
ζ	correction factor for mass transfer	–
ζ_F	frictional portion of friction parameter	–

Symbol	Name	SI-unit (as an example)
ζ_R	form component of friction parameter	–
η	dynamic viscosity	kg/s m = Pa s
η	dimensionless wall distance, $\eta = z/\delta$	–
η_δ^*	factor, $\eta_\delta^* = \delta(w_{G,x}/\nu_L \cdot x)^{1/2}$	–
ϑ	temperature	K
$\Delta\vartheta, \Delta\bar{\vartheta}$	temperature difference	K
$\Delta\vartheta_{max}$	temperature difference $\vartheta_w - \vartheta_s$ at peak heat flux	K
$\Delta\vartheta_u$	temperature difference in subcooled boiling	K
ϑ_∞	temperature far from the wall	K
$\Delta\vartheta_{id}$	"ideal" temperature difference in boiling of mixtures	K
$\Delta\tilde{\vartheta}$	liquid superheating in a binary mixture	K
$\bar{\vartheta}$	mean temperature	K
ϑ_B	adiabatic mixing temperature	K
θ	dimensionless temperature, Eq. (14.15)	–
κ	parameter, Eq. (4.39)	–
λ	thermal conductivity	W/K m
λ_t	turbulent thermal conductivity	W/K m
μ	chemical potential	J/kg or J/mole
ν	kinematic viscosity	m^2/s
ξ	mass fraction	–
ξ'	mass fraction in the liquid mixture	–
ξ''	mass fraction in the vapor mixture	–
ρ	density	kg/m^3
ϱ'	density of saturated liquid	kg/m^3
ϱ''	density of saturated vapor	kg/m^3
σ	surface tension, interfacial tension	N/m
σ_{LG}	interfacial tension, liquid–gas	N/m
σ_r	radiation coefficient of a black body, $\sigma_r = 5.67051 \cdot 10^{-8}\ W/m^2K^4$	W/m^2K^4
σ_{SG}	interfacial tension, solid–gas	N/m
σ_{SL}	interfacial tension, solid–liquid	N/m
τ	shear stress	N/m^2
τ	dimensionless time, $\tau = t/t^+$	–
τ_δ	shear stress at film surface	N/m
τ_δ^*	reduced shear stress, $\tau_\delta^* = \tau_\delta[\varrho_L g(\nu_L^2/g)^{1/3}]$	–
τ^+	reduced shear stress, $\tau^+ = \tau[\varrho_L g(\nu_L^2/g)^{1/3}]$	–
φ	factor for suction of condensate	–
φ	surface ratio of finned tubes	–
φ	circumferential angle	–
φ	exponent	–

Symbol	Name	SI-unit (as an example)
Ψ	angle of curvature	–
Φ	heat transfer rate	W
Φ_E	kinetic energy	W
Φ	multiplier for pressure drop in two-phase flow	–
ω	factor	–
ω^*	temperature factor	1/K

c) Subscripts

Symbol	Name
a	outside
b	bulk, in the core
cr	critical
G	gas or vapor
i	internal
I	at the interface
K	cooling fluid
l	laminar
L	liquid
R	fin
s	saturation
S	stratified, solid wall
t	turbulent
u	upper limit
w	wall, water
x	in x-direction
y	in y-direction

d) Dimensionless Groups

Group	Name
$Ar = \frac{g}{\nu_L^2}\left(\frac{\sigma}{\varrho' g}\right)^{3/2}$	Archimedes number
$Bo = q/(\dot{m}\Delta h_v)$	boiling number
$Fr = w^2/gz$	Froude number
$Fr = \dot{m}^2/(\varrho_L^2 gd)$	Froude number
$Fr_L = \dot{m}^2(1-x^*)^2/(\varrho_L^2 gd)$	Froude number of the liquid in a two-phase flow
$Ja = (c_L \Delta\vartheta \varrho')/(\varrho'' \Delta h_v)$	Jakob number
$Le = D/a$	Lewis number

Group	Name
$Nu = \alpha l/\lambda$	Nusselt number
$Nu = \alpha(\nu_L^2/g)^{1/3}/\lambda_L$	Nusselt number in condensation
Nu	average Nusselt number
$Pr = \nu/a = \eta c_p/\lambda$	Prandtl number
$Pr_t = \varepsilon/\varepsilon_q$	turbulent Prandtl number
$Re = wl/\nu$	Reynolds number
$Re = w\delta/\nu$	Reynolds number
$Re = \Gamma/\eta_L$	Reynolds number for condensation
$Re_{2\Phi} = Re F^{1.25}$	modified Reynolds number of a two-phase flow
$Re = \dot{m}(1 - x^*)d/\eta_L$	Reynolds number of the liquid in a two-phase flow
$Re_\tau = w_\tau(\nu_L^2/g)^{1/3}/\nu_L$	Reynolds number defined with shear velocity
$Sc = \nu/D$	Schmidt number
$Sh = \beta x/D$	Sherwood number
$We = w^2\varrho d/\sigma$	Weber number

A Heat Transfer in Condensation

1 Fundamentals

1.1 The Process of Condensation

Mass transfers from one aggregate state to another state serve in nature and in technology to separate or to transform substances. The condensation of vapors to be treated here is such a mass transfer. Thereby, the vapor of a substance is converted into the liquid phase.

In the theory of thermodynamic equilibrium between different phases, one postulates the original and the new phase as given and the phase transformation as complete. The formation of a new phase from the original one is, therefore, impossible under the conditions of equilibrium. Phase transformations presuppose, therefore, deviations from thermal, mechanical, or material equilibrium.

If the temperature of a wall surface is lower than the saturation temperature of the adjacent vapor, then there is no prevailing equilibrium and the vapor is liquefied. Thereupon a condensate is formed, which as a result of its contact with the wall surface is subcooled, so that a further vapor can condense upon the previously formed condensate. The condensation of vapors is, therefore, always associated with a mass transfer, during which the vapor flows to the phase interface and is converted into the liquid phase.

The course of this process can be subdivided into several steps, in which several resistances, which are connected in series, must be overcome. The portion of these resistances in the total resistance can be very different. First, as a result of flow (convective transport) and the molecular movement (diffusion transport), vapor arrives at the phase interface. In the next step, the vapor at the phase interface is condensed and, finally, the condensation enthalpy being released at the phase interface is transported by conduction and convection to the cooled wall. Accordingly, three connected resistances are to be overcome: the thermal resistance in the vapor phase, the thermal resistance during the conversion of the vapor into the liquid phase, and the thermal resistance in the liquid phase.

Of these resistances, the one in the liquid phase is generally controlling. The thermal resistance in the vapor is often negligible because of the good mixing, but it can be substantial during large superheating of a vapor as well as during the condensation of vapor mixtures or of mixtures of vapors with inert gases because of the impeding effect of the diffusion. For the conversion of the vapor into the liquid phase, a temperature drop directly at the interface between the vapor and the liquid is necessary. This temperature drop is, however, very small and amounts usually to only a few hundredths of a degree, if one disregards very low pressures, which for water are below 0.01 bar [1.1]. Correspondingly, the associated "molecular kinetic" resistance at the interface from vapor to liquid, which is equal to the inverse value of the heat transfer coefficient [1.2],

$$\alpha_I = \frac{2f}{2-f}\left(\frac{M}{2\pi RT}\right)^{1/2}\frac{\Delta h_v^2 pM}{RT^2} \tag{1.1}$$

is usually negligible.

In Eq. (1.1), f is the condensation coefficient representing the fraction of all occurring molecules held back by the condensate surface, M is the molecular weight (SI unit kg/mole), R the universal gas constant (SI unit J/mole K), T the absolute temperature (SI unit K), Δh_v the vaporization enthalpy (SI unit J/kg), and p the pressure (SI unit Pa). According to recent measurements, the condensation coefficient is somewhat smaller than unity. For the derivation of Eq. (1.1) one assumes that vapor acts as an ideal gas.

1.2 The Different Types of Condensation

Actual condensation can take place in very different ways. If the condensate forms a continuous film, Fig. 1.1, then one speaks of *film condensation*. The condensate film can be stagnant or flow in laminar or turbulent form. Its thermal resistance controls the velocity of condensation, if one can disregard the molecular kinetic resistance. For calculation, it is sufficient to determine this resistance as Nusselt [1.3] did for the first time for laminar film flow. Instead of in film form, condensate can occur in the form of drops, as Fig. 1.2 shows. This type of condensation is called *drop condensation*. Whether film or drop condensation prevails, depends upon whether the wall is completely or incompletely wetted. The mode depends upon the forces acting upon a liquid

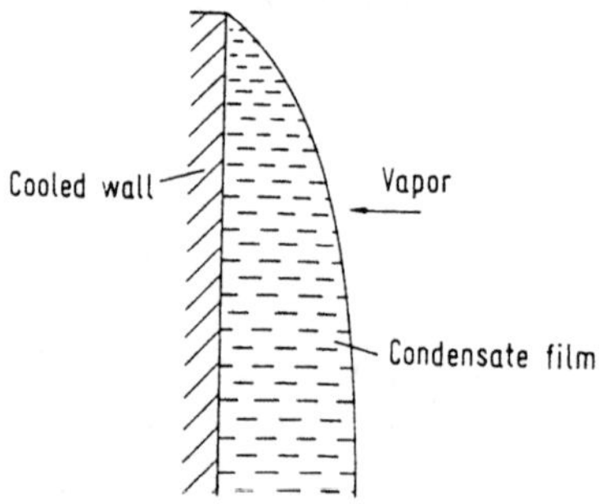

Fig. 1.1. Film condensation

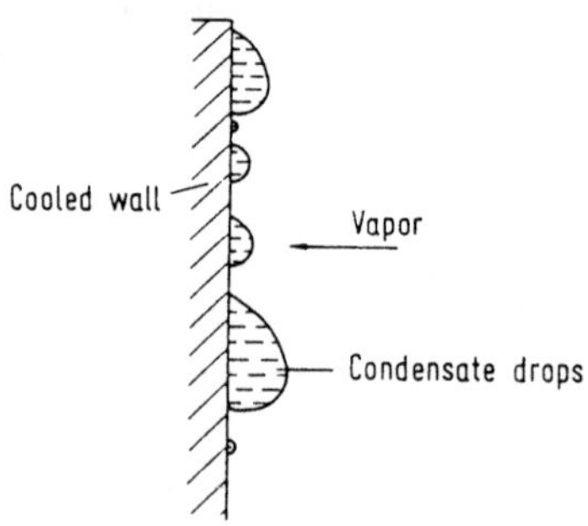

Fig. 1.2. Drop condensation

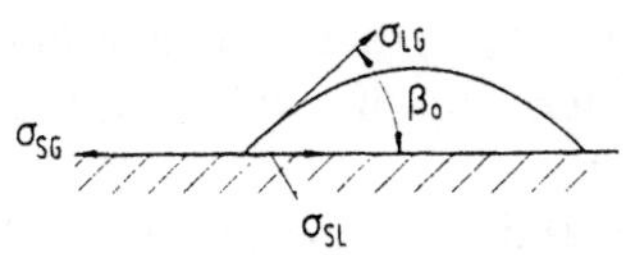

Fig. 1.3. Interfacial tensions at the drop edge during incomplete wetting. The indices S, L, G signify the solid, liquid, and the gaseous phase; β_0 is the wetting angle

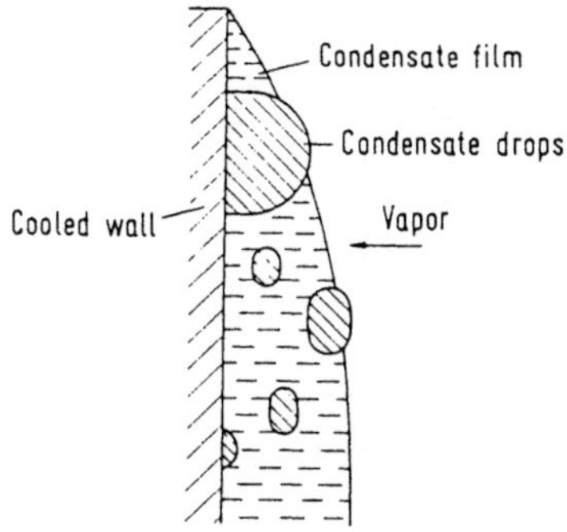

Fig. 1.4. Condensation of vapors of immiscible liquids

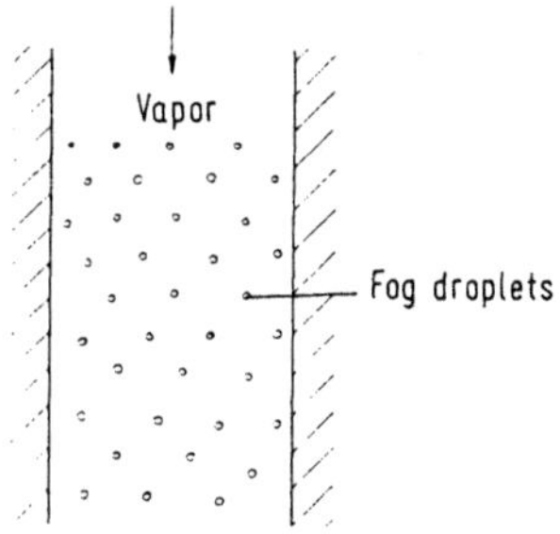

Fig. 1.5. Condensate formation as fog drops

drop, as shown in Fig. 1.3. If σ_{LG} signifies the interfacial tension (SI unit N/m) of the liquid (index L) against its own vapor (index G), σ_{SL} that of the stationary wall (index S) against the liquid, and σ_{SG} that of the stationary wall against the vapor, then in equilibrium a wetting or contact angle β_0 must occur according to the equation

$$\sigma_{SG} - \sigma_{SL} = \sigma_{LG} \cos \beta_0 . \tag{1.2}$$

Finite values of the wetting angle mean incomplete wetting and drop formation. If, on the other hand, there is $\beta_0 = 0$, then the drop spreads out over the entire wall. Additionally, if the so-called wetting tension is $\sigma_{SG} - \sigma_{SL} > \sigma_{LG}$, then equilibrium can no longer occur according to Eq. (1.2). There occurs complete wetting and thus film formation.

In practice, one often has to condense vapor mixtures whose liquid phases are immiscible. There then results a mixed form of drop and film condensation, as Fig. 1.4 shows. In an extended liquid film of the one phase, there occur rather large drops of the other phase, of which several extend to the wall, whereas others are inclosed in the liquid film or float upon it.

Finally, with sufficient supercooling of a vapor and with the presence of "condensation nuclei"—these are tiny particles on which condensate can be deposited—a fog can be formed, as Fig. 1.5 schematically shows.

2 Film Condensation of Stagnant Vapors

2.1 The Nusselt Film Condensation Theory

If a vapor condenses on a vertical or an inclined surface, a liquid film is formed, which flows off under the influence of gravity. If the vapor velocity is very low and the liquid film is thin, there is laminar flow in the condensate, and the heat is transferred to the wall chiefly by conduction through the film. The convective effect is negligibly small.

Already in 1916 Nusselt [1.3] proposed a simple theory for the calculation of heat transfer in the case of laminar film condensation on horizontal tubes and on vertical or inclined walls. This theory is known in the technical literature as the *Nusselt film condensation theory*. It is to be discussed in the following using the example of the condensation on a vertical wall, Fig. 2.1. As the diagram shows, saturated steam at temperature ϑ_s is condensing on a vertical wall whose temperature ϑ_w is constant and lower than the saturation temperature. There forms a continuous condensate film, which flows downward under the influence of gravity and whose thickness $\delta(x)$ constantly increases. The velocity profile $w(y)$ is obtained from a balance of forces. Assuming a steady flow, the forces exerted by the shear stresses are in equilibrium with gravity, and, in conformity with the sketch on the right in Fig. 2.1, one obtains

$$\varrho_L g \mathrm{d}V + \tau(y + \mathrm{d}y)\mathrm{d}x\mathrm{d}z = \tau(y)\mathrm{d}x\mathrm{d}z. \tag{2.1}$$

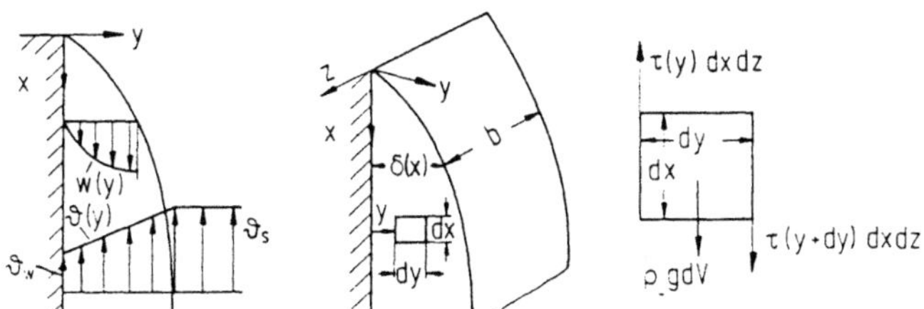

Fig. 2.1. Laminar condensate film on a vertical wall, including velocity and temperature profiles and balance of forces

With $\tau(y+\mathrm{d}y)-\tau(y)=(\partial\tau/\partial y)\mathrm{d}y$ and $\mathrm{d}V=\mathrm{d}x\mathrm{d}y\mathrm{d}z$ one obtains

$$\frac{\partial\tau}{\partial y}=-\varrho_{\mathrm{L}}g. \tag{2.2}$$

If the condensate is a Newtonian liquid, then there applies

$$\tau=\eta_{\mathrm{L}}\frac{\partial w}{\partial y}. \tag{2.3}$$

Assuming a temperature-independent dynamic viscosity, Eq. (2.2) converts into

$$\eta_{\mathrm{L}}\frac{\partial^2 w}{\partial y^2}=-\varrho_{\mathrm{L}}g, \tag{2.4}$$

from which, after integration, assuming constant density, one obtains a parabolic velocity profile in the y direction:

$$w=\frac{\varrho_{\mathrm{L}}g}{2\eta_{\mathrm{L}}}y^2+c_1y+c_0 \tag{2.5}$$

The coefficients c_1 and c_0 can still be dependent upon the x-coordinate. Two boundary conditions are available for their determination.

At the wall, $y=0$, the velocity is $w=0$, and if one further assumes that the vapor velocity is not very great and that, consequently, the shear exerted by the vapor upon the condensate film is small, then at the condensate surface $y=\delta$

$$\partial w/\partial y=0.$$

With these two boundary conditions,

$$c_0=0 \quad \text{and} \quad c_1=\varrho_{\mathrm{L}}g\delta/\eta_{\mathrm{L}},$$

so that one obtains as the velocity profile

$$w=\frac{\varrho_{\mathrm{L}}g}{\eta_{\mathrm{L}}}\delta^2\left(\frac{y}{\delta}-\frac{y^2}{2\delta^2}\right). \tag{2.6}$$

By integration, this gives the average velocity $\bar{w}(x)$ across the film thickness $\delta(x)$

$$\bar{w}=\frac{1}{\delta}\int_0^\delta w\mathrm{d}y=\frac{\varrho_{\mathrm{L}}g}{3\eta_{\mathrm{L}}}\delta^2. \tag{2.7}$$

The condensate mass flow rate is then

$$\dot{M}=\bar{w}\varrho_{\mathrm{L}}b\delta=\frac{\varrho_{\mathrm{L}}^2gb}{3\eta_{\mathrm{L}}}\delta^3 \tag{2.8}$$

and the change of this mass flow rate with the film thickness is

$$\frac{\mathrm{d}\dot{M}}{\mathrm{d}\delta}=\frac{\varrho_{\mathrm{L}}^2gb}{\eta_{\mathrm{L}}}\delta^2. \tag{2.9}$$

For the formation of the condensate mass flow rate $\mathrm{d}\dot{M}$, a heat transfer rate $\mathrm{d}\Phi = \Delta h_v \mathrm{d}\dot{M}$ is required, where Δh_v is the vaporization enthalpy. This energy is transferred, according to assumption, by pure heat conduction through the condensate film. The influence of convection on the condensate film is considered negligible. Assuming a linear temperature profile as plotted in Fig. 2.1, the heat transfer through a surface segment $b\mathrm{d}x$ will be

$$\mathrm{d}\Phi = \lambda_L \frac{\vartheta_s - \vartheta_w}{\delta} b\mathrm{d}x.$$

On the other hand $\mathrm{d}\Phi = \Delta h_v \mathrm{d}\dot{M}$, where the condensate mass flow rate is given by Eq. (2.9). The last equation thus converts into

$$\lambda_L \frac{\vartheta_s - \vartheta_w}{\delta} b\mathrm{d}x = \Delta h_v \mathrm{d}\dot{M} = \Delta h_v \frac{\varrho_L^2 g b}{\eta_L} \delta^2 \mathrm{d}\delta,$$

or it is

$$\delta^3 \frac{\mathrm{d}\delta}{\mathrm{d}x} = \frac{\lambda_L \eta_L}{\varrho_L^2 g \Delta h_v} (\vartheta_s - \vartheta_w),$$

from which, when considering $\delta(x=0) = 0$, one obtains by integration the film thickness as

$$\delta = \left[\frac{4\lambda_L \eta_L (\vartheta_s - \vartheta_w)}{\varrho_L^2 g \Delta h_v} x \right]^{1/4}. \tag{2.10}$$

The film thickness increases by the fourth root of the distance x from the leading edge.

Because the temperature profile was assumed to be linear, the local heat transfer coefficient α is given by

$$\alpha = \frac{\lambda_L}{\delta} = \left[\frac{\varrho_L^2 g \Delta h_v \lambda_L^3}{4\eta_L (\vartheta_s - \vartheta_w)} \frac{1}{x} \right]^{1/4}, \tag{2.11}$$

and the average heat transfer coefficient for a wall of height h is

$$\bar{\alpha} = \frac{1}{h} \int_0^h \alpha \mathrm{d}x = \frac{4}{3} \alpha(x=h) = 0.943 \left[\frac{\varrho_L^2 g \Delta h_v \lambda_L^3}{\eta_L (\vartheta_s - \vartheta_w)} \frac{1}{h} \right]^{1/4}. \tag{2.12}$$

The condensate properties are best evaluated at the mean temperature $\bar{\vartheta} = (\vartheta_w + \vartheta_s)/2$. In the unusual case where the vapor density ϱ_G is not negligible compared to the liquid density ϱ_L, the buoyant forces resulting from pressure gradients must be considered and one must replace the quantity ϱ_L^2 in the brackets by $\varrho_L(\varrho_L - \varrho_G)$.

With the aid of the energy balance for the condensate mass flow rate formed along the height h,

$$\dot{M} \Delta h_v = \bar{\alpha} (\vartheta_s - \vartheta_w) b h,$$

one can eliminate the temperature difference $\vartheta_s - \vartheta_w$. One then obtains the following expression equivalent to Eq. (2.12):

$$\frac{\bar{\alpha}}{\lambda_L}\left(\frac{\eta_L^2}{\varrho_L^2 g}\right)^{1/3} = 0.925\left(\frac{\dot{M}/b}{\eta_L}\right)^{-1/3}. \tag{2.12a}$$

As can be seen from the above equations, one achieves large heat transfer coefficients if the temperature difference $\vartheta_s - \vartheta_w$ and the wall height are small.

In both cases, the condensate film is thin, and thus its thermal resistance is small. The above results are valid also for the condensation of vapors on the inside and the outside of vertical tubes, if the tube diameter is large in comparison to the film thickness. The width $b = d\pi$ is then to be inserted into Eq. (2.12a).

The above derivation applies to a vertical wall or to a vertical tube of a not too small diameter. If the wall is inclined to the horizontal by the angle γ, then the gravitational acceleration g in the previous equations must be replaced by its component parallel to the wall $g \sin\gamma$. According to Eq. (2.11), the heat transfer coefficient α_γ is related to that for the vertical wall by the following equation:

$$\alpha_\gamma = \alpha(\sin\gamma)^{1/4} \quad \text{and correspondingly} \quad \bar{\alpha}_\gamma = \bar{\alpha}(\sin\gamma)^{1/4}. \tag{2.13}$$

These results cannot be applied to inclined tubes, because then the liquid film does not flow off evenly distributed over the circumference.

Nusselt also calculated the heat transfer for laminar film condensation on *horizontal tubes*. In this case, one obtains a differential equation for the film thickness, which can be integrated numerically. If one designates the outer tube diameter by d, the results for the mean heat transfer coefficient $\bar{\alpha}_{hor}$ over the circumference of the horizontal tube can be represented by

$$\bar{\alpha}_{hor} = 0.728\left(\frac{\varrho_L^2 g \Delta h_v \lambda_L^3}{\eta_L(\vartheta_s - \vartheta_w)}\frac{1}{d}\right)^{1/4} = 0.864\bar{\alpha}(h = d\pi/2), \tag{2.14}$$[1]

where $\bar{\alpha}$ is the average heat transfer coefficient on the vertical wall according to Eq. (2.12). With the aid of the energy balance for the condensate

$$\dot{M}\Delta h_v = \bar{\alpha}_{hor}(\vartheta_s - \vartheta_w)d\pi l$$

one can eliminate the temperature from the former equation and replace it by the condensate mass flow rate. One obtains then an equation equivalent to Eq. (2.14)

$$\frac{\bar{\alpha}_{hor}}{\lambda_L}\left(\frac{\eta_L^2}{\varrho_L^2 g}\right)^{1/3} = 0.959\left(\frac{\dot{M}/l}{\eta_L}\right)^{-1/3}. \tag{2.14a}$$

From a comparison of Eqs. (2.14) and (2.12), one can see that the average heat transfer coefficient on the tube in a horizontal position is related to that of the

[1] Nusselt obtained by graphical integration a slightly less accurate value of 0.725, which was adopted by all subsequent authors.

vertical tube by

$$\bar{\alpha}_{hor}/\bar{\alpha} = 0.772(h/d)^{1/4}. \tag{2.15}$$

If one chooses as an example a tube length of 3 m and a diameter of 0.029 m, one gets $\bar{\alpha}_{hor} = 2.46\bar{\alpha}$; on the tube in a horizontal position there condenses about $2\frac{1}{2}$ times as much vapor as on the one in the vertical position.

If tubes lie one beneath the other in a horizontal tube bank, the condensate film on the lower tubes will be thickened by the condensate from above, and the heat transfer coefficient will be diminished. On the other hand, the convection in the liquid film will be increased by the condensation occurring from above, and thus the effect of the higher thermal resistance will be reduced. As the mean value of the heat transfer coefficient $\bar{\alpha}_n$ for n tubes lying one over the other one obtains according to Nusselt $\bar{\alpha}_n/\bar{\alpha}_1 = n^{-1/4}$, if $\bar{\alpha}_1$ is the mean heat transfer coefficient for the top-most tube given by Eq. (2.14). Here, however, there is no consideration of the improvement of the heat transfer by the stronger convection produced by the occurring condensate. The heat transfer coefficient $\bar{\alpha}_n$ calculated in this manner is, therefore, somewhat too small.

More exact values are obtained according to Chen [2.1], if one additionally considers that the condensate film mixes with the cooler condensate trickling down from above and is thus even more subcooled, so that new vapor can condense. In this way, there results

$$\bar{\alpha}_n = \left[1 + 0.2\,\frac{c_{pL}(\vartheta_s - \vartheta_w)}{\Delta h_v}(n-1)\right] n^{-1/4}\bar{\alpha}_1, \tag{2.16}$$

where once again $\bar{\alpha}_1$ is the mean heat transfer coefficient for the top-most tube according to Eq. (2.14). This relation represents well the values measured on tubes arranged vertically in line in the range $c_{pL}(\vartheta_s - \vartheta_w)(n-1)/\Delta h_v < 2$.

2.2 Deviations from the Nusselt Film Condensation Theory

Deviations from the Nusselt film condensation theory result from the fact that the assumptions of the theory are not always fulfilled. One of the most important assumptions is that heat is only transferred by conduction through the condensate film. Accordingly, the total thermal resistance is supposed to be in the condensate film. Actually, there are additional heat resistances present. To these belong the initially mentioned resistance at the phase interface between condensate film and vapor, additional resistances in the presence of noncondensable gases, and possibly also a contact resistance between condensate film and cooling wall. These resistances are to be discussed individually in the following.

2.2.1 Consideration of the Resistance by the Condensate Film

An improved calculation of the thermal resistance in the condensate film goes back to Rohsenow [2.2]. It takes into consideration the fact that the pressure

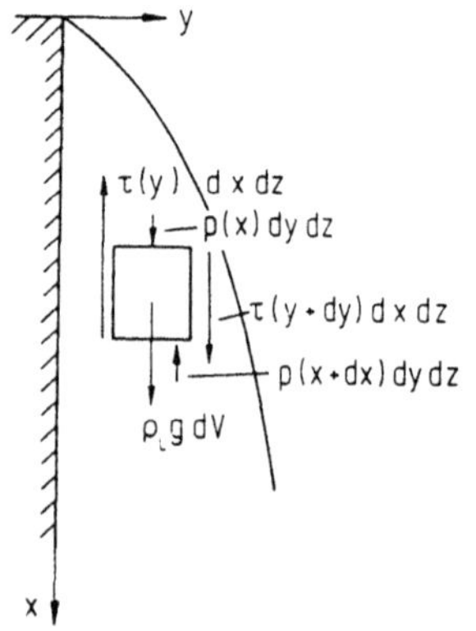

Fig. 2.2. Laminar condensate film on a vertical wall. Balance of forces, allowing for pressure differences

in the condensate and in the vapor space increases as a consequence of gravity. If one moves in Fig. 2.1 by a distance dx in the direction of the coordinate x, the pressure will increase by $\mathrm{d}p = \varrho_G g \mathrm{d}x$, whereby ϱ_G is the density of the vapor. In accordance with Fig. 2.2, the balance of forces now contains the weight $\varrho_L g \mathrm{d}V$ of the liquid volume dV, the forces exerted by the shear stresses, and the forces resulting from the pressure difference. It reads

$$\varrho_L g \mathrm{d}V + \tau(y+\mathrm{d}y)\mathrm{d}x\mathrm{d}z + p(x)\mathrm{d}y\mathrm{d}z = \tau(y)\mathrm{d}x\mathrm{d}z + p(x+\mathrm{d}x)\mathrm{d}y\mathrm{d}z.$$

With $\tau(y+\mathrm{d}y) - \tau(y) = (\partial\tau/\partial y)\mathrm{d}y$,

$$p(x)\mathrm{d}y\mathrm{d}z - p(x+\mathrm{d}x)\mathrm{d}y\mathrm{d}z = -(\mathrm{d}p/\mathrm{d}x)\mathrm{d}x\mathrm{d}y\mathrm{d}z$$

and $\mathrm{d}V = \mathrm{d}x\mathrm{d}y\mathrm{d}z$ one obtains

$$\frac{\partial\tau}{\partial y} = -\varrho_L g + \frac{\mathrm{d}p}{\mathrm{d}x}.$$

If one considers only the vapor space, the balance of forces there is considered to be

$$\frac{\mathrm{d}p}{\mathrm{d}x} = \varrho_G g.$$

Thus, the balance of forces is given by

$$\frac{\partial\tau}{\partial y} = -(\varrho_L - \varrho_G)g$$

or because of $\tau = \eta_L \partial w/\partial y$, in case one assumes as before a temperature-independent dynamic viscosity

$$\eta_L \frac{\partial^2 w}{\partial y^2} = -(\varrho_L - \varrho_G)g. \tag{2.17}$$

This equation takes the place of the earlier balance of momentum, Eq. (2.4), in which the vapor density ϱ_G was assumed to be small in comparison to the liquid density ϱ_L.

Thus, the term $\varrho_L g$ must be replaced by $(\varrho_L - \varrho_G)g$. According to Eq. (2.6), one then obtains for the velocity profile

$$w = \frac{(\varrho_L - \varrho_G)g}{\eta_L} \delta^2 \left(\frac{y}{\delta} - \frac{y^2}{2\delta^2} \right) \tag{2.18}$$

and, correspondingly, for the average velocity according to Eq. (2.7),

$$\bar{w} = \frac{(\varrho_L - \varrho_G)g}{3\eta_L} \delta^2 \tag{2.19}$$

and for the condensate mass flow rate according to Eq. (2.8),

$$\dot{M} = \frac{\varrho_L(\varrho_L - \varrho_G)gb}{3\eta_L} \delta^3. \tag{2.20}$$

In addition, in the energy balance the subcooling of the condensate is taken into consideration; the average enthalpy of the flowing condensate is therefore less than the enthalpy of the saturated liquid. At a cross section at position x, the enthalpy flow of the condensate is

$$\dot{M}(h'' - h_L) = \dot{M}\Delta h_v + \int_0^\delta \varrho_L w b c_{pL}(\vartheta_s - \vartheta)\mathrm{d}y, \tag{2.21}$$

if h_L is the specific average enthalpy of the liquid. If the liquid has the saturation temperature $\vartheta = \vartheta_s$ across a cross section, then $h'' - h_L = h'' - h'$. If one presupposes again a linear temperature profile

$$\vartheta_s - \vartheta = (\vartheta_s - \vartheta_w)\left(1 - \frac{y}{\delta}\right) \tag{2.22}$$

and also assumes temperature-independent properties in the condensate, then, after insertion of the temperature from Eq. (2.22) and the velocity from Eq. (2.18) into Eq. (2.21), one obtains as the result of integration

$$\dot{M}(h'' - h_L) = \dot{M}[\Delta h_v + \tfrac{3}{8}c_{pL}(\vartheta_s - \vartheta_w)] = \dot{M}\Delta h_v^*. \tag{2.23}$$

It follows from this fact, according to the assumptions made, the specific enthalpy h_L of the condensate flowing off is independent of the film thickness. The equation shows further that the vaporization enthalpy in the equations of the Nusselt film condensation theory must be replaced by the enthalpy difference Δh_v^*. Moreover, if one considers that the temperature profile in the condensate is slightly curved, then one obtains according to Rohsenow [2.2], instead of Eq. (2.23) for Δh_v^*, the more precise value

$$\Delta h_v^* = \Delta h_v + 0.68 c_{pL}(\vartheta_s - \vartheta_w). \tag{2.24}$$

From Eq. (2.12), one thus obtains for the average heat transfer coefficient the improved equation

$$\bar{\alpha} = 0.943\left[\frac{\varrho_L(\varrho_L - \varrho_G) g \Delta h_v^* \lambda_L^3}{\eta_L(\vartheta_s - \vartheta_w)}\frac{1}{h}\right]^{1/4}. \tag{2.25}$$

The equation holds for Prandtl numbers $Pr > 0.5$ and $c_{pL}(\vartheta_s - \vartheta_w)/\Delta h_v \leqq 1$. The best correlation with experiments resulted if one inserted the property values, especially the dynamic viscosity η_L, at a temperature $\vartheta_L = \vartheta_w + \frac{1}{4}(\vartheta_s - \vartheta_w)$.

A further, usually similarly slight deviation from the Nusselt film condensation theory results if the *vapor* is *superheated*. Besides the vaporization enthalpy, one must also eliminate the superheating enthalpy $c_{pG}(\vartheta_G - \vartheta_s)$ in order to cool the superheated vapor of the temperature ϑ_G down to the saturation temperature ϑ_s at the phase interface. In place of the enthalpy difference Δh_v^*, according to Eq. (2.24), one has to insert into Eq. (2.25) the enthalpy difference

$$\Delta h_{vu}^* = c_{pG}(\vartheta_G - \vartheta_s) + \Delta h_v + 0.68 c_{pL}(\vartheta_s - \vartheta_w). \tag{2.26}$$

Because the temperature difference through the condensate film is still $\vartheta_s - \vartheta_w$ there also results the heat flux as before from $q = \alpha(\vartheta_s - \vartheta_w)$ and the mass flux of the condensate from $\dot{M}/A = q/\Delta h_{vu}^*$.

In technical cases, however, the subcooling of the condensate, the buoyant forces, and the superheating of the vapor seldom reach such values that one must calculate according to the corrected Eq. (2.24), or as the case may be, Eq. (2.26).

The complete solution of the boundary layer equations for the laminar film condensation are found in Koh et al. [2.3]. Sparrow and Eckert [2.4] have investigated analytically the influence of superheated vapor on the heat transfer. Here, too, it was confirmed that the deviations from the Nusselt film condensation theory are slight.

2.2.2 Consideration of the Resistance at the Phase Interface Between Liquid and Vapor

As mentioned at the beginning, the temperature at the surface of the condensate film is slightly below the saturation temperature, so that vapor at the phase interface constantly condenses and is replaced by new vapor following behind. Accordingly, there is an additional resistance to be overcome at the phase interface, which is, however, of significance only at sufficiently small pressures, and whose magnitude was already given by Eq. (1.1).

2.2.3 Influence of Noncondensable Gases

If a vapor condenses in the presence of a noncondensable gas, it must diffuse through the gas to the phase interface. For this to happen, a partial pressure gradient toward the phase interface is necessary. As Fig. 2.3 shows, the partial pressure p_1 of the vapor falls from a constant value p_{1G} at a rather large distance

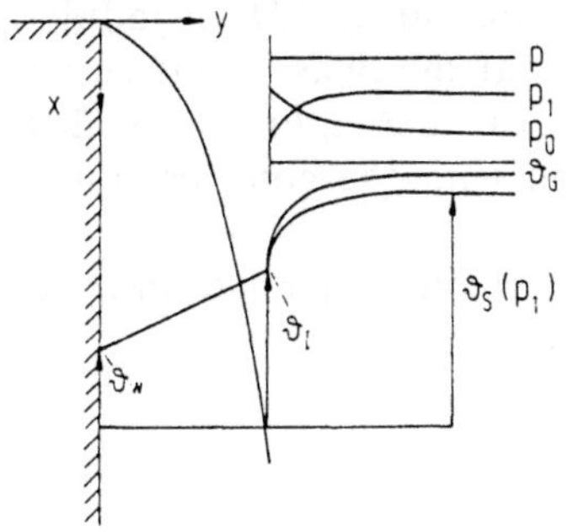

Fig. 2.3. Influence of inert gas on the profiles of partial pressure and of temperature. ϑ_G vapor temperature, ϑ_s saturation temperature, p_1 partial pressure of the condensing vapor, p_0 partial pressure of the inert gas, total pressure

$$p = p_1 + p_0$$

from the phase interface to a lower value p_{1I} at the phase interface. Correspondingly, the accompanying saturation temperature $\vartheta_s(p_1)$ also falls to the value ϑ_I at the phase interface. The pressure p_0 of the inert gas increases toward the phase interface such that $p_1 + p_0$ always equals the constant total pressure p.

Depending upon the gas content, the saturation temperature ϑ_I at the phase interface can be considerably below the saturation temperature corresponding to the pressure p which would result if no inert gas were present. The temperature difference between the phase interface and the wall is decreased by the inert gas, and, thus, the heat transfer coefficient is also diminished. In order to avoid this, one provides that the inert gas can be removed through valves. Large condensers are connected with a steam-jet apparatus which sucks off the gas. In some instances, for example in the case of condensation of water from mixtures of water vapor and air or in the case of the condensation of ammonia from its mixtures with air, inert gas is necessarily present; its influence on the heat transfer must therefore be considered.

In order to obtain a simple estimate of the influence of the inert gas, we will start with the energy balance

$$\Phi = \alpha_L A(\vartheta_I - \vartheta_w) = \dot{M}\Delta h_v + \alpha_G A(\vartheta_G - \vartheta_I). \tag{2.27}$$

The coefficient of heat transfer α_G signifies the heat transferred from the vapor-gas mixture to the phase interface. In a flowing gas, the value of α_G is determined by the properties of the gas and the velocity of the flow.

The mass flow rate $\dot{M}$ of the vapor transported to the surface of the condensate increases with the difference of the partial pressures $p_{1G} = y_{1G}p$ in the vapor space and p_{1I} at the phase interface.

According to the laws of mass transfer, there results

$$\dot{M} = \varrho_G \beta_{G0} A \ln \frac{p - p_{1I}}{p - p_{1G}}, \tag{2.28}$$

if ϱ_G is the density of the gas-vapor mixture at the pressure p and the temperature ϑ_G. Strictly speaking, this equation is valid only if the temperature is constant up to the phase interface. Except for lower temperatures, the assumption of a constant temperature will scarcely cause an error in the value of $\dot{M}$. In addition, p_{1I} is the partial pressure of the condensing vapor at the phase interface and p_{1G} that in the core of the gas-vapor flow. A more exact derivation of Eq. (2.28) will be discussed later in Sect. 6.4.3 in connection with the condensation of vapor mixtures.

According to the Lewis equation, one can approximately relate the mass transfer coefficient to the coefficient of heat transfer

$$\beta_{G0} = \frac{1}{\varrho_G} \frac{\alpha_{G0}}{c_{pG}}. \tag{2.29}$$

The index 0 in β_{G0} and α_{G0} denotes that we are dealing here with mass and heat transfer coefficients of a vapor flow, and that the condenser surface is regarded as a stationary solid wall. We place into Eq. (2.27) $\alpha_G = \alpha_{G0}\zeta$, in which the correction factor ζ allows for the fact that, in reality, the vapor does not flow along a solid wall, but rather that part of the flow disappears at the phase interface.

With the aid of Eqs. (2.28) and (2.29), one obtains from the energy balance Eq. (2.27)

$$\vartheta_I - \vartheta_w = \frac{\alpha_{G0}}{\alpha_L}\left[\frac{\Delta h_v}{c_{pG}} \ln \frac{p - p_{1I}}{p - p_{1G}} + \zeta(\vartheta_G - \vartheta_I)\right] \tag{2.30}$$

as the determining equation for the unknown temperature ϑ_I at the phase interface. This equation can be solved only iteratively. For small inert gas contents $p_0 \ll p$ or $p_1 \to p$, the quotient becomes $(p - p_{1I})/(p - p_{1G})$, and thereby the first term in the brackets becomes large. Equation (2.30) is then simplified to

$$\vartheta_I - \vartheta_w = \frac{\alpha_{G0} \Delta h_v}{\alpha_L c_{pG}} \ln \frac{p - p_{1I}}{p - p_{1G}}. \tag{2.31}$$

If one assumes that the condensate film has the thickness δ, at a definite location, then, in case no inert gas is present, the saturation temperature $\vartheta_s(p)$ would prevail at the surface of the condensate film and the heat flux would be

$$q = \frac{\lambda_L}{\delta}(\vartheta_s - \vartheta_w).$$

Given the presence of inert gas and an equal film thickness, a lower heat flux

$$q_G = \frac{\lambda_L}{\delta}(\vartheta_I - \vartheta_w)$$

is transferred. The ratio of the two heat fluxes is

$$\frac{q_G}{q} = \frac{\vartheta_I - \vartheta_w}{\vartheta_s - \vartheta_w} \leq 1. \tag{2.32}$$

With Eq. (2.31) there results the approximate equation

$$\frac{q_G}{q} = \frac{\alpha_{G0}\Delta h_v}{(\vartheta_s - \vartheta_w)\alpha_L c_{pG}} \ln \frac{p - p_{1I}}{p - p_{1G}}. \tag{2.33}$$

As can be recognized from the above, in the case of large temperature differences $\vartheta_s - \vartheta_w$, the velocity of flow should be made sufficiently large so that the heat transfer coefficient of the vapor is large, and thus the heat flux q_G is not too small.

As the result of a precise calculation, Figs. 2.4 and 2.5 show the ratio of the two heat fluxes q_G/q as a function of the temperature difference $\vartheta_s - \vartheta_w$ with the inert gas concentration as the parameter. Both figures hold for the condensation of water vapor from a mixture with air. Figure 2.4 holds for condensation in the case of forced convection, if the influence of gravity in relation to that of inertia can be neglected.

Figure 2.5 shows the reverse case, in which the forces of inertia are of negligible influence in relation to gravity. This corresponds to free convection.

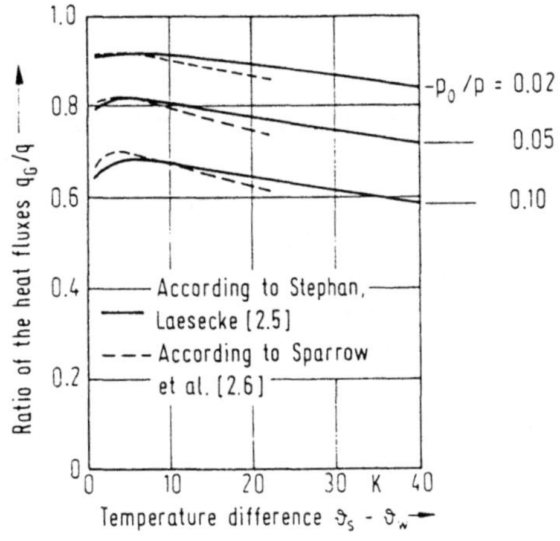

Fig. 2.4. Influence of inert gas on heat transfer during condensation in a forced flow of water vapor with air; q_G is the heat flux with and q without inert gas

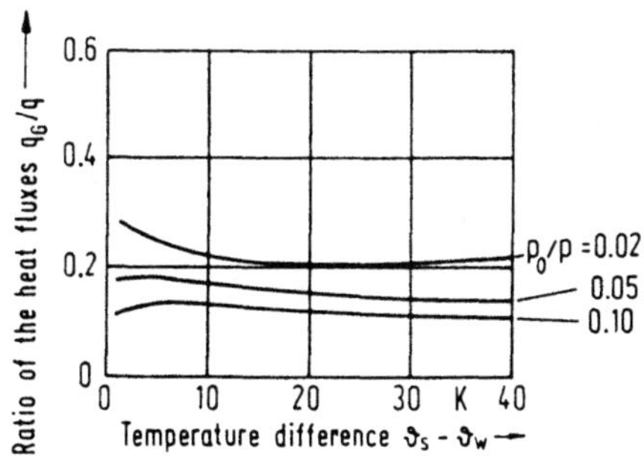

Fig. 2.5. Influence of inert gas on heat transfer during condensation in free convection of water vapor with air [2.5]; q_G is the heat flux with and q without inert gas

From both figures it can be seen clearly that the heat transfer diminishes with increasing inert gas concentration, and that during free convection the heat transfer is much more strongly reduced by the inert gas than during forced convection.

The calculation of heat exchanger surface area will be treated in more detail in Sect. 6.4 in connection with the condensation of mixtures of vapors.

2.2.4 Wave Formation on the Film Surface

The Nusselt film condensation theory assumed a steady film thickness increasing by means of further condensation. Experiments (among others [2.7–2.9]) showed, however, that during well-defined laminar film flow the film surface can be wavy. Such waves were observed not only on rough, but also on polished walls. Obviously, the small disturbances of velocity always present in such a flow are not dampened under certain circumstances, and waves are formed. They lead to an improvement of the heat transfer by 10 to 25% compared to the Nusselt film condensation theory. According to Grimley [2.10], waves and ripples appear at a critical Reynolds number of

$$Re = \frac{\bar{w}\delta}{\nu_L} = \frac{\dot{M}/b}{\eta_L} = 0.392\left[\left(\frac{\sigma}{\varrho_L g}\right)^{1/2}\left(\frac{g}{\nu_L^2}\right)^{1/3}\right]^{3/4}, \tag{2.34}$$

in which σ is the surface tension and $\nu_L = \eta_L/\varrho_L$ the kinematic viscosity. On the basis of a perturbation calculation, van der Walt and Kröger [2.11] have proven that for water and several other liquids, such as the refrigerant R-12, this equation represents a quite good approximation for the onset of waves.

In liquid sodium, however, the wave formation does not set in until critical Reynolds numbers, which in round numbers are five times larger than those calculated according to Eq. (2.34). If one disregards the wave formation and calculates according to the Nusselt film condensation theory, one obtains heat transfer coefficients that are too small. One, therefore, sizes the condensator too large. Unfortunately, there is at the moment no reliable theory for the calculation of the influence of wave formation upon heat transfer.

For practical purposes, one multiplies the heat transfer coefficients α according to the Nusselt film condensation theory with a correction factor f

$$\alpha_{\text{wavy}} = f\alpha, \tag{2.35}$$

which takes into account the influence of wave formation. Because the waves lead to a smaller time-average film thickness, the heat transfer is improved by the wave formation. The correction factor above is, therefore, larger than unity, and, according to experiments by van der Walt and Kröger [2.11], is to a large extent independent upon the quantity of condensate. A good mean value is $f = 1.15$.

2.2.5 Temperature-Dependent Property Values

In Eqs. (2.11) and (2.12) of the Nusselt film condensation theory, the property values of the condensate film are assumed to be temperature-independent. This assumption is well fulfilled if the temperature drop $\vartheta_s - \vartheta_w$ is sufficiently small. Otherwise, one has to consider that the dynamic viscosity, the thermal conductivity, and, to a lesser extent, also the density of the condensate film are variable with temperature. The momentum balance

$$\frac{\partial}{\partial y}\left[\eta_L(\vartheta)\frac{\partial w}{\partial y}\right] = -[\varrho_L(\vartheta) - \varrho_G]g, \tag{2.36}$$

then takes the place of Eq. (2.17), and the temperature profile in the condensate film results from the energy balance

$$\frac{\partial}{\partial y}\left[\lambda_L(\vartheta)\frac{\partial \vartheta}{\partial y}\right] = 0. \tag{2.37}$$

These equations are to be solved with the boundary conditions

$$w(y=0) = 0, \quad \frac{\partial w(y=\delta)}{\partial y} = 0,$$

$$\vartheta(y=0) = \vartheta_w \quad \text{and} \quad \vartheta(y=\delta) = \vartheta_s.$$

From the velocity profile, one obtains the condensate mass flow rate

$$\dot{M} = \int_0^\delta \varrho_L(\vartheta) w b \,\mathrm{d}y$$

and from this also its dependence on the film thickness $\mathrm{d}\dot{M}/\mathrm{d}\delta$. For the formation of this condensate mass flow rate, a heat transfer rate $\mathrm{d}\Phi = \Delta h_v \mathrm{d}\dot{M}$ is required. This corresponds to heat conduction in the condensate film:

$$\mathrm{d}\Phi = -\lambda_w\left(\frac{\partial \vartheta}{\partial y}\right)_{y=0} b\mathrm{d}x = \Delta h_v \mathrm{d}\dot{M}. \tag{2.38}$$

The thermal conductivity is to be evaluated at the wall temperature ϑ_w. With the help of the temperature profile, which is found by the solution of Eq. (2.37), Eq. (2.38) gives the as yet unknown film thickness δ. The heat transfer coefficient follows then by definition from

$$\alpha(\vartheta_s - \vartheta_w) = -\lambda_w\left(\frac{\partial \vartheta}{\partial y}\right)_{y=0}. \tag{2.39}$$

We shall dispense here with the solution of Eqs. (2.36) and (2.37) under the given boundary conditions. One can find the pertinent details in a work by Voskresenskij [2.12].

When further assuming that the density of the condensate film is only slightly dependent upon the temperature and is very much larger than that of the vapor $\varrho_L \gg \varrho_G$, one finds that the ratio of the heat transfer coefficient α to the heat transfer coefficient α_{Nu} predicted by the Nusselt theory can be represented by

$$\frac{\alpha}{\alpha_{Nu}} = f\left(\frac{\lambda_s}{\lambda_w}; \frac{\eta_s}{\eta_w}\right). \tag{2.40}$$

The index s shows that the properties in the condensate film have to be evaluated at the saturation temperature, and the index w shows that they have to be evaluated at the wall temperature. The heat transfer coefficient according to the Nusselt film condensation theory α_{Nu} is to be evaluated with the mean property values

$$\eta_L = \tfrac{1}{2}(\eta_s + \eta_w) \quad \text{and} \quad \lambda_L = \tfrac{1}{2}(\lambda_s + \lambda_w).$$

With property ratios

$$\eta^* = \eta_s/\eta_w \quad \text{and} \quad \lambda^* = \lambda_s/\lambda_w$$

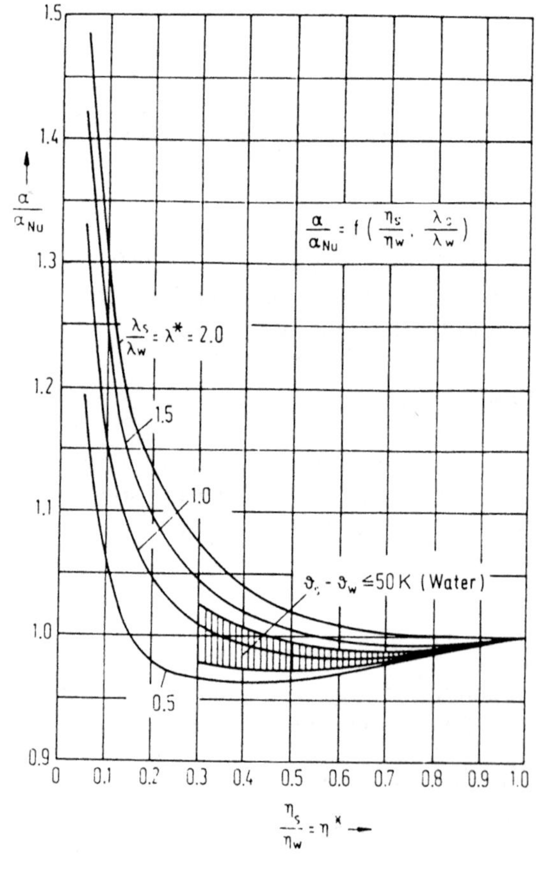

Fig. 2.6. Influence of temperature-dependent properties on heat transfer during film condensation [2.12]

Eq. (2.40) reads in its entirety

$$\frac{\alpha}{\alpha_{Nu}} = \frac{1+\eta^*}{10(1+\lambda^*)^3}\left[5+\lambda^*(14+11\lambda^*)+\frac{\lambda^*}{\eta^*}(1+4\lambda^*+5\lambda^{*2})\right]. \quad (2.41)$$

For the limiting case of $\lambda^* = \eta^* = 1$ one obtains $\alpha = \alpha_{Nu}$.

Equation (2.41) is depicted in Fig. 2.6. As one recognizes from the figure, the temperature dependence of the dynamic viscosity and of the thermal conductivity can have a substantial influence on the heat transfer, in so far as these property values change strongly with the temperature. For condensing water vapor, the value λ_s/λ_w changes at a temperature difference between saturation and wall temperature of $\vartheta_s - \vartheta_w \leqq 50$ K between 0.6 and 1.2, whereas for η_s/η_w the change is between 1 and 1.3. This region is shown as the hatched area in Fig. 2.6. As can be seen, within this area the deviations from the Nusselt theory are smaller than 3%.

2.3 Film Condensation with a Turbulent Film

The Nusselt film condensation theory presupposed a laminar film flow. Because the mass flow rate of condensate increases downstream, the Reynolds number based on the film thickness also increases steadily. The initially smooth film becomes wavy and changes finally into a turbulent film; the heat transfer is clearly better than with a laminar film.

Grigull [2.13] first calculated approximately the heat transfer in the case of turbulent film condensation by applying the Prandtl analogy for tube flow to the turbulent condensate film. Here, in addition to the values for the laminar film condensation, there also appears the Prandtl number as a new parameter. The results cannot be compactly depicted. In order to arrive at a clear representation, we shall first define the Reynolds number of the condensate film

$$Re = \bar{w}\delta/\nu_L = \bar{w}\delta\varrho_L b/(\nu_L\varrho_L b) = \dot{M}/(\eta_L b). \quad (2.42)$$

Within the framework of the Nusselt film condensation theory, this can be changed with the help of the mass flow rate of the condensate, Eq. (2.8),

$$\dot{M} = \bar{w}\varrho_L b\delta = \frac{\varrho_L^2 g b}{3\eta_L}\delta^3$$

into

$$Re = \frac{\dot{M}/b}{\eta_L} = \frac{\varrho_L^2 g}{3\eta_L^2}\delta^3 \quad \text{or} \quad (3Re)^{1/3} = \left(\frac{g}{\nu_L^2}\right)^{1/3}\delta.$$

If one here eliminates additionally the film thickness δ with the help of $\alpha = \lambda_L/\delta$, one obtains as another relation for the heat transfer equation for the Nusselt film condensation theory

$$Nu = \frac{\alpha(\nu_L^2/g)^{1/3}}{\lambda_L} = (3Re)^{-1/3}, \quad (2.43)$$

in which the Reynolds number is given by Eq. (2.42). In the case of a turbulent condensate film, there is now, as Grigull showed,

$$Nu = f(Re, Pr).$$

Figure 2.7 shows the results of an analytical solution of the differential equations for turbulent condensate film on a vertical tube [2.14]. Curve A represents the results of the Nusselt theory according to Eq. (2.43). Because of the mass balance $\dot{M} = \bar{w} d\pi\delta\varrho_L$ and with $\eta_L = \nu_L \varrho_L$, the Reynolds number can be written $Re = \dot{M}/(d\pi\eta_L)$ for the condensate film. Because the tube diameter d, which is the basis for the calculations of Fig. 2.7, was assumed to be much larger than the film thickness, the curvature of the condensate film has no influence on the heat transfer. The results are, therefore, also valid for condensation on a vertical smooth plate with the Reynolds number defined according to Eq. (2.42). In the transition zone between laminar and turbulent condensation, the condensate film is wavy. In Fig. 2.7 this transition range according to the results of Henstock and Hanratty [2.15] is drawn in with dashes.

One finds in the literature various statements concerning the value of the Reynolds number for the transition from laminar to turbulent flow. As Fig. 2.7 shows, one cannot clearly establish such a critical Reynolds number, as has been tried on occasion. Instead there is attached to the laminar range a transition region involving a wavy film, in which the Nusselt number deviates from the values of the Nusselt theory. This region begins at very small Reynolds numbers Re_u, if the Prandtl number is sufficiently large, whereas at small Prandtl numbers the laminar flow changes into a turbulent flow after a short transition region. The critical Reynolds number depends, therefore, upon the Prandtl number and can be far below a critical Reynolds number of approximately 400, at which the transition region ends.

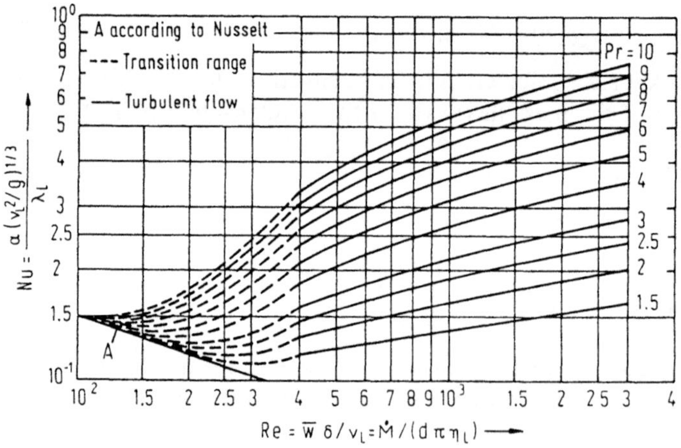

Fig. 2.7. Local Nusselt number dependence on the Reynolds number, according to [2.14], during condensation on a vertical tube or on a vertical flat wall

If one permits a deviation of 1% from the values of the Nusselt theory, then the theory will be valid up to a Reynolds number

$$Re_u = 256 Pr^{-0.47} \tag{2.44}$$

for $1 \leqq Pr \leqq 10$. In addition, for the application of the Nusselt theory, the shear stress exerted by the vapor must be sufficiently small. In the turbulent region, for Reynolds numbers greater than $Re_{cr} = 400$, the Nusselt numbers reach values far above those of the Nusselt theory.

In the transition region, the Nusselt number at first decreases with the Reynolds number and increases then after a minimum. The increase of film thickness with increasing Reynolds number causes at first a decrease of the heat transfer, but then the influence of the turbulence, increasing with the Reynolds number, prevails and leads to the improvement of the heat transfer.

One must notice that local Nusselt numbers are entered in Fig. 2.7. In a condenser, the Reynolds number based on the film thickness varies from zero at the beginning of the condensation up to a final value. In between, there can be in the transition region conditions with poor heat transfer, so that the mean coefficient of heat transfer can be smaller than the local one at the end of the condensing surface. Figure 2.8 shows the mean heat transfer coefficients associated with Fig. 2.7 according to

$$\bar{\alpha}(\vartheta_s - \vartheta_w) = \Delta h_v \dot{M}/A$$

whereby A is the pipe surface and $\dot{M}$ the condensate mass flow rate at the lower end of the tube.

As one recognizes from this, in order to achieve a good heat transfer at small Prandtl numbers, one should operate a condenser as much as possible in the region of the Nusselt theory, i.e., at small Reynolds numbers. This can be achieved by the utilization of short tubes. If, on the other hand, the Prandtl number is large, then good heat transfer is predominantly in the turbulent region, and can be achieved by utilization of sufficiently long tubes.

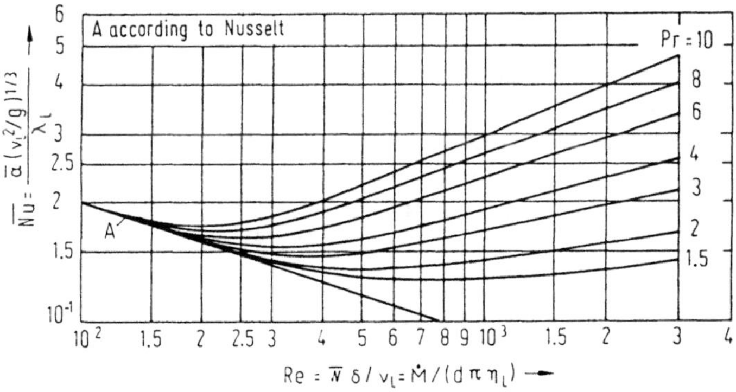

Fig. 2.8. Average Nusselt number dependence on the Reynolds number, according to [2.14], during condensation on a vertical tube or on a vertical flat wall

2.4 Practical Equations

2.4.1 Single Tubes

As already explained, the equations of the Nusselt film condensation theory are valid for laminar film condensation up to a Reynolds number

$$Re_u = 256 \cdot Pr^{-0.47} \quad \text{with} \quad 1 \leqq Pr \leqq 10,$$

if the actual heat transfer coefficients are not to deviate more than 1% from those of the Nusselt film condensation theory. Below this transition Reynolds number, the local heat transfer coefficient for vertical walls and tubes is calculated from Eq. (2.11); Eq. (2.13) applies for inclined walls or tubes. In Sect. 2.2, one will find corresponding equations for consideration of further deviations from the Nusselt theory.

At sufficiently large Reynolds numbers $Re \geqq 400$, the film is always turbulent. Starting from a solution of the momentum and energy equation for turbulent flow, Isashenko [2.16] developed for this region simple equations, which reproduce well the data known from the literature. According to this work, the local Nusselt number can be represented by

$$Nu = \frac{\alpha (\nu_L^2/g)^{1/3}}{\lambda_L} = 0.0325\, Re^{1/4}\, Pr^{1/2} \tag{2.45}$$

with the Reynolds number $Re = (\dot{M}/b)/\eta_L = \Gamma/\eta_L$ and the Prandl number Pr of the condensate. The equation is valid in the region $1 \leqq Pr \leqq 25$ and $400 \leqq Re \leqq 7 \cdot 10^5$. The Reynolds number of the condensate is obtained from

$$Re = \left[89 + 0.024 Pr^{1/2} \left(\frac{Pr}{Pr_w} \right)^{1/4} (Z - 2300) \right]^{4/3} \tag{2.46}$$

with

$$Z = \frac{c_{pL}(\vartheta_s - \vartheta_w)}{\Delta h_v} \frac{1}{Pr} \frac{x}{(\nu_L^2/g)^{1/3}}. \tag{2.46a}$$

In these equations, all the property values are to be evaluated at the saturation temperature, except for the Prandtl number Pr_w, which must be evaluated at the wall temperature. According to investigations of Labunzov [2.17], measured heat transfer coefficients deviated from the calculated ones by at most ± 12%.

Apparently without any knowledge of the work of Isashenko, Blangetti [2.18] derived an equation which deviates from Eq. (2.45). The local heat transfer coefficient is calculated according to

$$Nu = \frac{\alpha (\nu_L^2/g)^{1/3}}{\lambda_L} = 8.663 \cdot 10^{-3}\, Re^{0.382}\, Pr^{0.569}. \tag{2.47}$$

For the derivation of this equation, temperature-independent property values were assumed. In this equation, there still occurs the unknown condensate mass flow rate in the Reynolds number $Re = (\dot{M}/b)/\eta_L = \Gamma/\eta_L$. In order to calculate it, one must satisfy the energy balance at an arbitrary point x of the condensate film

$$\alpha(\vartheta_s - \vartheta_w) b \Delta x = \frac{d\dot{M}}{dx} \Delta x \Delta h_v \quad \text{or} \quad \alpha(\vartheta_s - \vartheta_w) = \frac{d\Gamma}{dx} \Delta h_v . \tag{2.48}$$

After the insertion of the heat transfer coefficient α from Eq. (2.47) into this equation, one obtains a common differential equation for the condensate mass flow rate per unit width $\Gamma(x)$. In order to integrate this equation, one must know an initial value $\Gamma_1(x_1)$. For this purpose, it will be expedient to assume that turbulent film condensation occurs only above a Reynolds number $Re = \Gamma/\eta_L$ of 400, so that therefore a lower limit of integration is $\Gamma_1 = 400\eta_L$. In the case of this Reynolds number, as Labunzov [2.17] found through the analysis of data, the value of the parameter defined by Eq. (2.46a) is

$$Z = Z_1 = 2300.$$

Thereby, the position x_1 for the initial value of the integration is also fixed. The solution of the differential equation resulting from Eq. (2.48) yields then as the defining equation for the Reynolds number $Re = \Gamma/\eta_L$

$$Re = [40.56 + 5.354 \cdot 10^{-3} Pr^{0.569} (Z - 2300)]^{1.618} . \tag{2.49}$$

Together with Eq. (2.47) of Blangetti, one thus has an equation from which is obtained in a simple manner the heat transfer coefficient during turbulent film condensation.

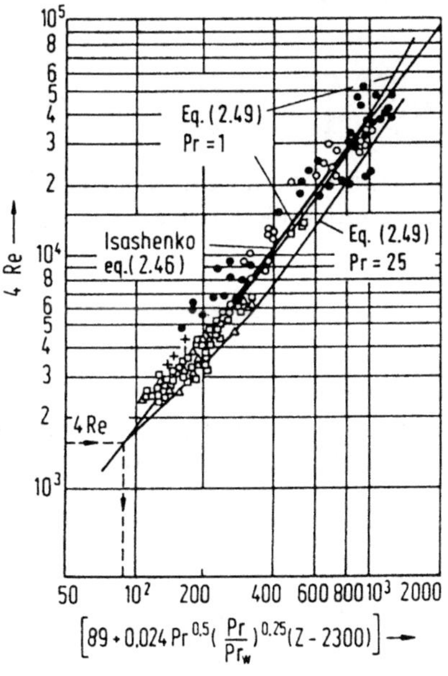

Fig. 2.9. Heat transfer during turbulent film condensation, comparing Eq. (2.46) of Isashenko [2.16] and Eq. (2.49) with data. Δ and + stand for measurements with water, $h = 3.66$ m, $Pr \approx 5.0$; □ water, $h = 6.1$ m, $Pr = 1.5$ to 2.5; ○ water, $h = 2.9$ m, $Pr = 1.5$ to 2.5; (●) diphenyl, $(h) = 3.65$ m, $Pr \approx 5.0$

Figure 2.9 shows a comparison of Eq. (2.46) of Isashenko with Eq. (2.49). In it are shown the test data with which Isashenko compared his correlation. As one sees, the agreement with the measurements is satisfactory. Equation (2.49) of Blangetti yields too low Reynolds numbers and thus also too low heat transfer coefficients, solely in the region of small values of the Prandtl number.

Thus, there are now available equations for the calculation of the heat transfer with laminar film condensation and with turbulent film condensation. For calculation of the heat transfer in the transition region between laminar and turbulent film condensation, empirical interpolation formulas have proven successful. One such formula is

$$\alpha = \sqrt[4]{(f\alpha_1)^4 + \alpha_t^4}. \tag{2.50}$$

The factor f takes into account the waviness of the laminar condensate film $f \approx 1.5$; α_1 is the heat transfer coefficient in the case of laminar film condensation from the Nusselt film condensation theory and α_t that in the case of turbulent film condensation, as results, for example, from Eq. (2.45) together with Eq. (2.46).

2.4.2 Tube Bundles

According to Nusselt, the mean heat transfer coefficient in the case of condensation on a single horizontal tube is given by [see Eq. (2.14)]:

$$\bar{\alpha}_{\text{hor}} = 0.728\left(\frac{\varrho_L^2 g \Delta h_v \lambda_L^3}{\eta_L(\vartheta_s - \vartheta_w)}\frac{1}{d}\right)^{1/4}. \tag{2.51}$$

The mean heat transfer coefficient $\bar{\alpha}_n$ of n tubes lying above one another was described by Chen [see Eq. (2.16)]

$$\bar{\alpha}_n = \left[1 + 0.2\,\frac{c_{pL}(\vartheta_s - \vartheta_w)}{\Delta h_v}(n-1)\right] n^{-1/4}\bar{\alpha}_1, \tag{2.52}$$

in which $\bar{\alpha}_1 = \bar{\alpha}_{\text{hor}}$ according to Nusselt. As an approximation, one also frequently uses in practice the empirical equations recommended by Kern [2.19]

$$\bar{\alpha}_n/\bar{\alpha}_1 = n^{-1/6} \tag{2.53}$$

and

$$\bar{\alpha}_n^*/\bar{\alpha}_1 = n^{5/6} - (n-1)^{-5/6} \quad \text{for} \quad n \geqq 10. \tag{2.53a}$$

$\bar{\alpha}_n$ is here the mean heat transfer coefficient of n tubes in a vertical row and $\bar{\alpha}_n^*$ the mean heat transfer coefficient of the n-th tube. Equation (2.53) is obtained by $\bar{\alpha}_n = \frac{1}{n}\sum_{k=1}^{n}\bar{\alpha}_k^*$, where $\bar{\alpha}_k^*$ is the mean heat transfer coefficient of a tube in the k-th row, which is obtained from Eq. (2.53a) for $n = k$.

Figure 2.10 shows $\bar{\alpha}_n/\bar{\alpha}_1$ as a function of n. Included in the figure are the values according to Eq. (2.53) of Kern and also experimental values for smaller

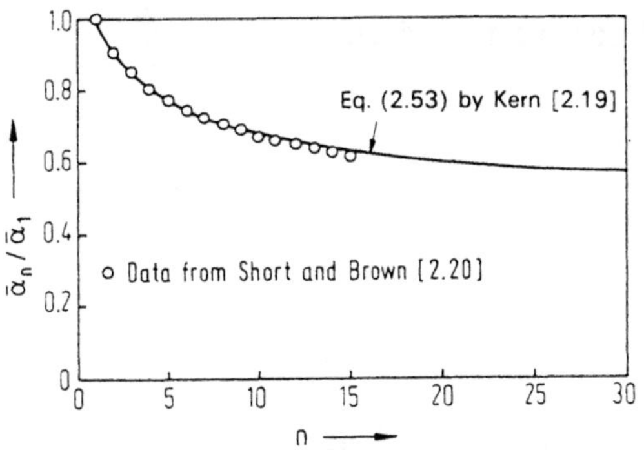

Fig. 2.10. Relative average heat transfer coefficient $\bar{\alpha}_n/\bar{\alpha}_1$ for n condenser tubes lying one above the other. $\bar{\alpha}_1$ is the average heat transfer coefficient of the top tube

values of n, which were calculated from the data of Short and Brown [2.20]. These equations are valid, however, only as long as the vapor velocity is so low that the shear stress exerted upon the liquid film is insignificant in comparison with the influence of gravity.

In tube bundle condensers, there are, however, frequently high vapor velocities around the pipes. Theoretical investigations of the heat transfer at very high vapor velocities, in which the influence of the shear stress was predominant and the vapor flowed over horizontal pipes, could be expressed by the simple Eq. [2.21]

$$Nu = \frac{\bar{\alpha} d}{\lambda_L} = 0.9\left(\frac{w_G d}{\nu_L}\right)^{1/2} \tag{2.54}$$

where w_G is the vapor velocity. This is ascertained from $w_G = w_{G0}/\varepsilon$ where w_{G0} is the vapor velocity, that would occur if no tubes were present. The value ε is the porosity. By this is meant the volume free of tubes relative to the actual volume available for vapor flow. In the commonly occurring range of vapor velocities, both the shear stress and gravity have an effect on the condensate film. The mean heat transfer coefficient on a single tube can be obtained here by superposition of the heat transfer coefficient $\bar{\alpha}_{hor} = \bar{\alpha}_{Nu}$ according to Nusselt, Eq. (2.51), and $\bar{\alpha} = \bar{\alpha}_\tau$ according to Eq. (2.54) [2.21]:

$$\bar{\alpha} = \left[\frac{\bar{\alpha}_\tau^2}{2} + \left(\frac{\bar{\alpha}_\tau^4}{4} + \bar{\alpha}_{Nu}^4\right)^{1/2}\right]^{1/2}. \tag{2.55}$$

This equation holds for the mean heat transfer coefficient on a single tube and does not yet include the influence of the condensate raining down from the tubes situated above. For practical purposes, this influence is taken into account by means of Eqs. (2.53) and (2.53a) of Kern, or by Eq. (2.52) of Chen, by replacing $\bar{\alpha}_1$ with the appropriate value $\bar{\alpha}_{hor}$ according to Eq. (2.51).

In condensers, the vapor-side pressure drop is often considerable. This leads to a decrease in the saturation temperature and also to a decrease of the vapor velocity along the path of flow. The heat transfer must, therefore, be calculated in stages.

3 Drop Condensation of Stagnant Vapors

If the condensate does not completely wet the wall, then, as was explained in Sect. 1.2, no cohesive condensate film is formed. Rather there are formed single drops of liquid. Heat transfer coefficients with drop condensation are distinctly larger than with film condensation. In the case of the condensation of water vapor, heat transfer coefficients larger by a factor of four to eight have been measured. To be sure, it has been shown that all the investigated substances, especially water, which condense on the heating surfaces commonly used, wet completely, as long as the material of the heating surface and the liquid have not been polluted. This also corresponds to the experience that a water film is a good sign of well cleaned laboratory apparatus.

The form of condensation is influenced chiefly by foreign substances adsorbed on the solid surface. They can lead locally to finite values of wetting angle and thus cause incomplete wetting. As foreign substances, there come into question organic materials added to water as antiwetting agents (injected materials, promoters). Occasionally, such foreign substances are unintentionally present in the vapor, for example, the lubricating oil of a boiler feed pump. Slight amounts still soluable in the condensate are sufficient to produce drop condensation. Also the addition of waxlike substances has been suggested [3.1]. For the maintenance of a stable drop condensation, however, a constant or periodic addition of the promoter is necessary, because it can be removed from the surface in the course of time by washing it off or by its being dissolved in the condensate. The formation of drops is further influenced by wall roughness, wall height, heat flux, and temperature difference.

Thin gold layers applied by electrolysis and noble metal platings of gold, rhodium, palladium, or platinum have proven to be effective as promoters [3.2]. As experiments [3.3] have shown, these must have a definite minimal thickness, which in the case of gold layers amounts in round numbers to 0.2 μm, in order to maintain a lasting drop condensation. Although this layer thickness is about half as large as the wave-length of the visible light, the quantity of gold needed is by no means inconsiderable. It amounts to 3.8 g per m^2 of heating surface. This means that even at low gold prices of roughly 15,000 US-$ per kilogram, the costs for the material for plating alone will amount to roughly 57 US-$ per m^2 of heating surface. On the other hand, there is a five to seven times greater heat transfer in the case of the condensation of water vapor [3.3]. Nevertheless, in the face of the high expense, gold plating will be used only in special cases. Otherwise, film condensation will be accepted.

Even if the type of condensation is not precisely known, one will assume film condensation for the calculation of condenser surfaces, because then the condenser surface is sufficiently dimensioned.

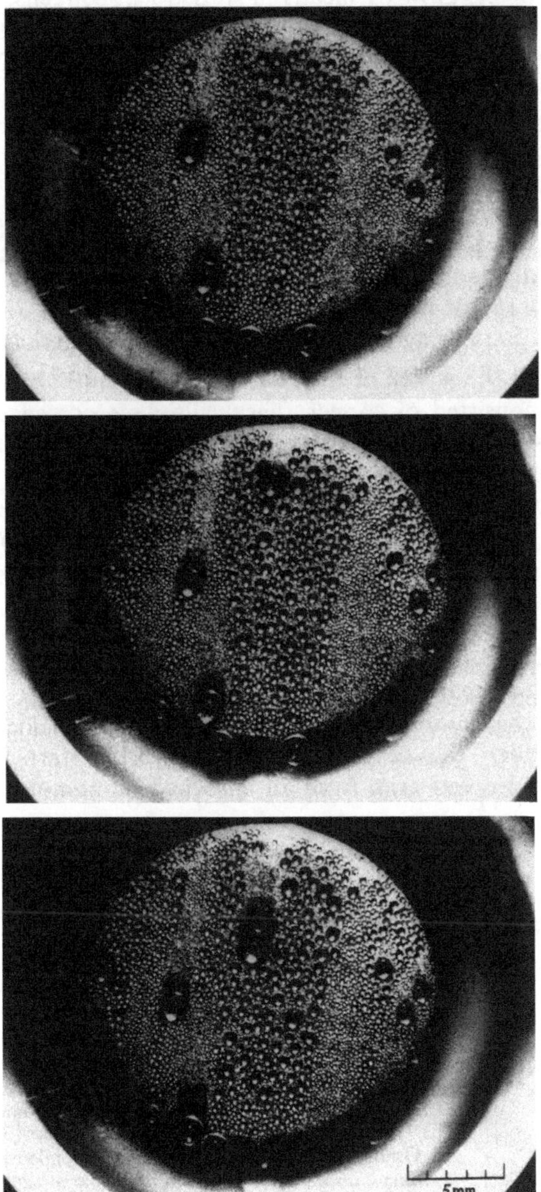

a
b
c

Fig. 3.1 a–c. Drop condensation on a vertical condensation surface, according to Krischer and Grigull [3.4]. **a** beginning of condensation; **b** condensation pattern at 16 s; **c** condensation pattern at 32 s. The tracks of the rolling or sliding drops can be clearly recognized. Above, in Fig. **b**, a drop just departing. $q = 0.12\ \mathrm{W/cm^2}$, $\Delta\vartheta = 0.05\ \mathrm{K}$. Diameter of condensation surface: 18 mm

Experiments for drop condensation are difficult because, for the determination of heat transfer coefficients, temperature differences of 1 K and less must be measured; furthermore, the wall temperature is variable with time and place. The cost of the testing technique for obtaining precise results is, therefore, considerable.

As an example for drop condensation Fig. 3.1 shows photos taken on a vertical surface. The little droplets grow by means of subsequent condensation out of the vapor space or by coalescing. As soon as a definite drop size is reached, the drop slides off and carries along the drops lying in its path. New drops are immediately formed behind the one that has departed. They are formed predominantly out of the water film left behind and at the scratches in the surface. According to experiments of Krischer and Grigull [3.4], the nucleation site density depends chiefly upon the subcooling of the heating surface. Experiments with water vapor condensing on copper and brass surfaces yielded upon subcooling of 0.1 K nucleation site densities of approximately $3 \cdot 10^3$ nuclei per mm^2 of heating surface, whereas at a subcooling of 0.4 K the nucleation site density lay at $15 \cdot 10^3$ nuclei per mm^2.

Figure 3.2 shows results concerning heat transfer with drop condensation, which were put together from the works of the last twenty years [3.4]. Most of the condensing surfaces consisted of copper. The walls were coated with various liquids that served as promoters. There were usually very different results depending upon the promoter and the constituent materials of the condensation surface. As one recognizes, individual test results vary from one another up to a factor of 30 in the heat flux. The curves 2, 5, and 6 of Wenzel [3.6], Le Fèvre and Rose [3.9], and Tanner et al. [3.10] show a weak dependence on the temperature rise and only slight deviations among one another.

In Fig. 3.3, the results from [3.10] and [3.12] together with those of Krischer and Grigull [3.4] are sketched separately.

Several *theories on drop condensation* have been built up for the calculation of heat transfer. One of the oldest, that developed by Eucken [3.13], starts with the conception that the first droplets arise from an adsorbed monomolecular

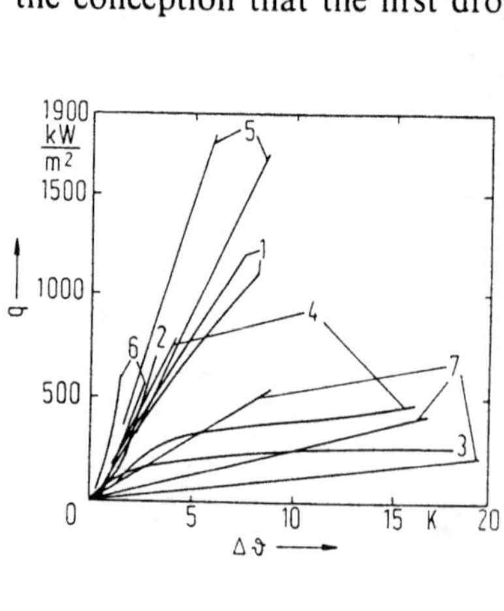

Fig. 3.2. Experimental results on drop condensation of water at approximately 1 bar pressure, according to [3.4]. *1* Hampson and Özisik 1952, curves for two different promoters [3.5]; *2* Wenzel 1957 [3.6]; *3* Welch and Westwater 1961 [3.7]; *4* Kast 1965, chrome-plated (upper curve) and plain copper surface [3.8]; *5* LeFèvre and Rose 1965, different promoters [3.9]; *6* Tanner et al. 1968, different promoters [3.10]; *7* Griffith and Lee 1967, gold-plated condensation surfaces: upper curve copper, middle curve zinc, and lower curve steel [3.11].

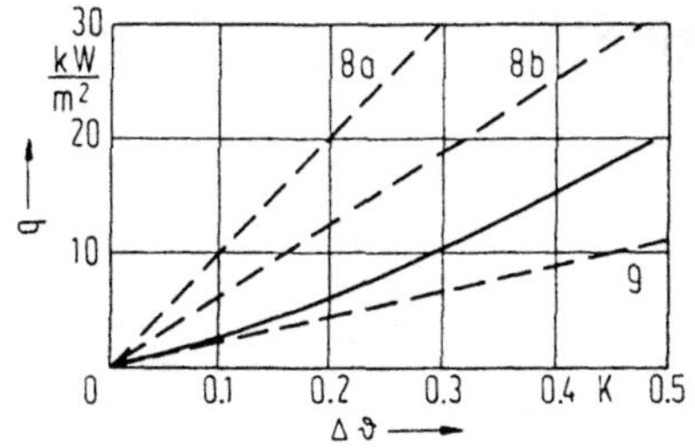

Fig. 3.3. Experimental results at approximately 0.03 bar pressure, according to [3.4]. Dashed curve: interpolated values *8* from data of Tanner et al. [3.10] a) promoter "Montan wax" and b) promoter "dioctadecyldisulfide"; *9* data of Brown and Thomas [3.12]

condensate layer, that the formation of the condensate layer is favored by nuclei, and that new condensate constantly flows to the droplets by means of surface diffusion primarily at the drop edge. This theory was later taken up by other authors and further developed [3.14, 3.15].

Other theories presuppose that between the drops there exists a thin, unstable water film, which bursts and disappears into the drop after reaching a critical thickness of several μm [3.16]. This conception has been disproved in the meantime by experiments [3.17, 3.18].

The conception that can be best brought into harmony with experiments is the idea that initially tiny droplets form at sites of nuclei, at depressions in the condensation surface, or at liquid remnants. The rate of their growth is determined by the resistance to heat conduction in the drops and also partially by the thermal resistance at the phase interface to the vapor. The rate of growth is thus dependent upon the specific drop radius and the driving temperature difference, a fact that has been proven experimentally [3.4].

A complete theory has not yet been successfully developed because the nucleus density is unknown and because one can predict only with difficulty the radius of the drops sliding off, since the radius is dependent upon the purity and smoothness of the condensation surface and upon the surface tension.

4 Condensation of Flowing Vapors

4.1 Laminar Film Condensation

Nusselt already extended the film condensation theory and took into consideration the fact that the vapor flowing along the condensate film influences the velocity in the condensate. The boundary condition to Eq. (2.5) then no longer reads $\partial w/\partial y = 0$ at $y = \delta$, but rather the velocity has a finite gradient at the free film surface corresponding to the shear stress exerted by the flowing vapor. Accordingly, in Eq. (2.5) for the velocity profile

$$w = -\frac{\varrho_L g}{2\eta_L} y^2 + c_1 y + c_0$$

the coefficients c_1 and c_0 are to be so determined that the boundary conditions

$$w(y=0) = 0 \quad \text{and} \quad \eta_L(\partial w/\partial y)_{y=\delta} = \pm \tau_\delta \tag{4.1}$$

are fulfilled, whereby the positive sign holds for downward flowing vapor and the negative one for vapor flowing upward. For the calculation of the shear stress at the phase interface, one assumes equality of pressure and friction forces in the vapor space. Let the pressure drop along the flow path dx be dp. Then

$$\tau_\delta d\pi = \frac{d^2\pi}{4}\frac{dp}{dx} \tag{4.2}$$

holds for the tube flow. On the other hand, there holds for the pressure drop

$$\frac{dp}{dx} = \zeta \frac{\varrho_G w_G^2}{d}. \tag{4.3}$$

Herewith we obtain

$$\tau_\delta = \zeta \frac{\varrho_G w_G^2}{4}. \tag{4.4}$$

Together with Eq. (2.5) one obtains the velocity profile

$$w = \frac{\varrho_L g}{\eta_L}\delta^2\left(\frac{y}{\delta} - \frac{y^2}{2\delta^2}\right) \pm \frac{\tau_\delta y}{\eta_L} \tag{4.5}$$

with τ_δ according to Eq. (4.4). One obtains the mean velocity as

$$\bar{w} = \frac{1}{\delta}\int_0^\delta w\mathrm{d}y = \frac{\varrho_L g}{3\eta_L}\delta^2 \pm \frac{\tau_\delta \delta}{2\eta_L} \tag{4.6}$$

and from this the mass flow rate of condensate, which flows off through a plane vertical to the wall

$$\dot{M} = \bar{w}\varrho_L b\delta = \frac{\varrho_L^2 g b}{3\eta_L}\delta^3 \pm \frac{\varrho_L \tau_\delta \delta^2 b}{2\eta_L}. \tag{4.7}$$

The film thickness is found, as was already shown in Sect. 2.1, with the aid of the energy balance

$$\lambda_L \frac{\vartheta_s - \vartheta_w}{\delta} b\mathrm{d}x = \Delta h_v \mathrm{d}\dot{M}$$

to be

$$\delta^4 \pm \frac{4}{3}\frac{\tau_\delta \delta^3}{\varrho_L g} = \frac{4\lambda_L \eta_L(\vartheta_s - \vartheta_w)}{\varrho_L^2 g \Delta h_v} x, \tag{4.8}$$

from which one gets the film thickness δ and then the heat flux $q = \lambda_L(\vartheta_s - \vartheta_w)/\delta$.

The shear stress τ_δ is given by Eq. (4.4). Whereas Nusselt here assumed a constant mean vapor velocity, Jakob et al. [4.1] allowed for its decrease along the cooled wall of a vertical pipe.

Figures 4.1 and 4.2 show data of Jakob et al. for saturated and superheated vapor compared with the predictions of the Nusselt theory according to Eq. (4.8) for downwards directed vapor flow. The heat flux is plotted against the difference between the saturation and wall temperature $\vartheta_s - \vartheta_w$. The negative temperature differences in Fig. 4.1 mean that the wall temperature lies above the saturation temperature, $\vartheta_w > \vartheta_s$, which means simply desuperheating of the

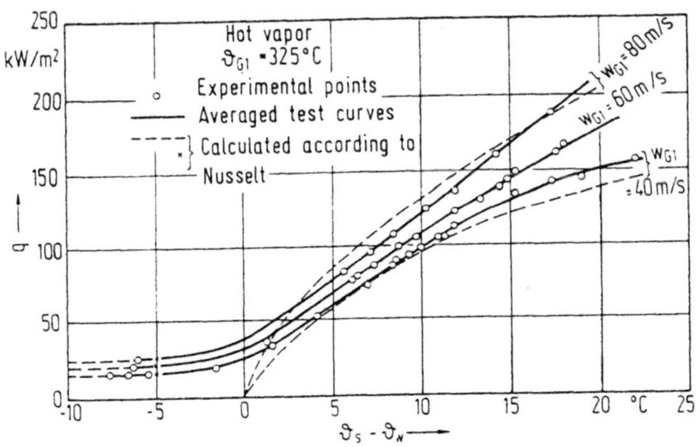

Fig. 4.1. Film condensation of superheated and saturated vapor flowing downward at 1.013 bar, according to [4.1]

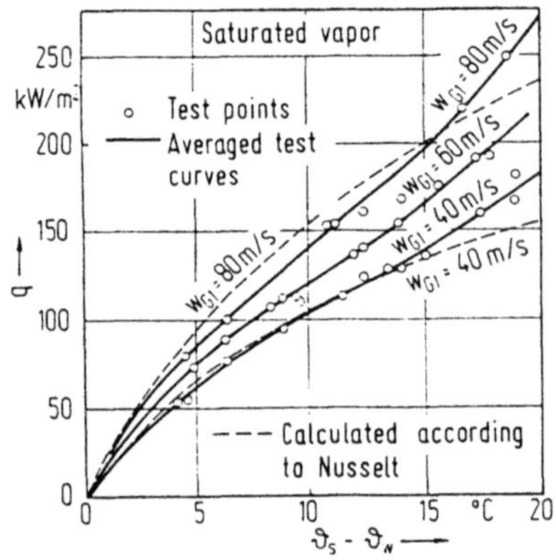

Fig. 4.2. Film condensation of saturated vapor flowing downward, according to [4.1]

vapor without condensation. The parameter w_{G1} is the entrance velocity of the vapor into the vertical tube (Fig. 4.2). If one also takes into consideration, according to Rohsenow et al. [4.2], the buoyancy effect occurring at vertical surfaces, as well as the subcooling of the condensate film and the superheating of the vapor, then one obtains, as was shown in Sect. 2.2, a somewhat different expression for the film thickness

$$\delta^4 \pm \frac{4}{3}\frac{\tau_\delta \delta^3}{(\varrho_L - \varrho_G)g} = \frac{4\lambda_L \eta_L(\vartheta_s - \vartheta_w)}{\varrho_L(\varrho_L - \varrho_G)g\Delta h^*_{vu}}x. \tag{4.9}$$

with Δh^*_{vu} according to Eq. (2.26). If the condensation surface is inclined by the angle γ toward the horizontal, then the gravitational acceleration g must be replaced by the normal component $g \sin\gamma$.

Rohsenow et al. [4.2] converted Eq. (4.9) to dimensionless form and then from the calculated film thicknesses determined the heat transfer coefficients $\alpha = \lambda_L/\delta$ and also the mean heat transfer coefficients $\bar{\alpha}$. Figure 4.3 shows as the result the average Nusselt number

$$\overline{Nu} = \frac{\bar{\alpha} l}{\lambda} \tag{4.10}$$

with the characteristic length

$$l = \left[\frac{\varrho_L \nu_L^2}{(\varrho_L - \varrho_G)g}\right]^{1/3}$$

plotted against the Reynolds number

$$Re = \frac{\bar{w}\delta}{\nu_L} = \frac{\dot{M}}{b\eta_L}, \tag{4.11}$$

if b is the width of the condensate film, which with condensation on a vertical

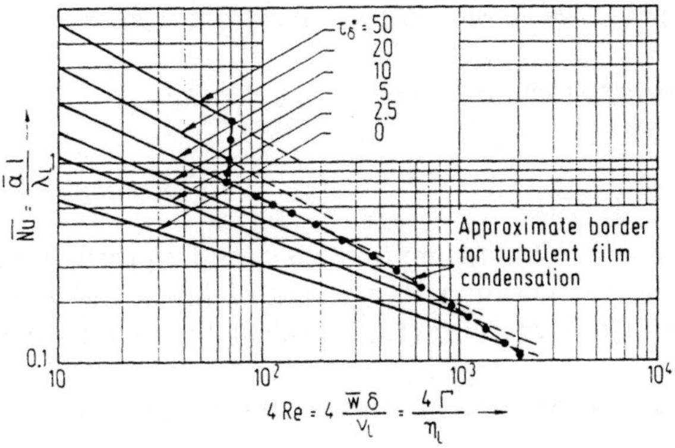

Fig. 4.3. Influence of shear stress of the vapor on laminar film condensation, according to [4.2]

tube equals $b = d\pi$. The dimensionless shear stress

$$\tau_\delta^* = \frac{\tau_\delta}{l(\varrho_L - \varrho_G)g} \tag{4.12}$$

is entered as a curve parameter. The results hold for vapor flowing downward. The dotted lines show the approximate boundary for turbulent film condensation.

4.2 Turbulent Film Condensation in Vertical Tubes

4.2.1 General Equations

The laws of conservation of mass, momentum, and energy apply to turbulent as well as to laminar film condensation. One needs, however, additional information about the mechanisms of the turbulent exchange of mass, momentum, and energy.

In the following, therefore, the general balance equations for the calculation of the heat transfer with turbulent film condensation will first be prepared and then their solution will be discussed.

4.2.1.1 The Mass and Momentum Balances

In a turbulent flow there occur as the consequence of turbulent fluctuations additional stresses

$$\tau_{xy} = -\overline{\varrho w_x' w_y'} = \tau_t,$$

whereby w_x' is the momentary velocity in the direction of the x-axis, w_y' that in the direction of the y-axis and the bar means the time average. In order to calculate the value of this turbulent stress, numerous solutions and calculation procedures have been suggested. The oldest solution dates from Boussinesq in the year 1877.

Relying on the Newton expression, he wrote

$$\tau_t = \varrho\varepsilon \frac{\partial \bar{w}_x}{\partial y}, \tag{4.13}$$

in which ε is the turbulent viscosity. This assumption has been frequently used with success for turbulent film condensation. The works of Jones and Renz [4.3] should be pointed out for other procedures for the calculation of turbulent film condensation.

The total shear stress is composed of laminar and turbulent portions

$$\tau = \varrho(\nu + \varepsilon)\frac{\partial \bar{w}_x}{\partial y}.$$

We will now consider a one-dimensional flow $w_x(y)$ and shall omit in the future the bar for the mean values. If for abbreviation one uses

$$1 + \frac{\varepsilon}{\nu} = \varepsilon^+, \tag{4.14}$$

then one obtains for the stress

$$\tau = \varrho\nu\varepsilon^+ \frac{\mathrm{d}w_x}{\mathrm{d}y}. \tag{4.15}$$

For the limiting case of pure laminar flow we have $\varepsilon^+ = 1$, and for completely turbulent flow, because $\varepsilon/\nu \gg 1$, $\varepsilon^+ = \varepsilon/\nu$.

With film flow in tubes the shear stress τ and the shear stress at the film surface τ_δ are connected with the pressure gradient $\mathrm{d}p/\mathrm{d}x$ in the tube. In keeping with Fig. 4.4, we consider the equilibrium between viscous force, pressure force, and gravitational force, which apply to an annular element of the condensate film and to the pertinent vapor volume between the tube cross sections at x

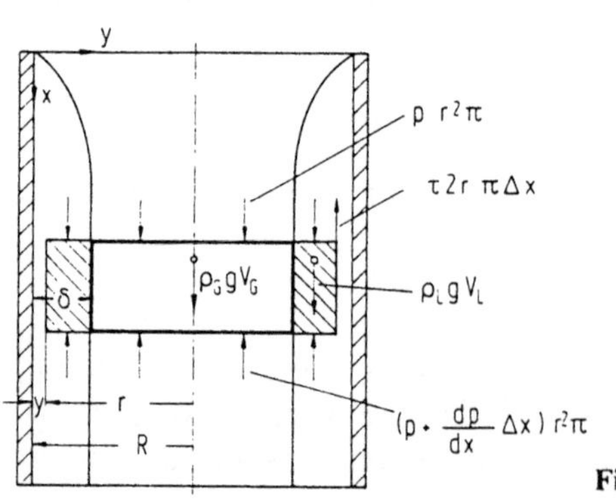

Fig. 4.4. Balance of forces on an annular element

and $x + \Delta x$. This gives

$$\tau 2r\pi\Delta x + \left(p + \frac{dp}{dx}\Delta x\right) r^2\pi = pr^2\pi + \varrho_G g V_G + \varrho_L g V_L$$

with $V_G = (R - \delta)^2 \pi \Delta x$ and $V_L = (R - y)^2 \pi \Delta x - V_G$.

With this is obtained

$$\tau = \frac{R-y}{2}\left(g\varrho_L - \frac{dp}{dx}\right) - g(\varrho_L - \varrho_G)\frac{(R-\delta)^2}{2(R-y)}. \tag{4.16}$$

If one stipulates the same balance of forces only for the vapor volume in heavy outline in Fig. 4.4, then there follows from this for the stress τ_δ at the film surface the expression

$$\tau_\delta 2(R-\delta)\pi\Delta x + \left(p + \frac{dp}{dx}\Delta x\right)(R-\delta)^2\pi = p(R-\delta)^2\pi + \varrho_G g V_G$$

or

$$\tau_\delta = \frac{R-\delta}{2}\left(g\varrho_G - \frac{dp}{dx}\right). \tag{4.17}$$

which also results from Eq. (4.16), if one sets there $y = \delta$. Together with Eq. (4.16) there follows from Eq. (4.17)

$$\tau = \frac{R-y}{R-\delta}\tau_\delta + g(\varrho_L - \varrho_G)\frac{2R(\delta - y) - \delta^2 + y^2}{2(R-y)}. \tag{4.18}$$

If the film is very thin in comparison to the tube radius, $\delta \ll R$, then Eqs. (4.17) and (4.18) are simplified, if one considers that $y < \delta \ll R$, to

$$\tau_\delta = \frac{R}{2}\left(g\varrho_G - \frac{dp}{dx}\right) \tag{4.17a}$$

and

$$\tau = \tau_\delta + g(\varrho_L - \varrho_G)(\delta - y). \tag{4.18a}$$

The wall shear stress resulting from Eq. (4.18a), valid for very thin films, is

$$\tau_w = \tau_\delta + g(\varrho_L - \varrho_G)\delta, \tag{4.18b}$$

or with Eq. (4.18a)

$$\tau = \tau_w - g(\varrho_L - \varrho_G)y, \tag{4.18c}$$

As Eq. (4.17a) shows, one must know only the pressure drop in order to calculate the stress at the film surface in the case of thin films, whereas for the determination of the wall shear stress according to Eq. (4.18b), the film thickness must also be known.

In the following it will now be assumed that the film is thin in comparison to the tube radius. The relation for the shear stress, Eq. (4.15), can then be written with Eq. (4.18c) as

$$\tau = \varrho_L \nu_L \varepsilon^+ \frac{dw_x}{dy} = \tau_w - g(\varrho_L - \varrho_G) y. \tag{4.19}$$

By integration the velocity profile is obtained:

$$w_x = \int_0^y \frac{1}{\varepsilon^+} \left[\frac{\tau_w}{\varrho_L \nu_L} - \frac{g}{\nu_L} \left(1 - \frac{\varrho_G}{\varrho_L} \right) y \right] dy. \tag{4.20}$$

For the calculation of the velocity profile one must, therefore, know the turbulent viscosity ε, or the value ε^+, and also the wall shear stress, which, in turn, is dependent upon the pressure drop and the local film thickness. In contrast to the "laminar" viscosity ν, the turbulent viscosity is not a pure property, but rather is strongly dependent upon the flow rate. In the limiting case of $\varepsilon^+ = 1$ for laminar film flow, there results from Eq. (4.20), upon consideration of the connection between wall shear stress and the shear stress at the film surface, Eq. (4.18b), the earlier calculated velocity profile for laminar flow for downward flowing vapor, Eq. (4.5).

One now introduces the following abbreviations, which have proven practical in the theory of turbulent flows:

the friction velocity	$w_\tau = \sqrt{\tau_w/\varrho_L}$,
the relative wall distance	$y^+ = (y w_\tau)/\nu_L$
and the relative velocity	$w^+ = w_x/w_\tau$.

Values designated by the superscript + are dimensionless here.

With such relations, one obtains the relative velocity profile

$$w^+ = \int_0^{y^+} \frac{1}{\varepsilon^+} \left[1 - \frac{g \nu_L}{w_\tau^3} \left(1 - \frac{\varrho_G}{\varrho_L} \right) y^+ \right] dy^+. \tag{4.20a}$$

For the calculation of the film flow rate, let us consider an annular element of the surface $dA = |2r\pi dr| = 2R\pi dy$. The mass flow rate through this is $d\dot{M} = \varrho_L w_x dA = \varrho_L w_x 2R\pi dy$. The mass flow rate at a film cross section then is found by integration to be

$$\dot{M} = 2R\pi \int_0^\delta \varrho_L w_x dy = 2R\pi \Gamma,$$

where Γ is the condensate flow rate per unit of circumference. It is, therefore,

$$\Gamma = \int_0^\delta \varrho_L \left\{ \int_0^y \frac{1}{\varepsilon^+} \left[\frac{\tau_w}{\varrho_L \nu_L} - \frac{g}{\nu_L} \left(1 - \frac{\varrho_G}{\varrho_L} \right) y \right] dy \right\} dy \tag{4.21}$$

or with the previous abbreviations

$$\Gamma = \int_0^\delta \varrho_L w_x dy = \int_0^\delta \varrho_L w_\tau w^+ dy = \int_0^{\delta^+} \varrho_L \nu_L w^+ dy^+. \tag{4.22}$$

If one assumes a constant density of the liquid, then there is

$$\frac{\Gamma}{\varrho_L} = \int_0^{\delta} w_x \mathrm{d}y = \bar{w}_x \delta,$$

where $\bar{w}_x$ is the mean velocity across the film, and the Reynolds number of the film is

$$\frac{\Gamma}{\varrho_L \nu_L} = \frac{\bar{w}_x \delta}{\nu_L} = Re. \tag{4.23}$$

Upon the assumption of a constant dynamic viscosity $\varrho_L \nu_L$, with Eq. (4.22),

$$Re = \frac{\Gamma}{\varrho_L \nu_L} = \int_0^{\delta^+} w^+ \mathrm{d}y^+. \tag{4.24}$$

The velocity profile $w^+(y^+)$ is given by Eq. (4.20a). In addition, we write there for the term

$$\frac{g\nu_L}{w_\tau^3} = \left(\frac{\nu_L}{w_\tau(\nu_L^2/g)^{1/3}}\right)^3$$

and to abbreviate, introduce the Reynolds number Re_τ formed with the friction velocity w_τ and the "length" $(\nu_L^2/g)^{1/3}$:

$$Re_\tau = \frac{w_\tau(\nu_L^2/g)^{1/3}}{\nu_L}. \tag{4.25}$$

Thus, after the insertion of the velocity w^+ from Eq. (4.20a), Eq. (4.24) reads

$$Re = \frac{\Gamma}{\varrho_L \nu_L} = \int_0^{\delta^+} \left\{ \int_0^{y^+} \frac{1}{\varepsilon^+} \left[1 - \frac{1}{Re_\tau^3} \left(1 - \frac{\varrho_G}{\varrho_L} \right) y^+ \right] \mathrm{d}y^+ \right\} \mathrm{d}y^+. \tag{4.26}$$

With a known film thickness, one can determine from this equation the pertinent film flow rate Γ or the Reynolds number of the liquid, or, conversely, one can calculate the film thickness for a given Reynolds number.

The calculation proceeds here in the following manner: From the pressure drop $\mathrm{d}p/\mathrm{d}x$ of the vapor, one obtains according to Eq. (4.17a) the shear stress τ_δ at the film surface. This together with the film thickness δ gives the wall shear stress according to Eq. (4.18b). Thereby $w_\tau = \sqrt{\tau_w/\varrho_L}$ and the Reynolds number according to Eq. (4.25) are established. With an assumption for ε^+ one can solve the integral of Eq. (4.26) and calculate the Reynolds number of the film. If one wants to determine the film thickness for a given Reynolds number, one first estimates the film thickness, then determines τ_w and Re_τ, and finally checks by integration of Eq. (4.26) whether the estimated value of the film thickness is correct. The estimated value is to be iteratively corrected until it coincides with the calculated value.

4.2.1.2 The Energy Balance

The Eq. (4.26) resulting from the consideration of momentum balance contains two unknowns, namely the film thickness δ and the associated flow rate Γ. The energy balance is required to complete the formulation. Upon assumption that the convective energy transport is negligibly small in comparison with the heat conduction, the energy equation reads

$$\frac{\partial q}{\partial y} = 0 \tag{4.27}$$

with

$$q = -(\lambda + \lambda_t)\frac{\mathrm{d}\vartheta}{\mathrm{d}y}, \tag{4.28}$$

where λ_t is the "turbulent thermal conductivity". Accordingly, one designates

$$\varepsilon_q = \frac{\lambda_t}{\varrho c_p}$$

as the "turbulent thermal diffusivity", and, analogously to the traditional molecular Prandtl number $Pr = \nu/a$, the quantity

$$\frac{\varepsilon}{\varepsilon_q} = Pr_t$$

as the "turbulent Prandtl number". With these quantities, Eq. (4.28) reads

$$q = -\lambda_L\left(1 + \frac{Pr}{Pr_t}\frac{\varepsilon}{\nu_L}\right)\frac{\mathrm{d}\vartheta}{\mathrm{d}y}$$

for the liquid film or upon consideration of $\varepsilon^+ = 1 + \varepsilon/\nu$:

$$q = -\lambda_L\left(1 + \frac{Pr(\varepsilon^+ - 1)}{Pr_t}\right)\frac{\mathrm{d}\vartheta}{\mathrm{d}y}. \tag{4.29}$$

By definition, the heat flux is, in turn, related to the heat transfer coefficient α according to

$$q = \alpha(\vartheta_w - \vartheta_s). \tag{4.30}$$

Insertion of this equation into Eq. (4.29) and separation of the variables yields

$$\frac{\mathrm{d}\vartheta}{\vartheta_s - \vartheta_w}\frac{\lambda_L}{\alpha} = \frac{\mathrm{d}y}{1 + \dfrac{Pr(\varepsilon^+ - 1)}{Pr_t}}$$

and after integration between the wall $y = 0$ and the film surface $y = \delta$,

$$\frac{\lambda_L}{\alpha} = \int_0^\delta \frac{\mathrm{d}y}{1 + \dfrac{Pr(\varepsilon^+ - 1)}{Pr_t}}.$$

After the insertion of a Nusselt number formed with the characteristic length $(\nu_L^2/g)^{1/3}$, one can write for this

$$Nu = \frac{\alpha(\nu_L^2/g)^{1/3}}{\lambda_L}$$

and with the dimensionless wall distance $y^+ = (w_\tau y)/\nu_L$ and the Reynolds number $Re_\tau = w_\tau(\nu_L^2/g)^{1/3}/\nu_L$, one can also write

$$\frac{1}{Nu} = \frac{1}{Re_\tau}\int_0^{\delta^+} \frac{dy^+}{1 + \dfrac{Pr(\varepsilon^+ - 1)}{Pr_t}}. \tag{4.31}$$

If one has followed the above procedure and determined the wall shear stress from the pressure drop and the film thickness, this equation will yield the requisite heat transfer coefficient. Together with Eq. (4.26) one can also determine the mass flow rate Γ or the Reynolds number Re_τ and also the heat transfer coefficient α for a given film thickness and for a known pressure drop. In order to establish at which point x the interrelated triple values δ, Γ, α occur, the energy balance at the film surface is available. It reads

$$\alpha 2R\pi\Delta x(\vartheta_s - \vartheta_w) = \frac{d\dot{M}}{dx}\Delta x\Delta h_v = 2R\pi\frac{d\Gamma}{dx}\Delta x\Delta h_v$$

or

$$\alpha(\vartheta_s - \vartheta_w) = \frac{d\Gamma}{d\delta}\frac{d\delta}{dx}\Delta h_v. \tag{4.32}$$

Here on the basis of the previous calculation α and $d\Gamma/d\delta$ are known as functions of the film thickness. The derivation $d\Gamma/d\delta$ results from Eq. (4.26) by differentiation, and the heat transfer coefficient α follows from Eq. (4.31). On the basis of the above equation, one can also express $d\delta/dx$ as a function of the film thickness δ. The integration then yields $\delta(x)$, with which the local heat transfer coefficient $\alpha(x)$ is also established. Generally speaking, this calculation can be carried out only with a computer.

4.2.2 Models for the Solution of the Equations

According to the previous considerations, when the film thickness δ is given, one obtains the flow rate Γ from Eq. (4.26) and the heat transfer coefficient α from Eq. (4.31). The solution of the integrals in Eqs. (4.26) and (4.31) is, however, possible only if the turbulent diffusivity parameter $\varepsilon^+(y^+)$ and the turbulent Prandtl number $Pr_t(y^+)$ are known. The theories of turbulent flows provide these parameters. The various solutions of the above-mentioned equations found in the literature differ only in the expressions for the turbulent diffusivity and the turbulent Prandtl number. Instead of the turbulent diffusivity parameter

ε^+, the velocity profile can be stated, because both quantities are connected with one another through Eq. (4.19). This is more clearly evident if one transforms Eq. (4.19) to

$$\frac{\tau}{\tau_w} = \varepsilon^+ \frac{dw^+}{dy^+} = 1 - g(\varrho_L - \varrho_G)\frac{y}{\tau_w}. \tag{4.33}$$

Rohsenow et al. [4.2] extended their investigations to turbulent film condensation by assuming in the condensate film the universal velocity profile according to Prandtl and Nikuradse, which is known from the turbulent tube flow, and also assuming constant properties and a turbulent Prandtl number $Pr_t = 1$. They subdivided the liquid film according to the procedure used for turbulent tube flow into three areas: the laminar sublayer near the wall, a transition layer, and the outer fully turbulent layer that extends to the film surface.

In the wall-contiguous laminar sublayer $0 \leqq y^+ \leqq 5$ there is a purely laminar flow. There is, therefore, $\varepsilon^+ = 1$ and, as one can recognize from Eq. (4.33), because y is very small,

$$\frac{dw^+}{dy^+} = 1 \quad \text{or} \quad w^+ = y^+.$$

In the bordering transition layer $5 \leqq y^+ \leqq 30$, the shear stress is said to be still practically equal to the wall shear stress $\tau = \tau_w$. Thus, there follows from Eq. (4.33)

$$\varepsilon^+ \frac{dw^+}{dy^+} = 1.$$

For the velocity profile, the Prandtl-Nikuradse relation is assumed

$$w^+ = -3.05 + 5.0 \ln y^+.$$

Thus, the turbulent diffusivity parameter

$$\varepsilon^+ = y^+/5$$

is fixed.

Finally, the velocity profile of turbulent tube flow $(30 \leqq y^+ \leqq \delta^+)$

$$w^+ = 5.5 + 2.5 \ln y^+,$$

is adopted in the adjoining fully turbulent region. When considering the previous definitions $y^+ = y w_\tau / \nu_L$ and $w_\tau = \sqrt{\tau_w / \varrho_L}$, one finds from Eq. (4.33) the turbulent diffusivity parameter

$$\varepsilon^+ = \frac{y^+}{2.5}\left[1 - \frac{g\nu_L}{w_\tau^3}\left(1 - \frac{\varrho_G}{\varrho_L}\right) y^+\right].$$

With these data, the temperature profile can be calculated. By integration of

Eq. (4.29), it is found to be

$$\vartheta - \vartheta_w = -\frac{q\delta}{\lambda_L}\int_0^{y/\delta} \frac{d(y/\delta)}{1+\frac{Pr(\varepsilon^+ - 1)}{Pr_t}}.$$

Furthermore, there is

$$\vartheta_s - \vartheta_w = -\frac{q\delta}{\lambda_L}\int_0^{1} \frac{d(y/\delta)}{1+\frac{Pr(\varepsilon^+ - 1)}{Pr_t}},$$

from which one obtains the dimensionless temperature:

$$\Theta = \frac{\vartheta - \vartheta_w}{\vartheta_s - \vartheta_w} = \frac{\int_0^{y/\delta} \frac{d(y/\delta)}{1+\frac{Pr(\varepsilon^+ - 1)}{Pr_t}}}{\int_0^{1} \frac{d(y/\delta)}{1+\frac{Pr(\varepsilon^+ - 1)}{Pr_t}}}. \tag{4.34}$$

During the integration, one must consider that the turbulent diffusivity parameter ε^+ changes in states. Accordingly, the integrals must be evaluated stepwise.

As a result of such a calculation, one obtains temperature profiles, as they are depicted in Fig. 4.5 for water at a presssure of 1 bar and for stagnant vapor ($\tau_\delta = 0$). The case of the laminar film, $\varepsilon^+ = 1$, as it is described by the Nusselt film condensation theory, is indicated by the dashed straight line. As one recognizes, the temperature profile become more and more pronounced with increasing Reynolds number. It is to be pointed out, however, that these profiles are only qualitatively correct and that they deviate numerically somewhat from the more recent values calculated with a more refined theory [4.5].

With the above assumptions, local heat transfer coefficients can also be determined as described. For practical purposes, the calculation required for

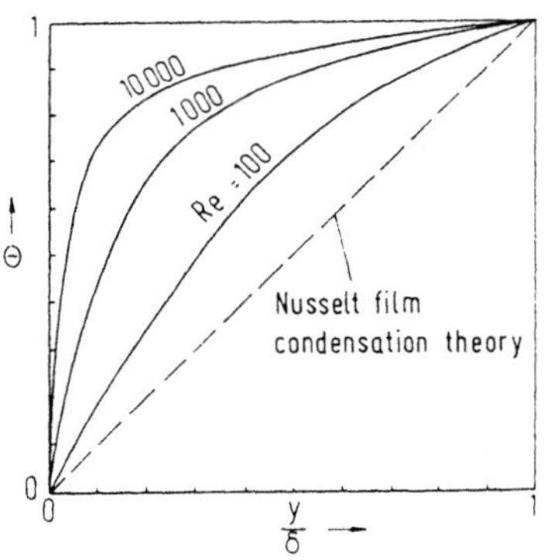

Fig. 4.5. Temperature profile during turbulent film condensation, according to Rohsenow

this is carried out on a computer. Rohsenow himself reported only mean Nusselt numbers. These could be rendered quite well in the region

$$5 \leqq \tau_\delta^* = \frac{\tau_\delta}{(v_L^2/g)^{1/3} \varrho_L g} \leqq 50$$

and $2 \leqq Pr \leqq 3$ by the equation found earlier by Carpenter and Colburn [4.4]

$$\overline{Nu} = \frac{\bar{\alpha}(v_L^2/g)^{1/3}}{\lambda_L} = 0.065\, Pr^{1/2} \tau_\delta^{*\,1/2} \tag{4.35}$$

Here, it has to be considered that in the calculations of Rohsenow the shear stress τ_δ was assumed to be constant; in addition $\varrho_G \ll \varrho_L$ was assumed, and therefore, the factor $1 - \varrho_G/\varrho_L = 1$ was set. Equation (4.35) is frequently used to obtain *approximate values* for the mean heat transfer coefficient. The shear stress τ_δ is calculated from

$$\tau_\delta = \zeta \frac{\varrho_G w_G^2}{2} = \zeta \frac{(\dot{M}_G/A)^2}{2\varrho_G}, \tag{4.36}$$

where A is the tube cross section, $\dot{M}_G$ the mass flow rate of vapor, and $\zeta(Re)$ the friction factor. Because the mass flux of the vapor decreases between the inlet cross section and the outlet cross section, Carpenter and Colburn recommend to calculate the shear stress τ_δ with a mean mass flux $\dot{m}_G$, which results from

$$\dot{m}_G = \frac{1}{\sqrt{3}} (\dot{m}_1^2 + \dot{m}_1 \dot{m}_2 + \dot{m}_2^2)^{1/2} \tag{4.37}$$

when assuming a linear velocity decrease of the vapor along the pipe, where $\dot{m}_1$ is the mass flux of the vapor in the inlet cross section and $\dot{m}_2$ that in the outlet cross section. With complete condensation of the vapor there would be $\dot{m}_2 = 0$ and $\dot{m}_G = 0.58\,\dot{m}_1$, i.e., the "mean" vapor quantity would be somewhat greater than half the quantity in the inlet cross section.

Another method for the calculation of heat transfer during turbulent film condensation has been worked out by Dukler [4.6]. It presupposes likewise a turbulent Prandtl number $Pr_t = 1$ and differs from the method of Rohsenow only in the assumptions for the turbulent diffusivity ε^+. In the layer near the wall

$$0 < y^+ \leqq 20$$

the assumption of Deissler [4.7, 4.8] for the turbulent diffusivity is adopted

$$\frac{\varepsilon}{\nu} = \varepsilon^+ - 1 = n^2 w^+ y^+ [1 - \exp(-n^2 w^+ y^+)], \tag{4.38}$$

in which the constant $n = 0.0124$ according to Deissler is replaced by the value $n = 0.01$, in order to achieve a better agreement with the experiments of Carpenter and Colburn [4.4]. In the region far from the wall,

$$20 \leqq y^+ \leqq \delta^+$$

the Kàrmàn assumption for turbulent boundary layers is assumed to be valid

$$\frac{\varepsilon}{\nu} \approx \varepsilon^+ = \kappa^2 \frac{(\mathrm{d}w^+/\mathrm{d}y^+)^3}{(\mathrm{d}^2 w^+/\mathrm{d}y^{+2})^2} \tag{4.39}$$

with $\kappa = 0.36$. In the liquid film, therefore, heat transfer is assumed only on the basis of turbulent fluctuations and not because of molecular conduction. By inserting these expressions into Eq. (4.20), while allowing for the assumption of $\varrho_G \ll \varrho_L$ and integrating, one obtains the velocity profile. From this results with Eq. (4.22) the flow rate Γ or the Reynolds number $Re = \Gamma/(\varrho_L \nu_L)$ of the condensate. With the aid of Eqs. (4.31) and (4.32), one finally obtains, as described before, the local heat transfer coefficients by the numerical solution of the equation.

As an example for the results obtained by Dukler, Fig. 4.6 shows the local Nusselt number as a function of the Reynolds number for a Prandtl number $Pr = 5.0$ and downwards directed vapor flow. As shown by a comparison with falling film flow [4.5], the results of Dukler agree fairly well only for $Pr \geqq 4$ with data [4.10, 4.11] and theory [4.5]. This is to be traced back to several weaknesses in the Dukler assumptions. As is seen upon close scrutiny, the turbulent diffusivity ε^+ at the point $y^+ = 20$ shows a steep increase, which is physically impossible. Because only heat transfer based upon the turbulent fluctuations was assumed in the liquid film, the results in the limiting case of small Reynolds numbers Re do not reduce to the values of the Nusselt film condensation theory. The theory of Dukler was later applied by Hewitt [4.12] also to the case of upward flowing vapor. Finally, Lee [4.13] also took into consideration the molecular viscosity in the assumptions.

The assumptions for the turbulent diffusivity ε^+ used by Rohsenow and Dukler yielded the greatest value of the fluctuations at the film surface, because they were taken over from the theory of turbulent tube flow. As Levich [4.14]

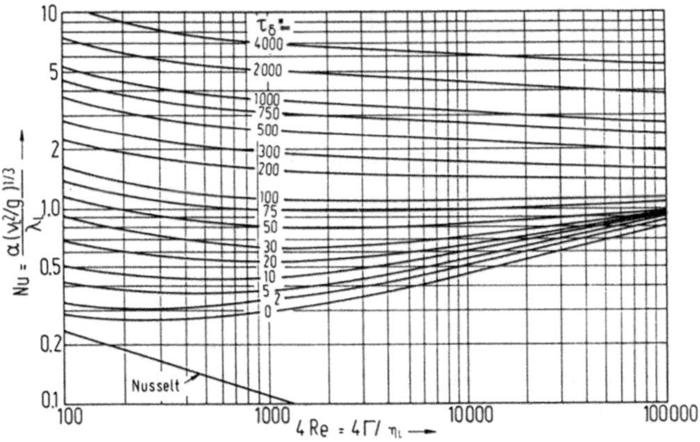

Fig. 4.6. Film condensation of saturated vapor flowing downward, according to [4.4], Prandtl number $Pr = 5.0$

recognized, however, surface waves produced by turbulence are damped under the influence of the surface tension. The latter exerts a "laminarizing" effect on the film surface. According to the conceptions of Levich, the greater the surface tension is the more the turbulent fluctuations will be damped. In addition, the turbulent oscillatory motion decreases by the square of the distance from the film surface. Therefore

$$\varepsilon \approx \frac{(\delta - y)^2}{\sigma}.$$

Lamourelle and Sandall [4.15] were the first to find an assumption for the influence of the surface tension on the turbulent shear stress. They based the assumption upon data on the absorption of gas flows of CO_2, O_2, H_2 and He at a turbulently flowing water film at 25 °C. Upon consideration of further experimental results, Mills and Chung [4.16] extended this assumption and represented it by dimensionless values. Accordingly, the coefficient ε_D for the turbulent mass transfer is

$$\frac{\varepsilon_D}{\nu_L} = \varepsilon_D^+ - 1 = 6.62 \cdot 10^{-3} Re^{1.678} \frac{g \varrho_L}{\sigma} (\delta - y)^2. \tag{4.40}$$

This assumption was obtained by measurements of the mass exchange; if, however, one assumes as is often the case for problems in mass exchange, a turbulent Schmidt number $Sc_t = \varepsilon/\varepsilon_D = 1$, where ε_D is also called the "turbulent diffusion coefficient", then one may adopt the above relation for the turbulent viscosity ε also. Then $\varepsilon_D = \varepsilon$ and $\varepsilon_D^+ = \varepsilon^+$. Allowing for the known definition $\delta^+ = \delta w_\tau/\nu_L$ and $y^+ = y w_\tau/\nu_L$ with $w_\tau = \sqrt{\tau_w/\varrho_L}$, Eq. (4.40) can also be written

$$\varepsilon^+ - 1 = 6.62 \cdot 10^{-3} Re^{1.678} \frac{g \varrho_L}{\sigma} (\delta - y)(\delta^+ - y^+) \frac{\nu_L}{w_\tau}. \tag{4.41}$$

This equation was obtained by experiments on the falling film. It is valid, therefore, only for vanishing shear stress at the phase interface. Allowing then for $\varrho_G \ll \varrho_L$, the wall friction is $\tau_w = \varrho_L g \delta$, as can be recognized from Eq. (4.18b). Thus, one can also write

$$w_\tau = \sqrt{\tau_w/\varrho_L} = \sqrt{g\delta}.$$

The thickness δ of the falling film in Eq. (4.40) was determined from an equation of Brötz [4.17], according to which

$$\delta = \left(\frac{\nu_L^2}{g}\right)^{1/3} 0.169 Re^{2/3}. \tag{4.42}$$

This equation holds, as was recently shown [4.5], in the range $590 \leqq Re \leqq 1700$. The measurements, upon which Eq. (4.40) is based, were carried out in somewhat the same range. From Eq. (4.42) one obtains for the friction velocity

$$w_\tau = g^{1/2} \left(\frac{\nu_L^2}{g}\right)^{1/6} 0.411 Re^{1/3}. \tag{4.43}$$

Insertion of this expression into Eq. (4.41) yields

$$\varepsilon^+ - 1 = 0.0161\, Re^{1.345} \left(\frac{\nu_L^2}{g}\right)^{1/3} \frac{g\varrho_L}{\sigma} (\delta - y)(\delta^+ - y^+). \tag{4.44}$$

Since this expression came from the transformation of the original Eq. (4.40), it is completely equivalent to that equation. It is, therefore, valid only for a falling film, which by definition is characterized by vanishing vapor shear stress $\tau_\delta = 0$. On the other hand, as has already been explained, Eq. (4.18a), allowing for $\varrho_G \ll \varrho_L$, the shear stress is connected with the vapor shear stress by

$$\tau = \tau_\delta + g\varrho_L (\delta - y).$$

The factor $g\varrho_L(\delta - y)$ in Eq. (4.44) can, therefore, be interpreted as the shear stress $\tau = g\varrho_L(\delta - y)$ of the falling film ($\tau_\delta = 0$). If one wishes to apply the equation also to the liquid film with finite vapor shear stress, it is obvious that the factor $g\varrho_L(\delta - y)$ must be replaced by the shear stress τ or by $\tau_\delta + g\varrho_L(\delta - y)$. One then obtains the modification suggested by Blangetti [4.18] for the assumption of Levich

$$\varepsilon^+ - 1 = 0.0161\, Re^{1.345} \frac{g\varrho_L l^2}{\sigma} \left[\frac{\tau_\delta}{g\varrho_L l} + \frac{1}{l}(\delta - y)\right](\delta^+ - y^+) \tag{4.45}$$

with the abbreviation $l = (\nu_L^2/g)^{1/3}$ for the characteristic length l. In the special case of the falling film $\tau_\delta = 0$, Eq. (4.45) converts into the previous Eq. (4.44). In Eq. (4.45), the dimensionless value

$$\left(\frac{g\varrho_L l^2}{\sigma}\right)^3 = Ka$$

is designated as the Kapitza number.

One must fully realize that the above assumptions can only be justified if one can successfully reproduce data with it. In actuality, measurements such as those made by Blangetti for $Re \lesssim 2000$, have proven the usefulness of the assumption.

Nevertheless, from a theoretical standpoint, several objections must be raised to the above assumptions. The concept of Levich, upon which they are based and according to which a purely laminar flow prevails at the film surface because of the attenuating effect of the surface tension, has until now been able to adequately describe only absorption measurements on falling films. By definition, however, no shear stress acts there on the film surface. On the contrary, it is quite conceivable that during the condensation of flowing vapors a sufficiently large vapor shear stress forces a turbulent flow at the film surface. Then, the turbulent fluctuation quantity $\varepsilon^+ - 1$ would not disappear at the film surface according to Eq. (4.45), but rather, being dependent upon the vapor shear stress, would assume a finite value. Whereas the earlier conceptions of Rohsenow, Dukler, and others surmised maximum turbulence on the film surface, a laminar flow is now presumed to be there according to the new assumption, Eq. (4.45).

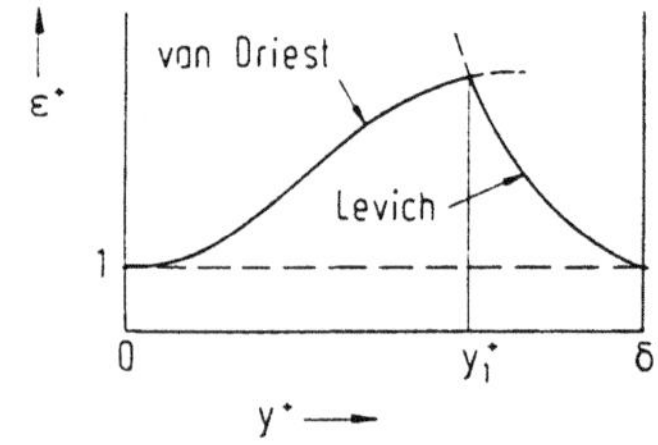

Fig. 4.7. Curve of the turbulent diffusivity ε^+

Reality might well lie between these extremes. Only a very precise investigation of the by no means simple turbulence mechanism at the film surface can clarify these still open questions.

Whereas Eq. (4.45) describes the course of the turbulent diffusivity in the vicinity of the film surface, one could use in the vicinity of the wall ($y^+ < 26$), in accordance with the suggestion of van Driest [4.19], a damping function

$$1 - \exp(-y^+/26)$$

as a factor in the expression for the turbulent diffusivity. The damping function causes a reduction of the diffusivity when approaching the wall. A modified form of the van Driest assumption [4.20] thus reads

$$\varepsilon^+ = \frac{1}{2}\left\{1 + \left[0.64 y^{+2}\tau^+\left(1 - \exp\left(-\frac{y^+\sqrt{\tau^+}}{26}\right)\right)^2\right]^{1/2}\right\}. \tag{4.46}$$

As Fig. 4.7 shows, the two assumptions according to van Driest, Eq. (4.46), and Levich in keeping with the suggestion by Blangetti, Eq. (4.45), intersect at a definite point y_1^+. Left of this point of intersection in the wall contiguous area, the assumption of van Driest is supposed to be valid, and to the right of it the expression near the film surface. Mills and Chung [4.16] have calculated the heat transfer on falling film according to this procedure, while in a corresponding manner with the aid of Eqs. (4.45) and (4.46) Blangetti treated the case of heat

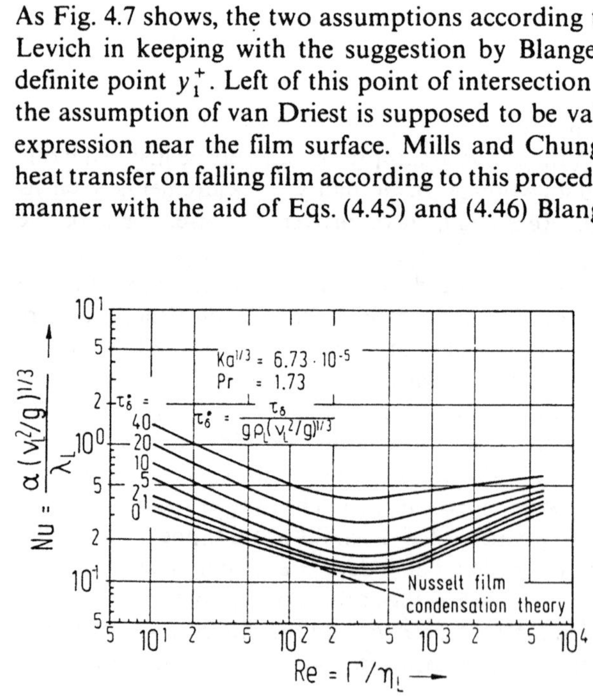

Fig. 4.8. Local Nusselt number during film condensation of water, according to [4.18]

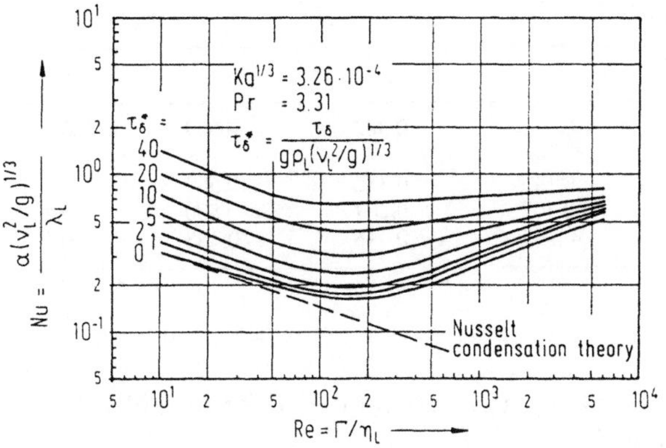

Fig. 4.9. Local Nusselt number during condensation of R12, according to [4.18]

transfer during the condensation of flowing vapors. The heat transfer coefficients result according to the previously discussed model from Eqs. (4.31) and (4.32).

Figures 4.8 and 4.9 reproduce for water and the refrigerant R-12 the variation of the local Nusselt number with the Reynolds number. Plotted, as the limiting case, is the Nusselt film condensation theory. The figures show that with small Reynolds numbers, one obtains the results of laminar film condensation. In the area of laminar flow, the Nusselt number decreases with increasing Reynolds number. After a minimum, the Nusselt number increases again in the area of the turbulent flow, as the observations about the film condensation of stagnant vapors have similarly shown. With increasing Prandtl number, the minimum is shifted to smaller Reynolds numbers.

4.2.3 Practical Equations

On the basis of the above considerations, Blangetti [4.18] calculated numerically the local heat transfer and described the numerical results by use of exponential equations. It was also shown that the Kapitza number is not needed for the reproduction of the results, although the conception of the damping effect of the surface tension on the film surface necessarily led to the Kapitza number. As an explanation, Blangetti suggested that the Kapitza number as well as the Prandtl number is a physical property, and that an inter-dependency exists between the two properties, so that the influence of the Kapitza number on the heat transfer can also be determined by the Prandtl number. The empirical correlation for the calculation of the local heat transfer during turbulent film condensation is

$$Nu = A Re^{n_1} Pr^{n_2}(1 + B\tau_\delta^{*n_3}) \tag{4.47}$$

with

$$Nu = \frac{\alpha(v_L^2/g)^{1/3}}{\lambda_L}, \quad Re = \frac{\Gamma}{\eta_L} \quad \text{and} \quad \tau_\delta^* = \frac{\tau_\delta}{\varrho_L g (v_L^2/g)^{1/3}}.$$

Table 4.1. Constants in Eq. (4.47)

τ_δ^*	A	B	n_1	n_2	n_3
0 (Falling film)	$8.663 \cdot 10^{-3}$	0	0.382	0.569	0
$0 < \tau_\delta^* \leqq 5$	$8.663 \cdot 10^{-3}$	0.145	0.382	0.569	0.541
$5 < \tau_\delta^* \leqq 10$	$2.700 \cdot 10^{-2}$	0.407	0.207	0.500	0.420
$10 < \tau_\delta^* \leqq 40$	$4.294 \cdot 10^{-2}$	0.647	0.096	0.458	0.473

The factors A, B and the exponents n_1, n_2, n_3 are tabulated for the different areas of the vapor shear τ_δ^* in Table 4.1.

This equation is valid only for turbulent film condensation. The local Nusselt number for both laminar and turbulent flows can be described approximately by means of a superposition [4.18]

$$Nu = [(f Nu_l)^4 + Nu_t^4]^{1/4} \tag{4.48}$$

where Nu_t is the local Nusselt number during turbulent film condensation, Eq. (4.47), and Nu_l that during laminar film condensation, which is obtained with the aid of Eq. (4.9).

In the area of the purely laminar flow, $Nu_t \ll Nu_l$; in pure turbulent flow, there is conversely $Nu_l \ll Nu_t$. Such an interpolation equation represents, therefore, the limiting cases of pure laminar and pure turbulent film condensation, and also yields the approximate values in the transition area. The factor $f = 1.15$ allows for the improvement of the heat transfer by the surface waves. The data upon which the equation is based extend only to Reynolds numbers Re of about 2000, but in the author's opinion [4.18] an extrapolation to higher Reynolds numbers of approximately 5000 may be made without hesitation. The equation holds further only for the condensation of nonmetallic fluids, $Pr \geqq 0{,}1$, and for the case where vapor and condensate are flowing downwards in a vertical tube.

Because one usually wishes to completely condense the vapor before it reaches the tube exit, the reverse case of an upward flowing vapor and a downward flowing condensate rarely occurs. Only during partial condensation of a mixture of vapors, is the vapor occasionally drawn off above, in order to be able to use the rectifying effect of the counterflow. The heat transfer during counterflow between vapor and condensate is less than during cocurrent flow, because the condensate film is retarded.

The shear stress τ_δ at the film surface in Eq. (4.47) can be determined in the following manner [4.21]. The friction factor

$$\frac{\zeta}{4} = \frac{\tau_\delta}{\varrho_G \bar{w}^2/2},$$

where $\bar{w}$ is the average vapor velocity through a cross section, is divided into two parts

$$\zeta = E\zeta_R + \zeta_F. \tag{4.49}$$

The parameter ζ_R designates the frictional portion, and ζ_F is a "form component" dependent chiefly upon the film flow. The correction factor E allows for the alteration of the velocity profile of the vapor by the condensate flow.

According to the film theory of mass transfer

$$E = \frac{\varphi}{e^{\varphi} - 1}$$

with

$$\varphi \approx \frac{q/\Delta h_v}{\zeta_R \varrho_G \bar{w}/8}.$$

As is recognized, in the case of small heat fluxes $q \to 0$, because of $\varphi \to 0$, the suction factor is $E \to 1$, because the influence of the vapor suction on the velocity profile disappears. In case the factor φ becomes small, $-0.1 < \varphi < 0$, the influence of E is negligible. The factors $\zeta_R/4$ and $\zeta_F/4$ are taken from a representation by Bergelin et al. [4.21], Fig. 4.10. The form component ζ_F in it depends upon the value

$$G = \frac{\Gamma \sigma_w}{\varrho_L \sigma},$$

where σ_w is the surface tension of the water and σ that of the condensate at the saturation temperature of the condensate.

To calculate local heat transfer coefficients from these equations, it is recommended that one proceeds as follows:

— Estimate at the end of the i-th section, whose length shall be Δx, the condensate mass flow rate relative to the circumference Γ_{i+1}. This mass flow rate is identical with that at the beginning of the $(i+1)$-th section.

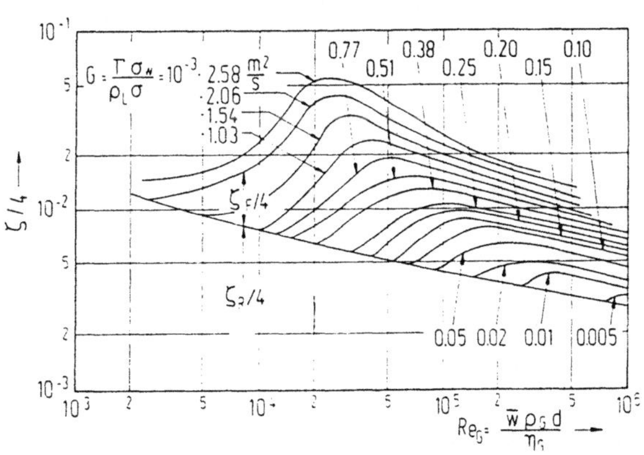

Fig. 4.10. Friction factor for vapors flowing in tubes

—On the basis of the estimated value, the heat flux

$$q = \frac{\mathrm{d}\Gamma}{\mathrm{d}x} \Delta h_v = \frac{\Gamma_{i+1} - \Gamma_i}{\Delta x} \Delta h_v$$

as well as the mean Reynolds number in the i-th section $Re = (\Gamma_{i+1} + \Gamma_i)/2\eta_L$ are known. With these, one can calculate the values q, E and τ_δ and obtain from Eq. (4.47) the mean heat transfer coefficient α of the i-th section.

—The original estimate must be corrected until the energy balance

$$\alpha(\vartheta_s - \vartheta_w) = \frac{\Gamma_{i+1} - \Gamma_i}{\Delta x} \Delta h_v$$

is fulfilled.

This calculation procedure is physically well founded, but it is only an approximation procedure that is quite time consuming and that, especially in the area of high vapor velocities, leads to inexact results, because additional pressure losses resulting from the entrainment of liquid droplets cannot be determined.

Shah [4.22] has reported another, and for practical purposes, simpler equation, which renders well the data of many authors. The equation starts with the well-known McAdams equation for heat transfer in single-phase turbulent tube flow, and expands it by a term that takes into consideration the phase transformation and the influence of the vapor flow on the condensate film:

$$Nu = 0.023 Re^{0.8} Pr^{0.4} \left\{ (1 - x^*)^{0.8} + \frac{3.8(1 - x^*)^{0.04} x^{*0.76}}{p^{*0.38}} \right\} \tag{4.50}$$

with

$$Nu = \frac{\alpha d}{\lambda_L}, \; Re = \frac{\bar{w} d}{\nu_L}, \; \bar{w} = \dot{M} \Big/ \left(\varrho_L \frac{\pi d^2}{4} \right), \quad Pr = \frac{\nu_L}{a_L} \;\text{ and }\; p^* = \frac{p}{p_{cr}}.$$

The value $x^* = \dot{m}_G/\dot{m}$ is the quality in a cross section of the tube. If one assumes a linear variation of the quality along a longitudinal section $L_2 - L_1$ of the tube, one obtains by integration the mean Nusselt number

$$\overline{Nu} = \frac{1}{L_2 - L_1} \int_{L_1}^{L_2} Nu \mathrm{d}L.$$

Let the quality at point L_2 be x_2^*, and that at point L_1 be x_1^*. Then, the average Nusselt number is

$$\overline{Nu} = \frac{0.023 Re^{0.8} Pr^{0.4}}{x_2^* - x_1^*} \left\{ -\frac{(1 - x^*)^{1.8}}{1.8} + \frac{3.8}{p^{*0.38}} \left(\frac{x^{*1.76}}{1.76} - \frac{0.04 x^{*2.76}}{2.76} \right) \right\}_{x^*_1}^{x^*_2} \tag{4.50a}$$

These equations represent measurements with water; with the refrigerants R-11, R-12, and R-113; and with methanol, ethanol, benzene, toluene, and trichloroethylene during condensation in vertical, horizontal, and inclined tubes with inner diameters of 7 to 40 mm at normalized pressures

$$0.002 \leqq p^* \leqq 0.44,$$

saturation temperatures

$$21\,°\mathrm{C} \leqq \vartheta_s \leqq 310\,°\mathrm{C},$$

vapor velocities

$$3\,\mathrm{m/s} \leqq w_G \leqq 300\,\mathrm{m/s},$$

mass fluxes

$$10.8\,\mathrm{kg/m^2s} \leqq \dot{m} \leqq 210.6\,\mathrm{kg/m^2s},$$

heat fluxes

$$158\,\mathrm{W/m^2} \leqq q \leqq 1.893 \cdot 10^6\,\mathrm{W/m^2},$$

Reynolds numbers

$$100 \leqq Re \leqq 63000$$

and Prandtl numbers

$$1 \leqq Pr \leqq 13.$$

The mean deviation from the experimental values is $\pm 15.4\%$.

For the practical calculation of heat transfer coefficients, it is recommended that one proceeds iteratively:

— Estimate the quality x_2^* at the end of a tube section and calculate with it from Eq. (4.50a) the mean heat transfer coefficient of this section.
— On the basis of the estimate,

$$x_2^* - x_1^* = \frac{\dot{M}_{G2} - \dot{M}_{G1}}{\dot{M}} = \frac{\dot{M}_{L1} - \dot{M}_{L2}}{\dot{M}}$$

is known and because of

$$\dot{M}_{L1} - \dot{M}_{L2} = 2R\pi(\Gamma_1 - \Gamma_2)$$

one also knows

$$\Gamma_2 - \Gamma_1 = \frac{\dot{M}}{2R\pi}(x_1^* - x_2^*).$$

— The calculation is to be repeated until the energy balance

$$\alpha(\vartheta_s - \vartheta_w) = \frac{\Gamma_2 - \Gamma_1}{L_2 - L_1}\Delta h_v \quad \text{or} \quad \alpha(\vartheta_s - \vartheta_w) = \frac{\dot{M}}{2R\pi}\frac{x_1^* - x_2^*}{L_2 - L_1}\Delta h_v$$

is fulfilled. If one changes the quality by less than 40% over a tube section, then, as Shah has shown, one does not make a large mistake in the calculation of the mean heat transfer coefficient if one assumes a linear decrease of quality over the respective tube section.

4.3 Film Condensation in Horizontal Tubes

Whereas in vertical tubes the condensate runs off mainly as a film, in horizontal tubes along with film or annular-dispersed flow, there can occur other flow patterns, which are depicted in Fig. 4.11. If the vapor shear stress on the condensate film is sufficiently large in comparison to the effect of gravity and the amount of condensate in a cross section is still small, then there develops an annular-dispersed flow, as is shown is the upper left corner of the diagram. Stratified flow results if the velocity of the vapor flow is moderate and, as a consequence, under the influence of gravity, the condensate can flow off around the circumference at the wall. An intermittent flow, which can consist of liquid with trapped vapor bubbles or of a liquid slug with vapor zones extending to the wall, forms at a low vapor velocity, but with a large liquid component. At a high vapor velocity and a large liquid component, one has a cohesive liquid zone with vapor bubbles distributed in it, as is shown in the upper right corner of Fig. 4.11.

Because the liquid component changes along a condenser tube, different flow patterns can also occur one after the other. Naturally, the heat transfer depends upon the phase distribution, and for the calculation of the heat transfer, one must first estimate which flow pattern is present. A starting point for this is given by the flow pattern chart, Fig. 4.12, which is based on an evaluation of the previously known data. In it, there is chosen as the ordinate a dimensionless vapor velocity introduced by Wallis [4.23]

$$j_G^* = \frac{x^* \dot{M}}{A[g d \varrho_G(\varrho_L - \varrho_G)]^{1/2}}.$$

Here x^* signifies the quality, $\dot{M}$ the mass flow rate in a tube cross section A, and d the tube diameter. The vapor volumetric content (void fraction) is $\varepsilon = V_G/V = A_G/A$, where V_G is the vapor volume and A_G the cross section occupied by the vapor. The value $(1-\varepsilon)/\varepsilon = A_L/A_G$ plotted as abscissa is,

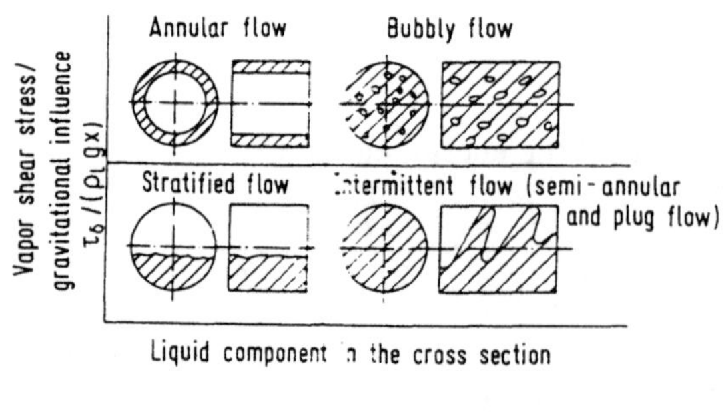

Fig. 4.11. Flow patterns during condensation in horizontal tubes

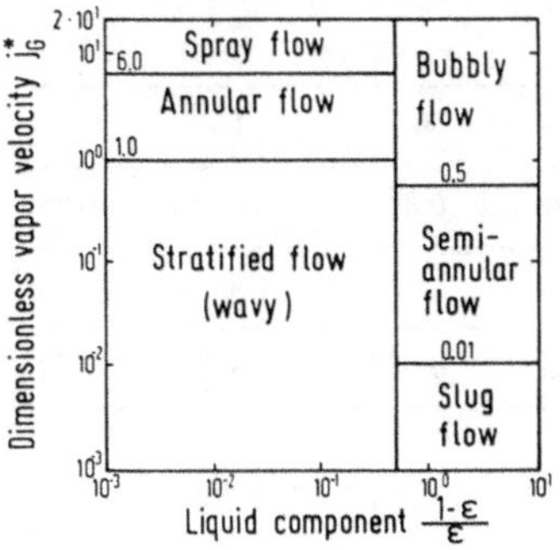

Fig. 4.12. Flow pattern map for condensation in horizontal tubes, according to [4.24]

therefore, the liquid volumetric fraction (holdup). The following reference values can be considered as approximate limits for the different forms of flow.

stratified flow	$j_G^* \leqq 1$ and $(1-\varepsilon)/\varepsilon \leqq 0.5$;
annular-dispersed flow (liquid phase)	$1 \leqq j_G^* \leqq 6$ and $(1-\varepsilon)/\varepsilon \leqq 0.5$;
spray flow	$j_G^* \geqq 6$ and $(1-\varepsilon)/\varepsilon \leqq 0.5$;
slug flow	$j_G^* \leqq 0.01$ and $(1-\varepsilon)/\varepsilon \geqq 0.5$;
semi-annular flow	$0.01 \leqq j_G^* \leqq 0.5$ and $(1-\varepsilon)/\varepsilon \geqq 0.5$;
bubbly flow	$j_G^* \geqq 0.5$ and $(1-\varepsilon)/\varepsilon \geqq 0.5$.

The most frequently occurring types of flow are annular-dispersed flow and stratified flow. Most measurements leading to the establishment of the flow pattern chart, Fig. 4.12, were collected for these patterns. Of these two types of flow, the annular-dispersed flow can occur also in vertical tubes, if the vapor velocity is so great that one can disregard the effect of gravity in contrast to the influence of the vapor shear stress on the film condensate, i.e., $(\varrho_L g l)/\tau_\delta \ll 1$. This holds also for the case of annular-dispersed flow in vertical tubes.

4.3.1 Heat Transfer During Stratified Flow

As is seen from the idealized sketch in Fig. 4.13, in a stratified flow, part of the heat is transferred by the condensate film flowing off at the circumference of the tube, and part by the condensate layer at the bottom of the tube. Accordingly, the entire heat transfer coefficient $\bar{\alpha}_{tot}$ averaged over the circumference is composed of the heat transfer coefficient $\bar{\alpha}_\varphi$ of the draining condensate, which

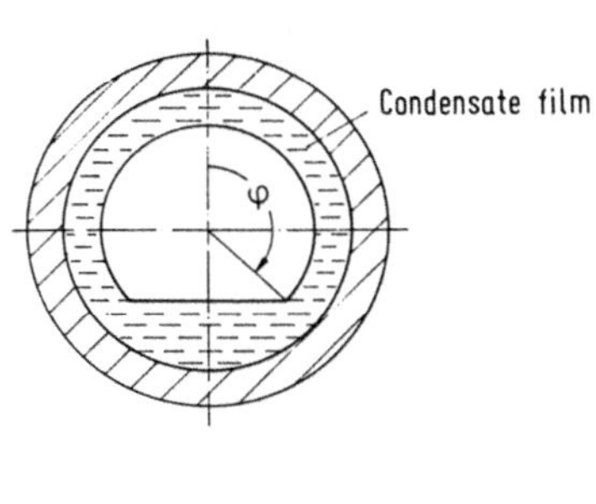

Fig. 4.13. Idealization of stratified flow during condensation in horizontal tubes

Table 4.2. Values of the function $F(\varphi)$, φ in degree

φ	0	10	20	30	40	50	60	70
$F(\varphi)$	–	0.903	0.902	0.899	0.896	0.892	0.887	0.881
φ	80	90	100	110	120	130	140	150
$F(\varphi)$	0.874	0.866	0.857	0.846	0.835	0.822	0.808	0.793
φ	160	170	180					
$F(\varphi)$	0.775	0.754	0.728					

extends to the circumferential angle φ, and of the heat transfer coefficient $\bar{\alpha}_B$ of the condensate layer on the bottom

$$\bar{\alpha}_{tot} = \frac{\varphi}{\pi}\,\bar{\alpha}_\varphi + \left(1 - \frac{\varphi}{\pi}\right)\bar{\alpha}_B. \tag{4.51}$$

Because the heat transfer coefficient $\bar{\alpha}_B$ of the condensate layer on the bottom is smaller, in general, in contrast to that of the film, and because the coefficient $(1 - \varphi/\pi)$ is also smaller than φ/π, because $\varphi > \pi/2$ is almost always valid, the above equation is simplified to

$$\bar{\alpha}_{tot} = \frac{\varphi}{\pi}\,\bar{\alpha}_\varphi. \tag{4.52}$$

The heat transfer coefficient $\bar{\alpha}_{tot}$ calculated from this is somewhat smaller than the actual one; hence, the heat exchange surface is dimensioned somewhat too large. One, therefore, has a certain safety reserve during the sizing of the exchange surface. In Eq. (4.52), the heat transfer coefficient $\bar{\alpha}_\varphi$ averaged over the circumference depends also upon the circumferential angle φ. For this, one can use the results of the Nusselt film condensation theory for the laminar film condensation on horizontal tubes. According to it, the heat transfer coefficient averaged over the circumferential angle φ is

$$\bar{\alpha}_\varphi = F(\varphi)\left[\frac{\varrho_L^2 g \Delta h_v \lambda_L^3}{\eta_L(\vartheta_s - \vartheta_w)}\,\frac{1}{d}\right]^{1/4}. \tag{4.53}$$

For the function $F(\varphi)$, one obtains from the numerical solution by Nusselt the numerical values put together in Table 4.2. These values can be described approximately according to a suggestion of Jaster and Kosky [4.25] by

$$F(\varphi) = 0.728\,\frac{\pi}{\varphi}\left[\frac{\varphi}{\pi} + \frac{\sin 2(\pi - \varphi)}{2\pi}\right]^{3/4}. \tag{4.54^1}$$

[1] With regard to the constant 0.728, see footnote on page 10.

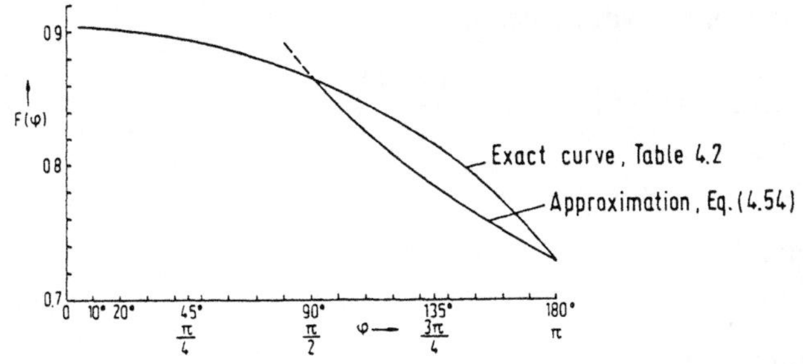

Fig. 4.14. Film condensation in a horizontal tube. Curve of function $F(\varphi)$

As Fig. 4.14 shows, this equation renders the more precise values according to Table 4.2 satisfactorily only as long as the draining condensate layer fills up no more than half of the tube cross section. In practice, however, this condition can be regarded as almost always fulfilled. The maximum error according to the approximation Eq. (4.54) in the range

$$\pi/2 \leqq \varphi \leqq \pi$$

is smaller then -4.3%; the average error amounts to roughly -2.4%.

The expression in the square brackets of Eq. (4.54) is so chosen that it can be easily connected with the void fraction. As noted earlier, the latter is defined by

$$\varepsilon = V_G/V = A_G/A, \tag{4.55}$$

if V_G is the vapor volume, V the entire volume, and A_G the cross section through which the vapor flows in the tube cross section A. If one regards the liquid film falling on the inner wall as very thin in comparison to the tube diameter, there results

$$A_G = r^2\pi - A_L,$$

whereby A_L results with the aid of Fig. 4.15 as the difference of the surface $r^2\beta$ and of the surface $r^2 \sin\beta \cos\beta$ of the triangle MAB

$$A_L = r^2(\beta - \sin\beta\cos\beta).$$

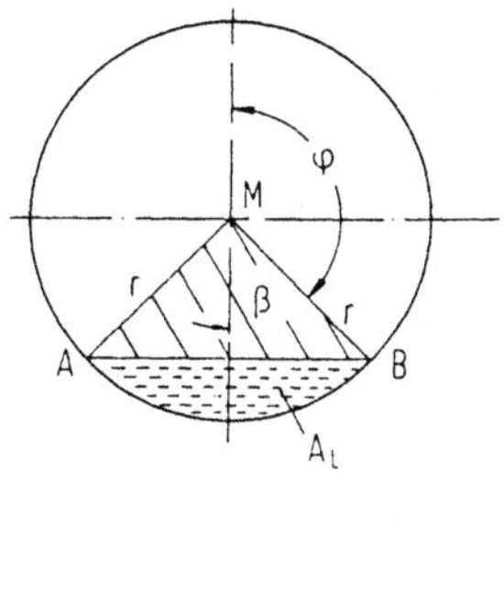

Fig. 4.15. Parameters for calculation of the void fraction $\varepsilon = A_G/A$

Thus, the flow cross section of the vapor becomes

$$A_G = r^2(\pi - \beta + \sin\beta\cos\beta) = r^2\left(\pi - \beta + \frac{\sin 2\beta}{2}\right)$$

and, thereby,

$$\varepsilon = \frac{A_G}{A} = 1 - \frac{\beta}{\pi} + \frac{\sin 2\beta}{2\pi}.$$

With $\beta = \pi - \varphi$, one obtains

$$\varepsilon = \frac{\varphi}{\pi} + \frac{\sin 2(\pi - \varphi)}{2\pi}. \tag{4.56}$$

Consequently, one can also write for the factor $F(\varphi)$, according to Eq. (4.54),

$$F(\varphi) = 0.728\frac{\pi}{\varphi}\varepsilon. \tag{4.57}$$

According to Eq. (4.53), the heat transfer coefficient of the draining condensate averaged over the circumference can thus be written approximately as

$$\bar{\alpha}_\varphi = 0.728\,\frac{\pi}{\varphi}\,\varepsilon\left[\frac{\varrho_L^2 g \Delta h_v \lambda_L^3}{\eta_L(\vartheta_s - \vartheta_w)}\,\frac{1}{d}\right]^{1/4}, \tag{4.58}$$

which is valid for $\frac{\pi}{2} \leqq \varphi \leqq \pi$, and one obtains for the entire heat transfer coefficient $\bar{\alpha}_{tot}$, according to Eq. (4.52),

$$\bar{\alpha}_{tot} = 0.728\varepsilon\left[\frac{\varrho_L^2 g \Delta h_v \lambda_L^3}{\eta_L(\vartheta_s - \vartheta_w)}\,\frac{1}{d}\right]^{1/4}. \tag{4.59}$$

In order to be able to work with this equation, one must know the quality ε. The explanations in the next section deal with this.

4.3.1.1 Relation Between Void Fraction and Quality in Stratified Flow

The volumetric vapor content (void fraction) in Eq. (4.59), which is defined by Eq. (4.55), depends on the vapor density, the liquid density, and the quality $x^* = \dot{M}_G/\dot{M}$. Usually, the volumetric vapor content is unknown, whereas the quality is known or can be calculated from the energy balance. Then, it is advantageous if the volumetric vapor content can be determined from the known quality by computation.

A relation between these values is found for an idealized flow in a paper of Zivi [4.26]. The wall friction is disregarded here, likewise the fluctuations of the vapor velocity, liquid velocity, pressure, and quality. The flow is assumed to be adiabatic, and the influence of gravity on the pressure drop is ignored. With these assumptions, it follows from the momentum equation that the work supplied to the flow serves to change the kinetic energy. Now, as is proven in the thermodynamics of irreversible processes, the velocity profile in a steady flow always adjusts itself so that the work fed to the flow or the kinetic energy of the flow assumes a minimum. If the velocity of the vapor is designated as w_G, and that of the liquid as w_L, then the mass flow rate of the vapor is

$$\dot{M}_G = w_G A_G \varrho_G = w_G \varepsilon A \varrho_G \tag{4.60}$$

and that of the liquid is

$$\dot{M}_L = w_L A_L \varrho_L = w_L (1 - \varepsilon) A \varrho_L . \tag{4.61}$$

With the aid of the quality $x^* = \dot{M}_G / \dot{M}$, one obtains from Eq. (4.60)

$$w_G = \frac{x^* \dot{M}}{\varepsilon A \varrho_G} . \tag{4.62}$$

Accordingly, there follows from Eq. (4.61), because of $1 - x^* = \dot{M}_L / \dot{M}$,

$$w_L = \frac{(1 - x^*) \dot{M}}{(1 - \varepsilon) A \varrho_L} . \tag{4.63}$$

The kinetic energy of the two-phase flow is given by

$$\Phi_E = \frac{w_G^2}{2} x^* \dot{M} + \frac{w_L^2}{2} (1 - x^*) \dot{M} . \tag{4.64}$$

So that the kinetic energy will reach a minimum as a function of the quality,

$$\frac{\partial \Phi_E}{\partial \varepsilon} = 0 = \dot{M} \left[w_G \frac{\partial w_G}{\partial \varepsilon} x^* + w_L \frac{\partial w_L}{\partial \varepsilon} (1 - x^*) \right] \tag{4.65}$$

must hold, in which because of Eqs. (4.62) and (4.63)

$$\frac{\partial w_G}{\partial \varepsilon} = - \frac{x^* \dot{M}}{\varepsilon^2 A \varrho_G} \quad \text{and} \quad \frac{\partial w_L}{\partial \varepsilon} = \frac{(1 - x^*) \dot{M}}{(1 - \varepsilon)^2 A \varrho_L} . \tag{4.66}$$

If one inserts these expressions and the velocities w_G and w_L, according to Eqs. (4.62) and (4.63), into Eq. (4.65) and solves the equation with respect to the void fraction, one finds

$$\varepsilon = \left[1 + \frac{1 - x^*}{x^*} \left(\frac{\varrho_G}{\varrho_L} \right)^{2/3} \right]^{-1} . \tag{4.67}$$

Corresponding to this is a ratio of vapor and liquid velocities, which is calculated from Eqs. (4.62) and (4.63) as

$$\frac{w_G}{w_L} = s = \frac{x^*}{1 - x^*} \frac{1 - \varepsilon}{\varepsilon} \frac{\varrho_L}{\varrho_G} . \tag{4.68}$$

The factor s states by how much the vapor flows faster than the liquid. It is designated as the slip or the slip factor. With the aid of Eq. (4.67), it is expressed as

$$s = \left(\frac{\varrho_L}{\varrho_G}\right)^{1/3}. \tag{4.69}$$

The slip depends, however, not only upon the density ratio of the two phases, but also upon the ratio of the inertia force to the friction force, and thus to the Reynolds number. Likewise the slip depends upon the ratio of inertia force to gravity, and thus to the Froude number. Equations that allow for these influences are found in the literature of two-phase flows [4.23, 4.27–4.29].

As an example can be mentioned an equation that allows for the influence of the Reynolds number and which has proven quite useful for rough calculations [4.29]

$$s = \left(\frac{\varrho_L}{\varrho_G}\right)^{0.205} \left(\frac{\dot{M} d}{A \eta_L}\right)^{-0.016}. \tag{4.70}$$

In this equation, $\dot{M}$ is the total mass flow rate of liquid and vapor in the tube cross section A, d the tube diameter, and η_L the dynamic viscosity of the liquid.

If one inserts Eq. (4.70) into Eq. (4.68) and solves the equation with respect to the quality, one obtains for this an equation that deviates somewhat from Eq. (4.67),

$$\varepsilon = \left[1 + \frac{1 - x^*}{x^*} \left(\frac{\varrho_G}{\varrho_L}\right)^{0.795} \left(\frac{\dot{M} d}{A \eta_L}\right)^{-0.016}\right]^{-1}. \tag{4.71}$$

For further assumptions for the relation between the void fraction and the quality, see the literature on two-phase flows [4.23, 4.27, 4.28].

4.3.2 Heat Transfer During Film Flow or Annular-Dispersed Flow

4.3.2.1 Mass and Momentum Balance

In film or annular-dispersed flow in a horizontal tube, the influence of gravity is negligible, and there is equilibrium between the forces of viscosity and pressure. As in Sect. 4.2 (see there Fig. 4.4), we shall again consider an annular element of the condensate film and of the associated vapor volume between the tube cross sections x and $x + \Delta x$. The equilibrium of forces requires

$$\tau 2 r \pi \Delta x + \left(p + \frac{dp}{dx} \Delta x\right) r^2 \pi = p r^2 \pi$$

or

$$\tau = -\frac{r}{2}\frac{dp}{dx} = -\frac{R - y}{2}\frac{dp}{dx}. \tag{4.72}$$

The same balance of forces for the vapor volume alone yields

$$\tau_\delta = -\frac{R-\delta}{2}\frac{dp}{dx}. \tag{4.73}$$

Thus, there is

$$\tau = \tau_\delta \frac{R-y}{R-\delta}$$

and

$$\tau_w = \tau_\delta \frac{R}{R-\delta}.$$

If the film is very thin in comparison to the tube radius $\delta \ll R$, then allowing for $y < \delta$ there result the approximations

$$\tau = -\frac{R}{2}\frac{dp}{dx} = \tau_\delta = \tau_w. \tag{4.72a}$$

With the aid of the assumption for the friction of turbulent flows

$$\tau = \varrho_L \nu_L \varepsilon^+ \frac{\partial w_x}{\partial y},$$

and because of $\tau = \tau_w$, there results the velocity profile in the condensate film:

$$w_x = \int_0^y \frac{1}{\varepsilon^+}\frac{\tau_w}{\varrho_L \nu_L}\,dy.$$

From this, one finds the mass flow rate of the condensate

$$\dot{M} = 2R\pi \int_0^\delta \varrho_L w_x dy = 2R\pi\Gamma$$

or the flow rate per unit of circumference

$$\Gamma = \int_0^\delta \varrho_L \left[\int_0^y \frac{1}{\varepsilon^+}\frac{\tau_w}{\varrho_L \nu_L}\,dy\right] dy. \tag{4.74}$$

4.3.2.2 The Energy Balance

If one disregards, just as in the case of the vertical tube, the convective energy transport in the flow direction, which is small in comparison to the heat conduction through the thin condensate film, then as in the case of condensation in vertical tubes, Sect. 4.2,

$$\frac{\partial q}{\partial y} = 0 \quad \text{with} \quad q = -(\lambda + \lambda_t)\frac{d\vartheta}{dy}$$

and

$$\lambda_t = \varrho_L c_p \varepsilon_q.$$

By integration, one finds Eq. (4.31), which was derived in Sect. 4.2. This equation serves for the calculation of the heat transfer coefficient, which can also be written in the form

$$\alpha = w_\tau \frac{\lambda_L}{\nu_L} Pr \frac{1}{T^+} = \left(\frac{\tau_w}{\varrho_L}\right)^{1/2} \varrho_L c_{pL} \frac{1}{T^+} \tag{4.75}$$

with

$$T^+ = T(\delta^+) = \int_0^{\delta+} \frac{dy^+}{\dfrac{1}{Pr} + \dfrac{\varepsilon^+ - 1}{Pr_t}}. \tag{4.75a}$$

For the calculation of T^+, as shown in Sect. 4.2, the turbulent diffusivity $\varepsilon^+(y^+)$ and the turbulent Prandtl number must be known. Assumptions have already been explained in Sect. 4.2.2. They lead, as explained there, to numerical values of the heat transfer coefficient or the Nusselt number, which can be rendered by the working formula of the type reported by Blangetti, Eq. (4.47). This equation is, therefore, also valid for a horizontal tube. Likewise, Eq. (4.50) by Shah also holds within the stated limits of validity. Butterworth [4.30] has found a somewhat different equation, which describes well in the range $Re > 1250$ the results of several authors. Accordingly, the value T^+ in Eq. (4.75) is given by

$$T^+ = 8.5 Re^{0.1} Pr^{0.57} \tag{4.76}$$

with $Re = \Gamma/\eta_L$.

4.3.3 The Practical Calculation of Heat Transfer Coefficients

As previously presented, the earlier working formulas for turbulent film condensation in vertical tubes are still valid, Sect. 4.2.3. Also the calculation procedure stated there can still be used as long as the Reynolds number Re of the liquid is sufficiently large, around $Re \gtrsim 1250$, because then the influence of gravity on the film thickness becomes negligible. Thus, the film thickness in the horizontal tube, which is obtained according to the assumption of Γ from Eq. (4.74), does not differ from that of the vertical tube, Eq. (4.21), given the same Reynolds number or the same condensate flow rate per unit of circumference $\Gamma = Re\eta_L$. Because, therefore, the mass balance and the energy balance for the vertical, horizontal, and also the inclined tube coincide, given a sufficiently large Reynolds number $Re \gtrsim 1250$, one may continue to use the calculation procedure explained previously. Especially here, one should refer to Eq. (4.50) by Shah [4.22], which is simple to use for practical purposes.

Equation (4.75) in conjunction with Eq. (4.76) leads also to rather accurate predictions. In order to determine from it the local heat transfer coefficients, one must proceed in the following manner:

—Again one first estimates the condensate flow rate per unit of circumference Γ_{i+1} at the end of the i-th section, whose length shall be Δx.

—Based upon this estimated value, the mean Reynolds number in the i-th section $Re = (\Gamma_{i+1} + \Gamma_i)/2\eta_L$ is known, as is the transmitted heat flux

$$q = \frac{d\Gamma}{dx}\Delta h_v = \frac{\Gamma_{i+1} - \Gamma_i}{\Delta x}\Delta h_v.$$

—In the next step, one calculates the mean wall shear stress τ_w in the i-th section. This wall shear stress is connected, according to Eq. (4.72a), with the pressure drop by

$$\tau_w = -\frac{d}{4}\frac{dp}{dx},$$

where d is the tube diameter. Because of the generally large Reynolds number of liquid and vapor, the pressure drop dp/dx is produced chiefly by friction and can be simply calculated from equations for the friction pressure drop of two-phase flows. Here one disregards the additional shear stress at the phase interface, which is produced by the condensation of the vapor. One thus acts as if one had a two-phase flow without condensation. The error caused by this leads only at very strong condensation to a slight elevation of the shear stress and thus of the pressure drop; however, with normal, not too extreme condensation, the error is negligible. Equations for the friction pressure drop dp/dx are found in the literature on two-phase flows [4.25, 4.27, 4.28]. As an example only the equation of Lockhart and Martinelli [4.31], which is often used for practical calculations, need be given. The friction pressure drop is traced back to that of the single-phase flow

$$\frac{dp}{dx} = \left(\frac{dp}{dx}\right)_L \Phi_L^2, \tag{4.77}$$

where $(dp/dx)_L$ is the pressure drop of the single-phase liquid, in our case that of the condensate in the i-th cross section, if it fills the entire tube. This pressure drop is calculated by using the familiar equation

$$\left(\frac{dp}{dx}\right)_L = \zeta \frac{1}{d} \varrho_L \frac{\bar{w}^2}{2},$$

where the velocity of the liquid is obtained from

$$\dot{M} = d\pi\Gamma = \varrho_L \bar{w} \frac{d^2\pi}{4}.$$

For expediency, for Γ in the i-th section, one puts $\Gamma = (\Gamma_{i+1} + \Gamma_i)/2$. The two-phase multiplier Φ_L, which is defined by Eq. (4.77), can be expressed by simple empirical equations, such as

$$\Phi_L^2 = 1 + \frac{C}{X_{tt}} + \frac{1}{X_{tt}^2}. \tag{4.78}$$

The constant C is dependent upon the character of the flow and is $C = 20$ for fully turbulent flow. For the value of C for other forms of flow refer to the literature on two-phase flow [4.23, 4.27, 4.28].

The parameter X_{tt} is the Martinelli parameter, which is given by

$$X_{tt} = \left(\frac{\varrho_G}{\varrho_L}\right)^{0.5} \left(\frac{\eta_L}{\eta_G}\right)^{0.1} \left(\frac{1-x^*}{x^*}\right)^{0.9}. \tag{4.78a}$$

The quality x^* in this equation can be easily expressed by the condensate flow rate per unit of circumference Γ, because by definition

$$x^* = \frac{\dot{M}_G}{\dot{M}} = 1 - \frac{\dot{M}_L}{\dot{M}} = 1 - \frac{2R\pi\Gamma}{\dot{M}}. \tag{4.78b}$$

With the aid of the wall shear stress, the local heat transfer coefficient α is found from Eq. (4.75).

The original estimate of Γ_{i+1} must be corrected until the energy balance

$$\alpha(\vartheta_s - \vartheta_w) = \frac{\Gamma_{i+1} - \Gamma_i}{\Delta x} \Delta h_v$$

is satisfied. To decide whether stratified flow or annular-dispersed flow is present, one can make use of the flow pattern map, Fig. 4.12. One should also refer to the descriptions in Sect. 4.2.

Between stratified flow and annular-dispersed flow there is a transition region, in which both forms of flow can alternately occur in a tube. In this region, neither the influence of gravity clearly predominates as with stratified flow nor that of friction, as with annular-dispersed flow. Rather gravity and the forces exerted by the shear stress are of equal order of magnitude. According to investigations by Palen et al. [4.32], the transition region lies between

$$0.5 < w_G^* = \left(\frac{\varrho_G w_G^2}{(\varrho_L - \varrho_G) g d}\right)^{1/2} < 1.5. \tag{4.79}$$

Below $w_G^* = 0.5$, there is pure stratified flow, and above $w_G^* = 1.5$, pure film flow.

The dimensionless vapor velocity w_G^* defined by the previous equation is closely connected to the dimensionless vapor velocity

$$j_G^* = x^* \dot{M}/A[g d \varrho_G(\varrho_L - \varrho_G)]^{1/2}.$$

depicted in Fig. 4.12. Because $x^*\dot{M} = \dot{M}_G = \varrho_G w_G \varepsilon A$, there results the relation $\varepsilon w_G^* = j_G^*$.

According to Fig. 4.12, stratified flows and annular-dispersed flows occur only if $(1-\varepsilon)/\varepsilon \leqq 0.5$, or if the void fraction is sufficiently large i.e., if $2/3 \leqq \varepsilon \leqq 1$. Equation (4.79) can thus be written

$$0.5 < j_G^*/\varepsilon < 1.5,$$

from which then results

$$1/3 < j_G^* < 1.5. \tag{4.80}$$

when allowing for the largest and the smallest value of ε.

It follows then that, within this range, stratified flow and annular-dispersed flow can occur alternately. Below $j_G^* = 1/3$ is the area of the pure stratified flow, and above $j_G^* = 1.5$ that of the pure annular-dispersed flow. In the transition region, one obtains the heat transfer coefficient by a superposition of the heat transfer coefficient α_S (subscript S = stratified) for the stratified flow with the value α_F (subscript F = film) for the annular-dispersed flow according to

$$\alpha = \alpha_F + (w_G^* - 1.5)(\alpha_G - \alpha_S), \tag{4.81}$$

where α_F and α_S are determined in the manner discussed previously.

Application equations for calculation of heat transfer with condensation on the shell side of a *tube bundle* were already presented in Sect. 2.4.

5 Condensation of Metal Vapors

Whereas heat transfer coefficients during film condensation of nonmetallic vapors can be determined quite precisely with the theories already treated, their application to metallic vapors results in deviations from measured values. As previous experiments showed, one obtains during the condensation of metal vapors considerably smaller heat transfer coefficients than predicted by the Nusselt film condensation theory. Thus, the ratio from measured heat transfer coefficients α and the values α_{Nu} calculated from the Nusselt film condensation theory equal approximately 0.03 to 0.2 for condensing potassium and 0.02 to 0.06 for condensing sodium vapors. Figure 5.1 gives a survey of measured values in comparison to the results of the Nusselt film condensation theory. The mean Nusselt number is obtained there from the well known equation for the film condensation on inclined vertical walls, Eqs. (2.12) and (2.13),

$$\overline{Nu} = \frac{\bar{\alpha} h}{\lambda_L} = 0.943 \left[\frac{\varrho_L(\varrho_L - \varrho_G) g \sin\gamma \Delta h_v}{\lambda_L \eta_L (\vartheta_s - \vartheta_w)} h^3 \right]^{1/4}, \tag{5.1}$$

which can be converted into

$$\overline{Nu}/0.943 \left[\frac{\varrho_L(\varrho_L - \varrho_G) g \sin\gamma Pr h^3}{\eta_L^2} \right]^{1/4} = \left[\frac{c_{pL}(\vartheta_s - \vartheta_w)}{\Delta h_v} \right]^{-1/4}, \tag{5.2}$$

where $Pr = c_{pL}\eta_L/\lambda_L$ is the Prandtl number of the condensate film. In Fig. 5.1, the left side of Eq. (5.2) is plotted as the ordinate, and the expression $c_{pL}(\vartheta_s - \vartheta_w)/\Delta h_v$ as the abscissa. If one chooses a double logarithmic plotting, then Eq. (5.2) is represented as a straight line. By this type of plotting one can also render well the heat transfer coefficients for small Prandtl numbers from the solution of the boundary layer equations [5.2, 5.3]. One reason for the small heat transfer coefficient may be the fact that the thermal conductivity of a liquid metal can be very large; consequently, the thermal resistance of the condensate film is small and, therefore, of the same order of magnitude as the otherwise negligibly small molecular kinetic thermal resistance at the phase interface. This thermal resistance increases with the saturation pressure or the vapor density, as is obvious from Eq. (1.1). Therefore, the heat transfer during the condensation of metal vapors is dependent upon the pressure. To be sure, this resistance is additionally determined by the condensation coefficient, i.e., the number of the condensing molecules in relation to all the molecules impacting the condensate

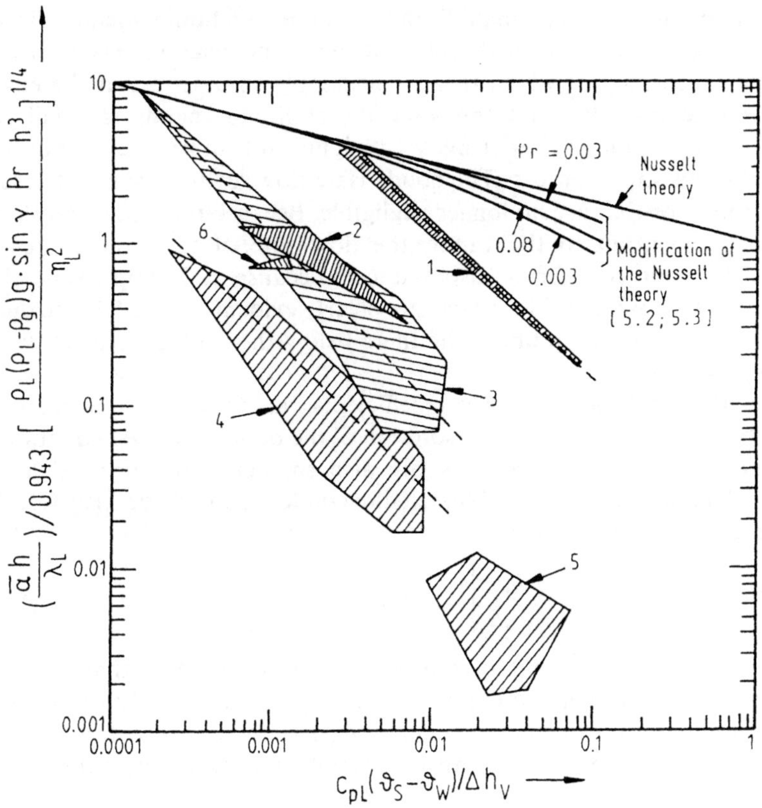

Fig. 5.1. Average Nusselt numbers during condensation of metal vapors, according to [5.1]. *1* to *3* mercury, *4* cadmium, *5* rubidium, *6* sodium

surface, about which there are no reliable data because of the experimental difficulties with liquid metals.

Many measurements [5.4–5.8] indicate that the condensation coefficient at sufficiently low pressures $p < 0.00384$ bar is approximately unity, and that at pressures in the range $0.00384\ \text{bar} < p < 1$ bar, it can be approximately described by

$$f = 0.062 p^{-0.5}.$$

However, there is considerable doubt whether the presently achievable accuracy in measurement permits such precise statements about the value of the condensation coefficient. The conjecture has therefore been made [5.9], that the condensation coefficient as stated by the above equation is based upon a false interpretation of insufficiently accurate results of measurement, and that the actual values lie close to unity. For a clarification of this question, further precise measurements of condensing metal vapors are needed.

Moreover, because of the small Prandtl number of liquid metals, $Pr = 10^{-2}$ to 10^{-4}, the assumption of the Nusselt film condensation theory is no longer tenable, according to which there is negligible shear stress at the edge of the condensate film. Because the viscosity of liquid metals is small in comparison to their thermal diffusivity, a significant momentum transfer between liquid and vapor occurs. The condensate flow leads to shear stresses at the phase interface that are no longer negligible. Because the liquid transfers a part of its momentum to the vapor, the flow rate of the condensate is diminished. The heat transfer is also reduced because it is approximately proportional to the flow rate. These facts are clearly evident in Fig. 5.1, because the curves for small Prandtl numbers lie there below those of the Nusselt film condensation theory.

Accordingly, local heat transfer coefficients, as they were calculated numerically by Koh et al. [5.3] from a solution of the boundary layer equations for the condensation on vertical walls, are also smaller than the local heat transfer coefficients according to the Nusselt film condensation theory, Eq. (2.11). If one writes Eq. (2.11) in the form

$$Nu_x \left[\frac{\varrho_L(\varrho_L - \varrho_G) g c_{pL} x^3}{4\eta_L \lambda_L} \frac{\Delta h_v}{c_{pL} \Delta \vartheta} \right]^{-1/4} = Y \tag{5.3}$$

with $Nu_x = \alpha x / \lambda_L$ and $\varrho_L(\varrho_L - \varrho_G)$ instead of ϱ_L^2, because we admit that there is not always $\varrho_G \ll \varrho_L$, then there would have to be $Y = 1$, if the conditions of the Nusselt film condensation theory are fulfilled. As Koh et al. showed, however, during the condensation of vapors with very small Prandtl numbers there is

$$Y = Y\left(Pr, \frac{c_{pL} \Delta \vartheta}{\Delta h_v} \right) < 1. \tag{5.3a}$$

In addition, Y is still a weak function of

$$\frac{(\varrho\eta)_L}{(\varrho\eta)_G}.$$

However, the influence of this value proved negligible in the range of values that occur for liquid metals. Table 5.1 gives typical values for Y.

Local heat transfer coefficients of metal vapors condensing in laminar flow on vertical walls can be calculated with the aid of these values according to the following method. The heat flow has to overcome two resistances connected one after the other, namely the molecular kinetic resistance and the resistance of the liquid film. We assume that the vapor is saturated and we indicate the temperature at the phase interface by ϑ_I. This gives then for the transferred heat flux

$$q = \alpha_I (\vartheta_s - \vartheta_I) \tag{5.4}$$

and

$$q = \alpha (\vartheta_I - \vartheta_w). \tag{5.5}$$

Table 5.1. Value of Y in Eq. (5.3) calculated by Koh et al. [5.3] for $[(\varrho\eta)_L/(\varrho\eta)_G]^{1/2} = 600$

Pr	$\frac{c_{pL}\Delta\vartheta}{\Delta h_v}$	Y
1	0.07704	0.9951
	0.3447	0.9893
	0.5231	0.9909
	0.7478	0.9967
0.03	$4.79 \cdot 10^{-5}$	0.9995
	$2.41 \cdot 10^{-4}$	0.9979
	$1.774 \cdot 10^{-3}$	0.9864
	$9.499 \cdot 10^{-3}$	0.9386
	0.03215	0.8510
	0.06831	0.7741
	0.1487	0.7100
0.008	$4.728 \cdot 10^{-4}$	0.9862
	$5.059 \cdot 10^{-3}$	0.8923
	0.01526	0.7849
	0.06579	0.6110
	0.1012	0.5639
0.003	$7.497 \cdot 10^{-5}$	0.9940
	$1.907 \cdot 10^{-3}$	0.8931
	$8.927 \cdot 10^{-3}$	0.7305
	0.02828	0.5867
	0.08746	0.4722

This gives

$$\frac{\alpha}{\alpha_I} = \frac{\vartheta_s - \vartheta_I}{\vartheta_I - \vartheta_w}, \tag{5.6}$$

where the heat transfer coefficient α_I is given by Eq. (1.1) and α by Eq. (5.3) together with the value Y from Table 5.1. Equation (5.6) serves for the determination of the temperature ϑ_I at the phase interface. Because the heat transfer coefficient α according to Eq. (5.3) is, in turn, dependent upon the temperature difference $\Delta\vartheta = \vartheta_I - \vartheta_w$, one can iteratively solve Eq. (5.6) for practical purposes. With the temperature ϑ_I the heat transfer coefficient α and also the entire thermal resistance are fixed. It is customary to introduce an overall heat transfer coefficient k by means of

$$\frac{1}{k} = \frac{1}{\alpha} + \frac{1}{\alpha_I}. \tag{5.7}$$

The transferred heat flux can then be calculated from

$$q = k(\vartheta_s - \vartheta_w). \tag{5.8}$$

The uncertainty of this calculation procedure is produced above all by the

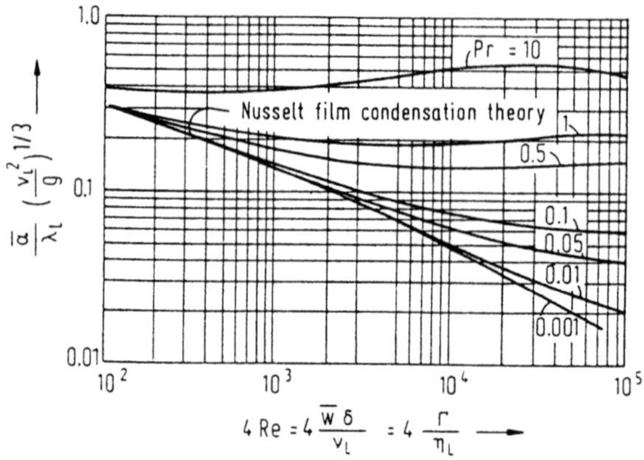

Fig. 5.2. Average Nusselt numbers for turbulent film condensation of metal vapors on a vertical plate, according to [5.10]

condensation coefficient, which is not known precisely enough. However, one can set a value of the coefficient close to unity, according to the presentation of Wilcox and Rohsenow [5.9].

The preceding equations for the heat transfer coefficient α hold for laminar film condensation. Lee [5.10] has analytically treated the case of turbulent film condensation for low Prandtl numbers, as they occur for liquid metals, by assuming in accordance with Nusselt, pure heat conduction in the liquid film and also equilibrium between gravity and friction forces. The momentum equation then contains the turbulent viscosity and the energy equation contains the turbulent thermal conductivity or the turbulent Prandtl number, in contrast to the Nusselt theory. For the turbulent viscosity, Lee used empirical equations of Deissler [5.11] and of von Kàrmàn [5.12]. The turbulent Prandtl number was considered to be constant. Because the equations correspond otherwise to those of the Nusselt theory, they need not be presented here in detail. Because of the complicated assumptions for the turbulent viscosity, a closed solution is not possible; only a numerical one is feasible. There results the average Nusselt numbers reproduced in Fig. 5.2. The figure was calculated with the assumption of a turbulent Prandtl number $Pr_t = 1$. As the investigations of Lee showed, the curves for small Prandtl numbers, i.e., in the range of liquid metals, are little influenced by the turbulent Prandtl number.

6 Condensation of Vapors of Miscible Liquids

6.1 Fundamentals. Influence of Mass Transfer

In the process industry, vapor mixtures are frequently liquefied, or the vapors contain inert gases, which by definition do not condense under the prevailing conditions. Their influence on the heat transfer during condensation was treated together with the deviations from the Nusselt film condensation theory in Sect. 2.2.3. There is yet to be considered how the heat transfer is changed if the condensate consists of a homogeneous phase and all components are essentially liquefied. Depending upon the purpose of the application, the vapor can be completely or partially condensed, so that vapor of a generally different composition than that of the entering vapor leaves the condenser. Such apparatus is designated as partial condenser or in rectification columns as dephlegmator. Because during the condensation of vapor mixtures as a rule the components with the higher boiling point condense first, and because the residual vapor thus becomes poor in these components, there is formed a concentration profile that is variable with the flow path and which influences considerably the heat exchange.

Whereas heat transfer during the condensation of binary mixtures has been frequently investigated, only little is known about the heat transfer during the condensation of mixtures with more than two components. Because the fundamental processes for these take place in a manner similar to those for binary mixtures, the characteristics of the latter during liquefaction will be discussed first in the following.

If a binary mixture, whose boiling point (bubble point) and dew point lines are depicted in Fig. 6.1a, condenses on a cooled wall with the temperature ϑ_w, there is formed a condensate, Fig. 6.1b, on which a vapor boundary layer is established. At the phase interface, a temperature ϑ_I results which lies between the temperature ϑ_G of the vapor far from the wall and the wall temperature ϑ_w. If the vapor is saturated, its temperature equals $\vartheta_G = \vartheta_s$, corresponding to Fig. 6.1a. The temperature distribution in the vapor and condensate is shown in Fig. 6.1c.

The component with the higher boiling point condenses more readily. Consequently, the vapor mixture at the phase interface generally contains more of the more volatile component than it does at a greater distance from it. A concentration profile builds up, as is depicted in Fig. 6.1d.

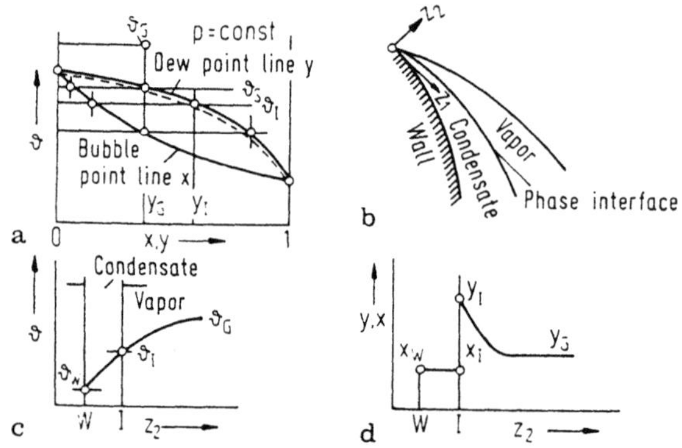

Fig. 6.1 a–d. Diagram for temperature and concentration for a binary mixture. Indices: *W* cold wall, *I* phase interface, *G* core flow in the vapor (*G* Gas). **a** profile of the boiling and dew point line, temperatures; **b** condensate and vapor boundary layer; **c** temperature profile; **d** concentration profile

The concentration of the more volatile component increases toward the phase interface. In the stationary state, the molecules of the more volatile component that are not condensed at the condensate surface are transported back into the vapor by diffusion. In the condensate, momentum, heat, and mass transfer by convection usually can be neglected compared to conduction and diffusion, unless the Prandtl and Schmidt numbers are very small. Therefore, in the condensate film, the Nusselt assumptions continue to hold, according to which the flow is determined only by the forces of viscosity and gravity, whereas the temperature profile is determined chiefly by heat conduction. The concentration is constant across a cross section of the condensate film, because the wall is impermeable for matter and convective mass transport is negligible in the liquid film. This is easily understood, because under the mentioned assumptions, the equation for diffusion, the so-called second law of Fick, reads

$$\frac{\partial}{\partial z_2}\left(D\frac{\partial c}{\partial z_2}\right)=0,$$

where z_2 is the coordinate vertical to the wall, D the diffusion coefficient, and $c = n/V$ the concentration of one of the two components of the binary mixture. If this equation is integrated according to the limiting conditions,

$$\left(\frac{\partial c}{\partial z_2}\right)_{z_2=0}=0 \quad \text{and} \quad c(z_2=\delta)=c_{\mathrm{I}},$$

there follows directly $c = c_w = c_{\mathrm{I}} = \text{const}$. Therefore, it is $x = x_w = x_{\mathrm{I}} = \text{const.}$, as shown in Fig. 6.1d.

The concentration profile is influenced considerably by the flow and the type of flow arrangement. During the usual cocurrent flow of vapor and condensate shown in Fig. 6.2a, the higher boiling component condenses first. The fraction of the lower boiling component, therefore, increases downstream in the vapor from the original composition y_α to the final composition y_ω. The condensate quantity accumulating in the inlet cross section is large and, consequently, much of the higher boiling component is withdrawn from the

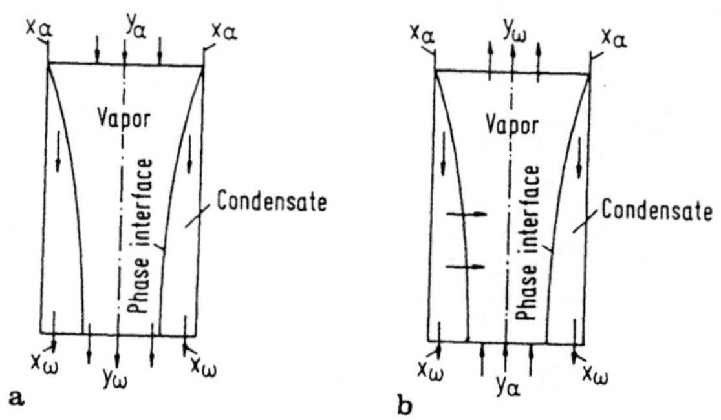

Fig. 6.2 a, b. Condensation during **a** cocurrent and **b** countercurrent flow. In case **a**, the falling liquid contains more volatile constituents (LB) than in **b**. In case **b**, LB-molecules pass over into the vapor, because rising less volatile constituents condense and thus drive out LB-molecules

vapor. The conditions are different during countercurrent flow, Fig. 6.2b, which occurs in falling film towers or in the reflux condensers of rectifier systems. There the vapor with initial composition y_α encounters a thicker condensate film. The condensing mass is smaller, and less of the higher boiling component is withdrawn from the vapor than is the case with cocurrent flow. The rising vapor now comes into contact with an increasingly thinner condensate film of the liquid trickling down. The accumulated condensate mass increases and the higher boiling components of the vapor condense preferentially. The condensation enthalpy being released effects simultaneously an evaporation of the low boiling components, which is indicated by the horizontal arrows in Fig. 6.2b. A mass exchange takes place between the liquid and the vapor by "rectification", whereby the exiting vapor contains more of the more easily volatile constituents and the condensate emerging below fewer, than is the case in cocurrent flow according to Fig. 6.2a. If one wants to produce a condensate more strongly enriched in substances that are difficult to vaporize, then the countercurrent flow conduction is preferable, an effect which Claude [6.1] pointed out more than fifty years ago. The flowing vapor contains during countercurrent flow more easily volatile substances than during cocurrent flow. As a result of the shear stress exerted by the vapor, the condensate is dammed up during countercurrent flow and the heat transfer is impaired.

Obviously, the behavior of the flow is decisive for the mass transfer, and thus, also for the heat transfer. It is generally true that a mass transfer resistance is formed as a consequence of the concentration field at the phase interface and that this resistance obstructs the mass flow in the direction of the phase interface. As a result, the heat transfer of condensing vapor mixtures for equal driving temperature gradients is less than that of condensing pure vapors. This decrease can assume considerable values. As an example of this, Fig. 6.3a shows the relative heat flux q/q_0 plotted versus the temperature ϑ_∞, which in this case is assumed equal to the saturation temperature along the dew point line. As is recognized, the actual heat flux q is clearly smaller than the heat flux q_0, which would be encountered if no resistance for the mass transfer in the vapor had to be overcome, and, therefore, the saturation temperature prevailed at the condensate surface.

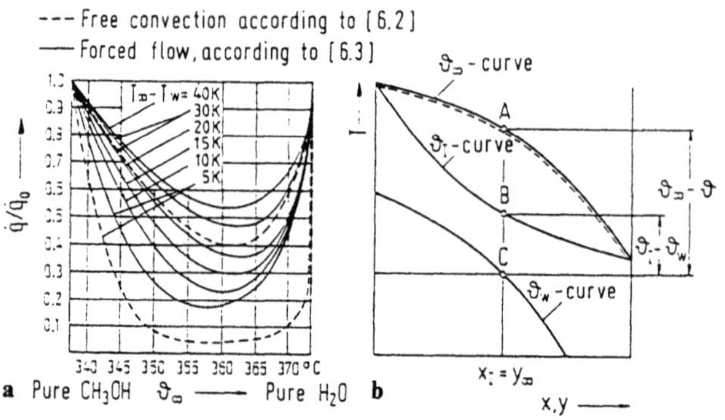

Fig. 6.3 a, b. Reduction of heat flow rate during condensation of methanol/water. **a** relative heat flux q/q_0 as a function of temperature ϑ_∞; **b** boiling diagram

The solid lines represent the condensation of a vapor mixture of methanol (CH_3OH) and water (H_2O) on a horizontal plate, for example, if the influence of gravity is negligible, or if the vapor velocity is sufficiently large and thus the *Froude number* $Fr = w_\infty^2/gz$ very large. The dashed curves are for the condensation in the case of small Froude numbers, for example, on a vertical wall with free convection of the vapor.

As in the case of the Nusselt film condensation theory, during the condensation of mixed vapors the heat flux increases with the driving temperature difference $\vartheta_\infty - \vartheta_w$. According to the Nusselt theory, the heat transfer coefficient decreases with the driving temperature difference corresponding to $\alpha \sim (\vartheta_\infty - \vartheta_w)^{-1/4}$, Eq. (2.11). The heat flux increases according to $q \sim (\vartheta_\infty - \vartheta_w)^{3/4}$. In addition, one recognizes in Fig. 6.3a a minimum of the heat flux at a definite temperature ϑ_∞. This results from the fact that the temperature difference $\vartheta_I - \vartheta_w$ between the condensate surface and the wall, which is decisive for the heat transfer, also assumes a minimum at a definite saturation temperature ϑ_∞, which can be explained with the aid of the boiling diagram, Fig. 6.3b. We assume here a sufficiently large temperature difference $\vartheta_\infty - \vartheta_w$, so that the vapor of the initial state A condenses completely and a condensate results that is designated by point B in Fig. 6.3b. The temperature ϑ_I at the phase interface is then equal to the boiling temperature of the liquid mixture and the composition of the accumulating condensate is identical with that of the vapor. This is called *local total condensation.* Let the wall temperature, assumed as constant, be designated by point C. The segment BC corresponds then to the temperature difference $\vartheta_I - \vartheta_w$, which is decisive for the heat flux q. If the Nusselt film condensation theory could hold also for the condensation of mixed vapors, there would be $q \sim (\vartheta_I - \vartheta_w)^{3/4}$. If one keeps $\vartheta_\infty - \vartheta_w$ constant and raises the temperature ϑ_∞, i.e., if one follows in Fig. 6.3a the path of a curve $\vartheta_\infty - \vartheta_w = \text{const}$ in the direction of increasing temperature ϑ_∞, one can again read off the segment BC in Fig. 6.3b, when one lays a parallel to the dew point line through point C. The segment $AC = \vartheta_\infty - \vartheta_w$ thus remains presumably unchanged, whereas *BC* at first grows smaller and then gets larger. The relative heat flux q/q_0 is also changed in the same proportion. If, on the other hand, one keeps the vapor temperature ϑ_∞ constant and diminishes the temperature difference $\vartheta_\infty - \vartheta_w$ by raising the wall temperature, then in Fig. 6.3b the ϑ_w-curve going through point *C* must be shifted upward. Thus, smaller values for the heat flux q and for the relative heat flux q/q_0 are obtained.

6.2 The Different Types of Condensation of Vapor Mixtures

If one observes a condensate film, then one sees that its surface is not smooth, but rather as a consequence of ever present mechanical, thermal, or chemical disturbances it is subjected to small variations in the thickness, temperature, and composition. Also the momentum exerted by the condensing vapor on the condensate film produces slight variations of the film thickness. Such variations lead, of necessity, to local changes of the surface tension. In thermodynamics, the surface tension is interpreted as free energy related to the unit area of the surface phase. On the basis of the second law of thermodynamics, every boundary surface is striving for a condition with a smaller surface tension. This happens by enlarging surfaces with smaller surface tension. Local changes of the surface tension can be equalized by convection, in that areas with a smaller surface tension expand at the cost of those with a greater surface tension. Thus, a marked flow detectable by the eye can occur on liquid surfaces. This phenomenon, which occurs both in completely miscible liquids and solutions as well as in only weakly miscible liquids, is called the *Marangoni Effect*. Local variations of the surface tension also cause variations of the thickness of condensate films, which either can increase further or can disappear again. If a thin condensate film were to be formed as the result of such a disturbance at a definite point and if it simultaneously had a greater surface tension than its neighboring thicker layers, then, because of its smaller surface tension, the thicker film would spread out into the area of the thinner film, until the films would again be equally thick. If on the other hand, the thicker condensate film had a greater surface tension than the thinner one, then the thinner one would continue to expand. Random disturbances would thus be increased. An unevenly thick film could be formed. For example, channels could be formed alongside thin liquid layers, even next to dry spots. These observations lead to the conclusion that the stability in respect to the variations of the surface tension depends upon the sign of the change of the surface tension σ with the film thickness δ. The condensate film is stable for

$$\frac{d\sigma}{d\delta} \leqq 0, \tag{6.1}$$

it is unstable for

$$\frac{d\sigma}{d\delta} > 0. \tag{6.2}$$

As Ford and Missen [6.4] have shown, the condition of Eq. (6.1) is necessary and sufficient as a criterion of stability; on the other hand, the condition of Eq. (6.2) is only a necessary criterion, i.e., if a film becomes instable, Eq. (6.2) is necessarily fulfilled. However, no instability need be present, if Eq. (6.2) is fulfilled. Experiments [6.5] have shown that instabilities are less marked or even disappear, if the condensation rate and thus the film thickness are great or if the composition approaches that of pure substances or that of an azeotropic point, at which mixtures in general condense as a cohesive film.

The surface tension of a liquid, which is in equilibrium with the vapor, is a function of the temperature and the composition of the mixture

$$\sigma = \sigma(\vartheta, x_1, x_2, \ldots, x_{K-1}), \tag{6.3}$$

where x_i is the mole fraction in the liquid phase and K the number of components of the complex mixture. This gives

$$\mathrm{d}\sigma = \left(\frac{\partial\sigma}{\partial\vartheta}\right)_{x_j} \mathrm{d}\vartheta + \sum_{k=1}^{K-1} \left(\frac{\partial\sigma}{\partial x_k}\right)_{\vartheta, x_{j \neq k}} \mathrm{d}x_k \tag{6.4}$$

and further

$$\frac{\mathrm{d}\sigma}{\mathrm{d}\delta} = \left(\frac{\partial\sigma}{\partial\vartheta}\right)_{x_j} \frac{\mathrm{d}\vartheta}{\mathrm{d}\delta} + \sum_{k=1}^{K-1} \left(\frac{\partial\sigma}{\partial x_k}\right)_{\vartheta, x_{j \neq k}} \frac{\mathrm{d}x_k}{\mathrm{d}\delta}. \tag{6.5}$$

The partial derivatives are to be calculated here at the temperature of the phase interface ϑ_I. These derivatives are given by the thermodynamic characteristics of the fluid, whereas the total derivatives $\mathrm{d}\vartheta/\mathrm{d}\delta$ and $\mathrm{d}x_k/\mathrm{d}\delta$ are fixed by the heat and mass transfer during the condensation. For binary mixtures, Eq. (6.5) converts into

$$\frac{\mathrm{d}\sigma}{\mathrm{d}\delta} = \left(\frac{\partial\sigma}{\partial\vartheta}\right)_{x_1} \frac{\mathrm{d}\vartheta}{\mathrm{d}\delta} + \left(\frac{\partial\sigma}{\partial x_1}\right)_{\vartheta} \frac{\mathrm{d}x_1}{\mathrm{d}\delta}, \tag{6.6}$$

where x_1 designates the mole fraction of the more easily volatile component in the liquid. Now, with condensing vapors, an increase of the temperature ϑ at the phase interface has as a result also a growth of the film thickness δ. It is

$$\mathrm{d}\vartheta/\mathrm{d}\delta > 0. \tag{6.7}$$

This follows already approximately from the Nusselt film condensation theory, Eq. (2.11), according to which there is $\vartheta_s - \vartheta_w \sim \delta^4$. Simultaneously, an increase in the film thickness, if one disregards azeotropic mixtures, causes an enrichment of the less volatile component in the condensate. In this, with increasing film thickness, the portion of the more easily volatile component decreases, and, therefore,

$$\mathrm{d}x_1/\mathrm{d}\delta < 0. \tag{6.8}$$

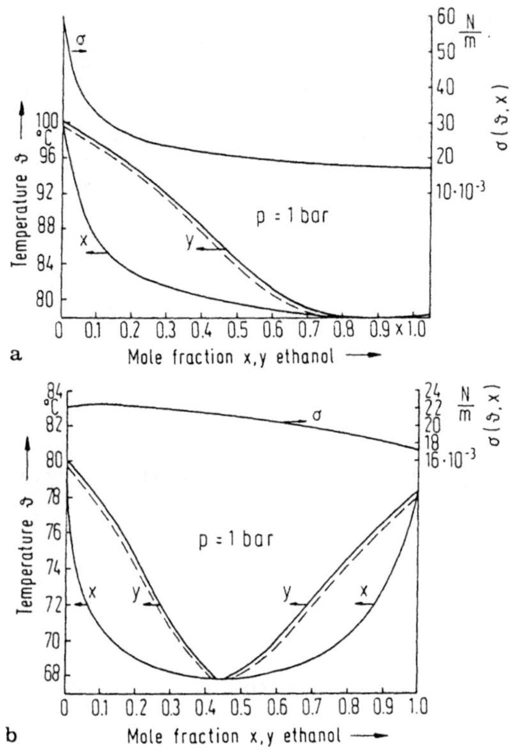

Fig. 6.4 a, b. Bubble point line (x), dew point line (y) and surface tension (σ) of the mixtures **a** ethanol/water and **b** ethanol/benzene

Furthermore, the more easily volatile component frequently also has the smaller surface tension. An increase in the mole faction x_1 of this component usually leads, therefore, to a decrease of the surface tension. Figure 6.4 shows as an example the surface tension of ethanol-water and ethanol-benzene as a function of the mole fraction x_1 of ethanol. In most cases, this gives

$$\left(\frac{\partial \sigma}{\partial x_1}\right)_{\vartheta} < 0. \tag{6.9}$$

The second term in Eq. (6.6) is thus positive, and, because of Eq. (6.7), the sign of the first term depends on the sign of $(\partial \sigma / \partial \vartheta)_{x_1}$. If one excludes the few mixtures with relatively extreme values of $\sigma(\vartheta)$, this sign is determined by the sign of $\Delta\sigma/\Delta\vartheta$, where $\Delta\sigma$ is the difference of the surface tensions of the pure components and $\Delta\vartheta$ the difference of their boiling temperatures. If

$$\Delta\sigma/\Delta\vartheta < 0, \tag{6.10}$$

then the first term in Eq. (6.6) also becomes negative. Because the first term is usually greater in value than the second term, then

$$\mathrm{d}\sigma/\mathrm{d}\delta \leqq 0.$$

Table 6.1. Condensation forms of several binary mixtures

Mixture (more volatile component named first)	Sign of $\Delta\sigma/\Delta\vartheta$	Condensation form: f = film nf = non-film	Azeotrop at 1 bar
n-Hexane/Benzene	+	nf	no
n-Pentane/n-Hexane	+	nf	no
n-Pentane/Benzene	+	nf	no
2,3-Dimethylbutane/Benzene	+	nf	no
2,3-Dimethylbutane/Toluene	−	f	no
n-Pentane/Methanol	+, −[a]	nf, f	yes
Methylene chloride/Methanol	−, +	f, ?	yes
n-Pentane/Methylene chloride	+, −	nf, f	yes
Methanol/n-Hexane	−, +	f, nf	yes
Methanol/Water	+	nf	no
Acetone/Methanol	−	f	no
Ethanol/Water	+, −	nf, ?	yes

[a] The first sign indicates the slope of $\Delta\sigma/\Delta\vartheta$ until the azeotropic point, the second sign from the azeotropic point until the more volatile component $x_1 = 1$.

According to Eq. (6.1), a stable condensate film is to be expected, if the condition of Eq. (6.10) is fulfilled. Conversely, if

$$\Delta\sigma/\Delta\vartheta > 0, \tag{6.11}$$

then the first term in Eq. (6.6) becomes positive. Because the second term is positive in most cases,

$$d\sigma/d\delta > 0.$$

Instabilities can occur, if the condensation rate is not very large and thus the film is not thick enough. The data which have been assembled in Table 6.1 based upon the work of Ford and Missen [6.4] and of Tamir [6.6] are in agreement with this.

The different condensation forms of mixtures with three components were investigated by Tamir [6.6]. He also showed that, corresponding to the previous considerations, the condensate can accumulate as a film, or that, when instabilities occur, the condensate can run off in the form of streamers, as drops, in mixed forms as film-streamers, or as a streamer-drop mixture.

Figures 6.5 through 6.8 show several of the different forms of condensate formation. The mixture composition and the pressure are practically the same for all the pictures. Only the driving temperature difference was changed. With large temperature differences or a composition with a preponderant portion of one component, film condensation occurs for the most part, whereas with smaller temperature differences, streamer and drop formation are the prevailing condensation forms. The measured heat transfer coefficients are larger up to a

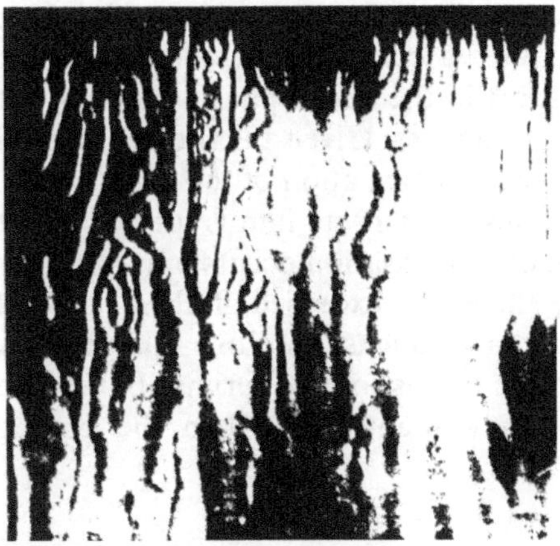

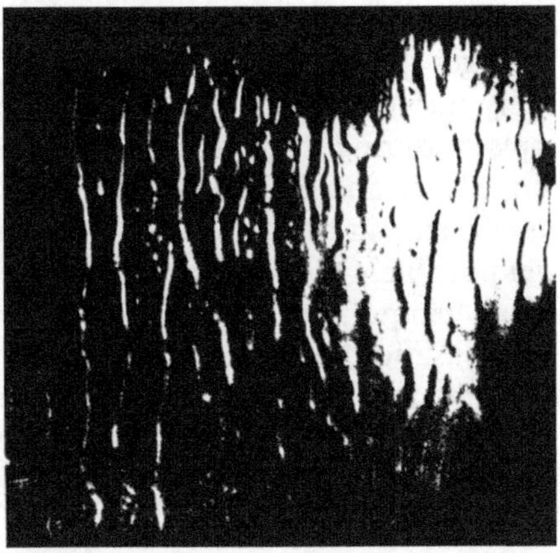

Fig. 6.5. **Fig. 6.6.**

Fig. 6.5. Streamer formation [6.6] during condensation of a mixture of acetone, $\xi_1 = 0.0837$, methanol $\xi_2 = 0.781$, and water $\xi_3 = 0.136$ at 0.986 bar. $\vartheta_\infty - \vartheta_\infty = 15\,°C$, $\sigma = 26.10^{-3}$ N/m at 25 °C

Fig. 6.6. Streamer and drop formation [6.6] during condensation of the mixture as in Fig. 6.5. $\vartheta_\infty - \vartheta_w = 10.9\,°C$; $\sigma = 26 \cdot 10^{-3}$ N/m at 25 °C

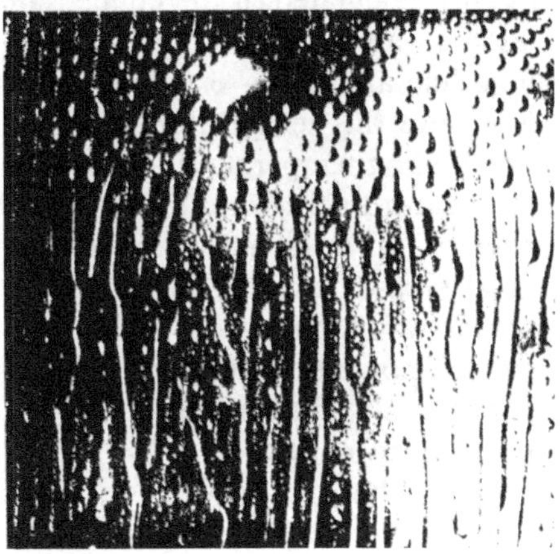

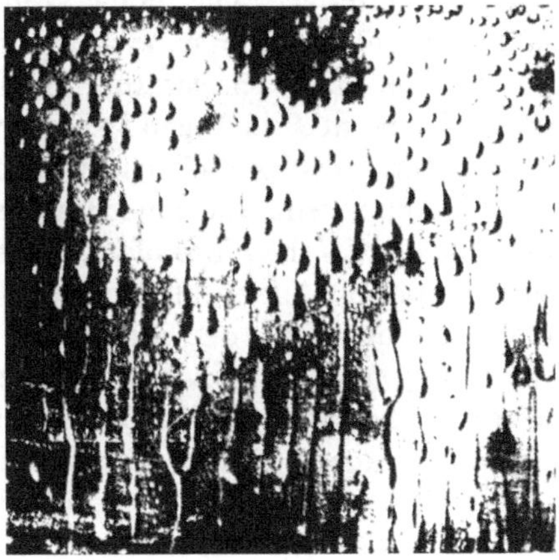

Fig. 6.7. **Fig. 6.8.**

Fig. 6.7. Streamer and drop formation [6.6] during condensation of the mixture as in Fig. 6.5. $\vartheta_\infty - \vartheta_w = 16.6\,°C$; $\sigma = 29.33 \cdot 10^{-3}$ N/m at 25 °C

Fig. 6.8. Drop formation [6.6] during the condensation of the same mixture as in Fig. 6.5. $\vartheta_\infty - \vartheta_w = 12\,°C$; $\sigma = 29.33 \cdot 10^{-3}$ N/m at 25 °C

factor of four than those of the pure film condensation. Such an improvement of the heat transfer is to be expected because the average heat resistance of a non-continuous film is in general smaller than that of a continuous film. To be sure, with mixtures, the increase of the heat transfer is by no means as large as with the change from film to drop condensation of pure substances, where the heat transfer coefficient increases by a factor of ten or more. Obviously, there emerges during the condensation of mixtures a considerable, additional heat resistance conditioned by the mass transport in the vapor phase, which is not present during the condensation of pure substances. Given the present state of knowledge and owing to the few experiments, a calculation of the heat transfer coefficients in the cases where no continuous condensate film is formed is not yet possible. On the other hand, this form of condensation occurs only at rather small temperature differences $\vartheta_\infty - \vartheta_w$; these are exceeded if there is total condensation in the condensers. With partial condensation, one must reckon with the fact that no continuous condensate film is formed, primarily because the coolant temperature there is often only slightly less than the boiling temperature of the most volatile component.

6.3 The Temperature at the Phase Interface

For the calculation of the heat transfer during film condensation of mixtures the temperature ϑ_I at the phase interface must be known. We shall start with the fact that there is thermodynamic equilibrium between liquid and vapor at the phase interface. The temperature ϑ_I is easily calculated, if the condensate at that point has the same composition as the vapor. It agrees, then, with the corresponding temperature on the boiling line and is, for example, determined by point B in Fig. 6.3b. In order to achieve this temperature, the wall temperature must be sufficiently far below the boiling temperature of the most volatile component. The condensation rate must accordingly be sufficiently large, and one speaks therefore of local total condensation. In technical condensers, this condition is often fulfilled; but there are also cases, such as during partial condensation, where one consciously chooses the wall temperature higher than the boiling temperatures of the more volatile components, so that these occur in the condensate in only slight amounts or not at all. The temperature at the phase interface is then also higher than the temperature at the boiling line.

In order to show how the temperature at the phase interface is obtained, we shall consider a binary mixture and establish for it the mass balance at the condensate surface. The mass flow rate $\dot{M}_G$ of the vapor flowing vertically to the phase interface is condensed there and is drawn off as condensate $\dot{M}_L$. It is

$$\dot{M}_G = \dot{M}_L = \dot{M}.$$

Let the more volatile component have at the phase interface the mass fraction ξ_I'' in the vapor mixture. A portion $\dot{M}\xi_I''$ of the mass flow rate is conveyed to the condensate surface by convection, and the portion $j_G A$ is supplied by diffusion and is drawn off in the condensate as the mass flow rate $\dot{M}\xi_I'$. The mass balance of the more volatile component at the surface A of the condensate film reads, therefore,

$$\dot{M}\xi_I' = \dot{M}\xi_I'' + j_G A, \tag{6.12}$$

where the diffusion flux is given by

$$j_G = -\varrho_G D\left(\frac{\partial \xi}{\partial z_2}\right)_I.$$

Let the coordinate z_2 be directed vertically from the condensate surface into the vapor. After the introduction of a mass transfer coefficient β_G, one can also write the diffusion flux as

$$j_G = \varrho_G \beta_G(\xi_I'' - \xi_G). \tag{6.13}$$

The mass fraction of the more volatile component in the gas at a great distance from the condensate surface is designated as ξ_G. With this, Eq. (6.12) converts into

$$\frac{\dot{M}}{A} = \dot{m} = -\varrho_G \beta_G \frac{\xi_I'' - \xi_G}{\xi_I'' - \xi_I'}. \tag{6.14}$$

The negative sign derives from the fact that the coordinate z_2 is by definition directed vertically away from the condensate surface and that the condensing vapor is flowing in opposite direction to this coordinate. In Eq. (6.14), there are now contained two limiting cases:

a) During local total condensation, the composition of the liquid ξ_I' is identical to that of the vapor $\xi_G = \xi_I'$. $\dot{M}/A = -\varrho_G \beta_G$ is then the mass flux of the vapor flowing to the condensate surface. In order to achieve local total condensation, the wall temperature must be sufficiently far below the boiling temperature of all components, so that this condenses all of them and that the condensate shows the same composition as the vapor. The mass flux $\dot{M}/A$ is correspondingly large.
b) For vanishingly small mass fluxes of the condensate $\dot{M}/A \to 0$, the vapor composition is $\xi_G = \xi_I''$ according to Eq. (6.14). Practically no condensate is formed and, consequently, no concentration profile is developed in the vapor space.

Both limiting cases are plotted in Fig. 6.9. Actual condensation rates lie between the two extreme cases. The temperature ϑ_I at the phase interface lies, as seen in Fig. 6.9, between the temperature of the dew point line (case b) and the temperature of the boiling line (case a). Accompanying vapor and liquid compositions are read off of the abscissa, points A and B in Fig. 6.9.

For the calculation of the temperature ϑ_I at the phase interface, one also needs as a further balance equation the energy equation at the condensate surface

$$q_L = q_G + \dot{m}\Delta h_v, \tag{6.15}$$

where q_L is the heat flux removed by conduction in the condensate, q_G the heat flux supplied by convection of the vapor to the condensate surface, and Δh_v the vaporization enthalpy of the mixture. We have

$$q_L = -\alpha_L(\vartheta_I - \vartheta_w). \tag{6.16}$$

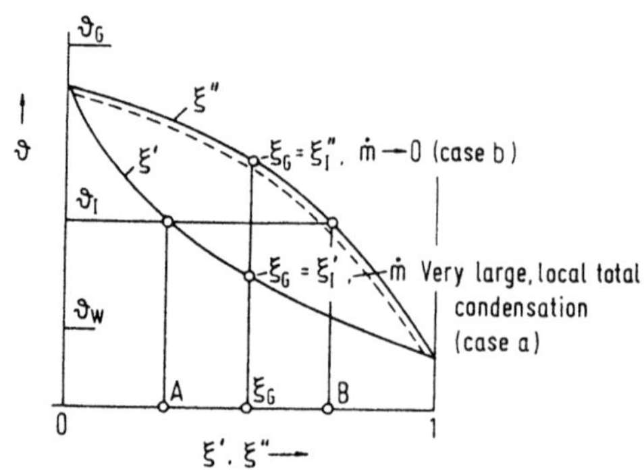

Fig. 6.9.

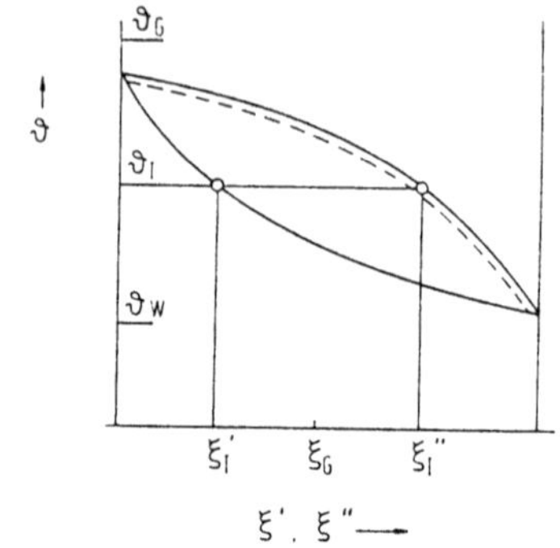

Fig. 6.10.

Fig. 6.9. Limiting cases of condensation

Fig. 6.10. Mass fractions and temperature at the condensation surface

On the other hand, this heat flux is transmitted by the wall to the coolant, whose temperature shall be ϑ_K. Let the heat transfer coefficient between wall and coolant be k'. Then,

$$q_L = -k'(\vartheta_w - \vartheta_K). \tag{6.17}$$

With the aid of Eq. (6.16) one can eliminate ϑ_w from this equation, and one then obtains

$$q_L = \frac{-\alpha_L}{(\alpha_L/k') + 1}(\vartheta_I - \vartheta_K). \tag{6.18}$$

In the case of a very large heat transfer coefficient $k' \to \infty$, the wall temperature ϑ_w is constant and is equal to the coolant temperature ϑ_K.

The heat flux from the vapor to the condensate surface is

$$q_G = -\alpha_G(\vartheta_G - \vartheta_I). \tag{6.19}$$

The negative sign in Eqs. (6.16) to (6.19) stems from the fact that the heat flow rates are directed against the coordinate z_2. If one inserts into the energy balance, according to Eq. (6.15), the heat flux q_G according to Eq. (6.19), together with the heat flux q_L, according to Eq. (6.18), and considers that the mass flux of the condensate is given by Eq. (6.14), then one obtains for the energy balance

$$\frac{\alpha_L}{(\alpha_L/k') + 1}(\vartheta_I - \vartheta_K) = \alpha_G(\vartheta_G - \vartheta_I) + \varrho_G\beta_G\frac{\xi_I'' - \xi_G}{\xi_I'' - \xi_I'}\Delta h_v. \tag{6.20}$$

From this equation can be calculated the unknown temperature ϑ_I. One must, however, consider that the mass fractions ξ_I' and ξ_I'' are dependent upon the temperature ϑ_I, as shown in Fig. 6.10 for a binary mixture. In addition, the heat transfer coefficient α_L is dependent upon the temperature ϑ_I. Because at least free convection occurs in the vapor, the term $\alpha_G(\vartheta_G - \vartheta_I)$ is often not negligible, but rather can be of an equal order of magnitude as the other expressions of Eq. (6.20), especially if the temperature of the phase interface ϑ_I is close to the wall temperature, so that the driving temperature gradient $\vartheta_I - \vartheta_w$ in the condensate is small, while that in the vapor $\vartheta_G - \vartheta_I$ is large.

The *practical calculation* of the temperature ϑ_I at the condensate surface is quite laborious and, even in the case of the binary mixture discussed here, can scarcely be achieved without a computer. In order to clarify the basic procedure, we shall simplify Eq. (6.20) by writing

$$\alpha_L(\vartheta_I - \vartheta_w) = \alpha_G(\vartheta_G - \vartheta_I) - \frac{\dot{M}}{A}\Delta h_v. \tag{6.21}$$

Here, $\dot{M}$ is the vapor mass flow rate condensing on a surface $A = b\Delta z$. Its sign is negative, in keeping with our system of signs. It flows perpendicular to the condensate surface and causes an increase in the condensate mass flow rate Γb, if we designate by Γ, as before, the condensate mass flow rate per unit of length; if the length of flow is designated by z,

$$\frac{\mathrm{d}(\Gamma b)}{\mathrm{d}z}\Delta z = -\dot{M} \quad \text{and, therefore,} \quad \frac{\mathrm{d}\Gamma}{\mathrm{d}z} = \frac{-\dot{M}}{b\Delta z} = \frac{-\dot{M}}{A} = -\dot{m}.$$

Based upon the earlier observations about the heat transfer during laminar or turbulent film condensation, one can also write for this

$$-\dot{m} = \frac{\mathrm{d}\Gamma}{\mathrm{d}\delta}\frac{\mathrm{d}\delta}{\mathrm{d}z}$$

so that Eq. (6.21) is converted into

$$\alpha_L(\vartheta_I - \vartheta_w) = \alpha_G(\vartheta_G - \vartheta_I) + \frac{\mathrm{d}\Gamma}{\mathrm{d}\delta}\frac{\mathrm{d}\delta}{\mathrm{d}z}\Delta h_v. \tag{6.22}$$

In this equation, the heat transfer coefficient α_L in the condensate film is only dependent upon the film thickness. According to the Nusselt film condensation theory for laminar film condensation, $\alpha_L = \lambda_L/\delta$, whereas one obtains for turbulent film condensation the heat transfer coefficient α_L from the earlier equations of Sect. 4.2.3. The heat transfer coefficient α_G is determined by the type of vapor flow. The differential quotient $\mathrm{d}\Gamma/\mathrm{d}\delta$ can also be expressed as a function of the film thickness δ, and, as has been shown earlier, Eq. (2.9), is during laminar film condensation

$$\frac{\mathrm{d}\Gamma}{\mathrm{d}\delta} = \frac{\varrho_L^2 g}{\eta_L}\delta^2. \tag{6.23}$$

On the other hand, one obtains during turbulent film condensation the expression $\mathrm{d}\Gamma/\mathrm{d}\delta$ by differentiation from Eq. (4.26) as a function of the film thickness. Equation (6.22) is, therefore, a differential equation of the form

$$f\left(\delta, \frac{d\delta}{dz}\right) = 0,$$

from which one gets by integration the film thickness

$$\delta = \delta(z, \vartheta_I). \tag{6.24}$$

With this one can also express the differential quotient $d\Gamma/d\delta$, for example, according to Eq. (6.23) as a function of the length of flow z and the temperature ϑ_I at the condensate surface as

$$\frac{d\Gamma}{d\delta} = f_1(z, \vartheta_I).$$

Likewise, there is also

$$\frac{d\delta}{dz} = f_2(z, \vartheta_I)$$

and thus, also, when allowing for Eq. (6.22),

$$\frac{d\Gamma}{d\delta}\frac{d\delta}{dz} = \frac{d\Gamma}{dz} = f_3(z, \vartheta_I) = -\dot{m}.$$

On the other hand, because of Eq. (6.14),

$$-\varrho_G \beta_G \frac{\xi_I'' - \xi_G}{\xi_I'' - \xi_I'} = \dot{m} = -f_3(z, \vartheta_I),$$

is also known. From this equation is obtained the unknown temperature ϑ_I of the condensate surface at a point z, because the boiling line $\xi_I'(\vartheta_I)$ and the dew point line $\xi_I''(\vartheta_I)$ are assumed to be given. One can calculate the mass transfer coefficient β_G approximately by using the analogy between heat and mass exchange, according to which

$$\varrho_G \beta_G = \frac{\alpha_G}{c_{pG}} Le^{2/3} \tag{6.25}$$

is valid with the Lewis number $Le = D/a$, in which D is the diffusion coefficient of the vapor and a its thermal diffusivity. For many gas mixtures, the Lewis number does not deviate greatly from unity. The mass transfer coefficient thus calculated needs an additional correction, which allows for the mass transport directed convectively toward the phase interface [6.7]. As can be seen from Sect. 2.2.4, the actual heat and mass transfer coefficients should be enlarged by a factor $f \approx 1.15$, which covers the waviness of the film surface.

6.4 The Practical Calculation of Heat Transfer Coefficients

6.4.1 The Approximation Procedure of Silver

Because it is difficult to calculate the temperature at the phase interface, and thus the heat transfer, especially during the condensation of mixtures with more than two components, one frequently makes use of simple but rather imprecise methods of calculation. Silver [6.8] and later Bell and Ghaly [6.9] described a procedure which has proved itself where there are not all too high requirements for precision. It starts from the following considerations. The entire heat flow, which is to be conducted to the cooling agent, is composed of three parts: the condensation enthalpy Φ_K, the heat flow Φ_L for the cooling of the condensate, and the heat flow Φ_G for the cooling of the gas. Temperature changes in the wall of the condenser shall be negligible. The heat flow

$$d\Phi = d\Phi_K + d\Phi_L + d\Phi_G$$

is conducted along one small surface section of the condenser to the cooling agent. This can also be expressed as

$$d\Phi = k' dA_0 (\vartheta_I - \vartheta_K), \tag{6.26}$$

where k' indicates the heat transfer coefficient from the condensate surface to the cooling agent, dA_0 a surface element of the condensate surface, ϑ_I its temperature, and ϑ_K the local temperature of the cooling agent. The area A_0 is practically equal to the surface area of the condenser.

The thermal resistance $1/k' dA_0$ is composed of the individual resistances from the condensate surface to the cooling agent. It is

$$\frac{1}{k' dA_0} = \frac{1}{\alpha dA_0} + \frac{r_{si}}{dA_0} + \frac{\delta_w}{\lambda_w dA_w} + \frac{r_{sa}}{dA_a} + \frac{1}{\alpha_K dA_a}$$

or

$$\frac{1}{k'} = \frac{1}{\alpha} + r_{si} + \frac{\delta_w dA_0}{\lambda_w dA_w} + r_{sa}\frac{dA_0}{dA_a} + \frac{dA_0}{\alpha_K dA_a}. \tag{6.27}$$

Here α designates the heat transfer coefficient of the condensate film, dA_0 a surface element, r_{si} a resistance for fouling deposits on the inner side of the cooling surface, δ_w the thickness of the wall, λ_w the wall thermal conductivity, dA_w the size of the wall surface – in the case of a tube, of course, the logarithmic mean between the inner and outer surface is to be used, r_{sa} a resistance for fouling deposits on the outer side of the cooling surface dA_a, and α_K the heat transfer coefficient of the cooling agent. In order to be able to eliminate the temperature ϑ_I at the condensate surface, which is still unknown in Eq. (6.26), one introduces the heat flow Φ_G for the cooling of the vapor

$$d\Phi_G = \alpha_G dA_0 (\vartheta_G - \vartheta_I) \tag{6.28}$$

and obtains

$$d\Phi = k' dA_0 \left(\vartheta_G - \vartheta_K - \frac{d\Phi_G}{\alpha_G dA_0} \right). \tag{6.29}$$

The heat flow withdrawn from the gas causes a drop of the gas temperature by $d\vartheta$; $d\vartheta$ is negative, and according to Eq. (6.28), $d\Phi_G$ is positive. It is given by

$$d\Phi_G = -\dot{M}_G c_{pG} d\vartheta_G. \tag{6.30}$$

The entire mass flow rate $\dot{M}$ of vapor and condensate undergoes an enthalpy change $-\dot{M}dh = d\Phi$; dh is negative, and $d\Phi$, according to Eq. (6.26), positive. Thus one obtains

$$d\Phi_G = \frac{\dot{M}_G c_{pG} d\vartheta_G}{\dot{M} dh} d\Phi = x^* c_{pG} \frac{d\vartheta_G}{dh} d\Phi \tag{6.31}$$

with the quality $x^* = \dot{M}_G/\dot{M}$. Insertion into Eq. (6.29) then gives

$$d\Phi = k' dA_0 (\vartheta_G - \vartheta_K) - \frac{k'}{\alpha_G} x^* c_{pG} \frac{d\vartheta_G}{dh} d\Phi. \tag{6.32}$$

Solving for dA_0 and integrating gives the necessary surface area

$$A_0 = \int_0^{\Phi_0} \frac{1 + \dfrac{k'}{\alpha_G} x^* c_{pG} \dfrac{d\vartheta_G}{dh}}{k'(\vartheta_G - \vartheta_K)} d\Phi, \tag{6.33}$$

where Φ_0 is the entire heat flux. In this equation, the expression $(k' x^* d\vartheta_G)/(\alpha_G dh)$ represents the surface area additionally needed to cool off further the vapor which is changeable in its composition along the flow path. If one had a pure vapor at saturation temperature, then there would be $d\vartheta_G/dh = 0$; if a gas were cooled without condensing, then $x^* = \dot{M}_G/\dot{M} = 1$ and $dh/d\vartheta_G = c_{pG}$; thus $x^* c_{pG} d\vartheta_G/dh = 1$. Equation (6.33) is then transformed into the known equation

$$A_0 = \int_0^{\Phi_0} \frac{\dfrac{1}{k'} + \dfrac{1}{\alpha_G}}{\vartheta_G - \vartheta_K} d\Phi,$$

or in differential form

$$d\Phi = \frac{1}{\dfrac{1}{k'} + \dfrac{1}{\alpha_G}} (\vartheta_G - \vartheta_K) dA_0. \tag{6.33a}$$

For the calculation of the condensing surface A_0 according to Eq. (6.33), a series of assumptions are still necessary. The most far reaching one assumes for the determination of $x^* d\vartheta_G/dh$ a thermodynamic equilibrium between vapor and liquid. One thus determines $x^* d\vartheta_G/dh$, as if liquid and vapor had at every point a uniform temperature $\vartheta = \vartheta_G$ and, with respect to mass exchange, were in equilibrium with one another. Then, the mass balance for a component yields

$$\dot{M}_G\xi_I'' + \dot{M}_L\xi_I' = \dot{M}\xi_G$$

or after division by the entire mass flow rate $\dot{M}$

$$x^*\xi_I'' + (1 - x^*)\xi_I' = \xi_G.$$

Thus

$$x^* = \frac{\xi_G - \xi_I'}{\xi_I'' - \xi_I'} \tag{6.34}$$

is representable as a function of the temperature ϑ_G, which is set according to the assumptions as equal to the equilibrium temperature.

Likewise the enthalpy can be formed as a function of the equilibrium temperature because of

$$\dot{M}_G h_G + \dot{M}_L h_L = \dot{M}h$$

or

$$x^* h_G + (1 - x^*)h_L = h. \tag{6.35}$$

The surface area calculation proceeds through the following steps:

—One first calculates the quality x^* and the enthalpy h of vapor and condensate as a function of the temperature ϑ_G. Figure 6.11 shows a typical example of the functions $h(\vartheta_G)$ and $x^*(\vartheta_G)$.

—Because of the energy balance

$$\mathrm{d}\Phi = -\dot{M}\mathrm{d}h = \pm \dot{M}_K c_{pK} \mathrm{d}\vartheta_K,$$

in which on the right side the positive sign is for cocurrent flow, and the negative sign for countercurrent flow of condensate and coolant, one can assign to every enthalpy change $\mathrm{d}h$ a change of the coolant temperature and thus calculate its dependence on the temperature ϑ_G. This follows from

$$\mathrm{d}\vartheta_K = \frac{-\dot{M}\mathrm{d}h}{\pm \dot{M}_K c_{pK}}$$

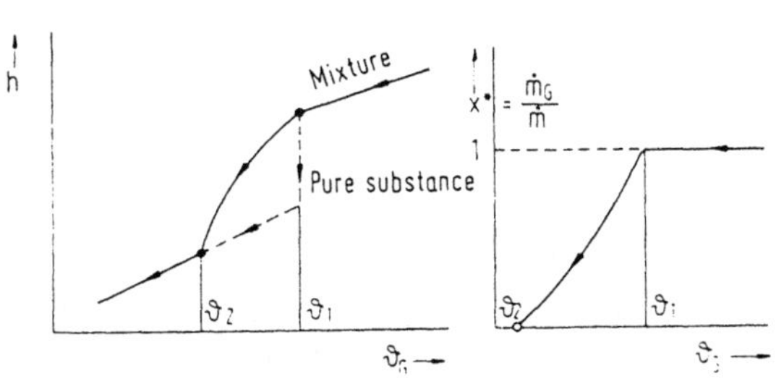

Fig. 6.11. Enthalpy and quality of a condensing mixture. ϑ_1 is the dew point temperature at the beginning of condensation; ϑ_2 is the bubble point temperature at the end of condensation

or

$$\vartheta_K - \vartheta_{K0} = \int_{\vartheta_{G0}}^{\vartheta_G} \frac{-\dot{M}\mathrm{d}h/\mathrm{d}\vartheta_G}{\pm \dot{M}_K c_{pK}} \mathrm{d}\vartheta_G, \tag{6.36}$$

in which ϑ_{K0} and ϑ_{G0} are the temperatures in the inlet cross section.

—Likewise, one can now calculate the heat transfer rate

$$\mathrm{d}\Phi = -\dot{M}\mathrm{d}h$$

or, after integration,

$$\Phi = -\int_{\vartheta_{G0}}^{\vartheta_G} \dot{M} \frac{\mathrm{d}h}{\mathrm{d}\vartheta_G} \mathrm{d}\vartheta_G \tag{6.37}$$

as a function of the temperature ϑ_G.

—As the next parameter, one calculates the overall heat transfer coefficient according to Eq. (6.27). Strictly speaking, it should be ascertained anew for each section of the condenser by calculating anew the heat transfer coefficient α of the condensate film for each specific section. This is for laminar film condensation according to Eq. (2.11) of the Nusselt film condensation theory, allowing for the corrections described in Sect. 2.2, or for turbulent film condensation according to Eq. (4.50) of Shah. In most cases, however, a change of the heat transfer coefficient α exerts little influence on the overall heat transfer coefficient k'; to a good approximation this remains constant.

—Because one knows the mass flow rate of the gas $\dot{M}_G = x^*\dot{M}$ because $x^*(\vartheta_G)$ is known as a function of the gas temperature ϑ_G, one can also calculate the heat transfer coefficient on the vapor side as a function of the gas temperature ϑ_G. For this purpose, one uses equations for the convective heat transfer on solid walls, and disregards, therefore, the suctioning off effect of the condensation and also the relative movement of gas and condensate.

—After one establishes in this manner the individual terms of the integrand in Eq. (6.33) as a function of the temperature ϑ_G (the heat flow is also a function of this temperature), the integrand itself can be represented as a function of the heat flow, and, thus, the requisite condenser surface area can be calculated.

The procedure described here has been modified by various authors [6.10–6.13], which, however, has scarcely led to higher accuracy or to simplifications in calculation. Depending upon the properties of the mixture, the assumption of a thermodynamic equilibrium between liquid and vapor for the calculation of $x^*\mathrm{d}\vartheta_G/\mathrm{d}h$ can yield too small as well as too large condenser surfaces. For this reason, one has developed the more precise methods to be discussed in the following section.

6.4.2 Binary Mixtures. The General Procedure

Using the example of binary mixtures, there shall first be a discussion of how one basically proceeds. As a further equation for a surface element ΔA, there is available the energy balance at the phase interface (see Eq. (6.15)),

$$\alpha_L(\vartheta_I - \vartheta_w) = \alpha_G(\vartheta_G - \vartheta_I) - \frac{\Delta \dot{M}}{\Delta A}\Delta h_v, \tag{6.38}$$

with the entire condensate mass flow accumulating at the phase interface, Eq. (6.14)

$$\frac{\Delta \dot{M}}{\Delta A} = -\varrho_G \beta_G \frac{\xi_I'' - \xi_G}{\xi_I'' - \xi_I'}. \tag{6.39}$$

This equation is valid, if both components in the condensate are present. If one of the components is a noncondensable inert gas, one gets $\xi_I' = 1$, because the liquid phase then consists only of the condensing components.

Before we consider the solution of Eqs. (6.38) and (6.39), we still need to discuss how one determines the heat and mass exchange coefficient α_G and β_G. When determining these values one must allow for the influence that the vapor flow directed vertically to the film surface has on the heat and mass transfer coefficients. In order to determine this influence, one can correct, as shown by Stephan and Laesecke [6.14], the heat and mass transfer coefficients by pure conduction or diffusion, respectively, by a suction term

$$\alpha_G = \alpha_{G0} + k|\dot{m}|c_{pG} \tag{6.40}$$

$$\varrho_G \beta_G = \varrho_G \beta_{G0} + k|\dot{m}|, \tag{6.41}$$

where the factor k considers that the suction itself is influenced by the vapor boundary layer. It is $k < 1$, if $\alpha_{G0}, \beta_{G0} \neq 0$, and it is $k = 1$, if no vapor boundary layer exists; thus $\alpha_{G0} = \beta_{G0} = 0$. Based upon the calculations of Sparrow et al. [6.15] on heat and mass transfer, the above expressions, according to Stephan and Laesecke [6.14], can be given the following forms for the case of laminar film condensation on a horizontal plate with parallel flow:

$$\frac{Nu}{Re_x^{0.5}} = 0.332 Pr + k \cdot R \cdot E \cdot Pr \frac{1}{\eta_\delta^*} \tag{6.42}$$

$$\frac{Sh}{Re_x^{0.5}} = 0.332 Sc + k \cdot R \cdot E \cdot Sc \frac{1}{\eta_\delta^*} \tag{6.43}$$

with the Nusselt number $Nu = \alpha_G x/\lambda_G$, the Reynolds number $Re_x = w_{G,x} x/\nu_G$, the Prandtl number $Pr = \nu_G/a_G$, the Sherwood number $Sh = \beta x/D$, the Schmidt number $Sc = \nu_G/D$, and the abbreviations

$$R \cdot E = \left(\frac{\varrho_L \eta_L}{\varrho_G \eta_G}\right)^{0.5} \frac{\lambda_L(\vartheta_I - \vartheta_w)}{\cdot\ \eta_L \Delta h_v}, \tag{6.44}$$

$$\frac{1}{\eta_\delta^*} = \frac{1}{\delta}\left(\frac{\nu_L x}{w_{G,x}}\right)^{1/2}. \tag{6.45}$$

The value η_δ^* depends in turn on $R \cdot E$ and, based on the calculations of Sparrow et al. [6.15], is tabulated in [6.14]. In the work of Sparrow is also shown how one forms the heat and mass transfer coefficients in the case of free convection or superimposed free and forced convection.

Heat and mass transfer coefficients can be calculated in a much easier manner, if the accumulated condensate mass is small. If, for example, there is $|\dot{m}|/\varrho_G \beta_{G0} \leqq 0.1$, then the mass transfer coefficient is

$$\varrho_G \beta_G = \varrho_G \beta_{G0} + k|\dot{m}| \leqq \varrho_G \beta_{G0}(1 + 0.1k) \quad \text{or} \quad \varrho_G \beta_G = \varrho_G \beta_{G0} \zeta$$

in which $k \approx 1$ is assumed. Thus, it is sufficient to choose for the mass transfer coefficient β_{G0} that of the vapor flow without condensation and to apply a small correction $\zeta > 1$. Similar considerations are valid for the heat transfer coefficient. If, for example, $|\dot{m}| c_{pG}/\alpha_{G0} \leqq 0.1$, one has

$$\alpha_G = \alpha_{G0} + k|\dot{m}| c_{pG} \leqq \alpha_{G0}(1 + 0.1k) \quad \text{or} \quad \alpha_G = \alpha_{G0} \zeta_0 .$$

The correction factor ζ_0 in

$$\alpha_G = \alpha_{G0} \zeta_0 \tag{6.46}$$

is given by the Ackermann correction [6.16]:

$$\zeta_0 = \frac{\varphi}{e^\varphi - 1} \quad \text{with} \quad \varphi = \frac{-|\dot{m}| c_{pG}}{\alpha_{G0}}. \tag{6.47}$$

One obtains the mass transfer coefficient from

$$\beta_G = \beta_{G0} \zeta \tag{6.48}$$

with the Stefan correction

$$\zeta = \frac{\Phi}{e^\Phi - 1} \quad \text{and} \quad \Phi = \frac{-|\dot{m}|}{\varrho_G \beta_{G0}}. \tag{6.49}$$

Besides the energy and mass balance at the phase interface, Eqs. (6.38) and (6.39), one must also consider another relation. Along the path of flow, the vapor is cooled according to

$$-\dot{M}_G c_{pG} \mathrm{d}\vartheta_G = \alpha_G \mathrm{d}A(\vartheta_G - \vartheta_I) - |\dot{m}| \mathrm{d}A c_{pG}(\vartheta_G - \vartheta_I).$$

If one solves this equation for the surface area under consideration and replaces

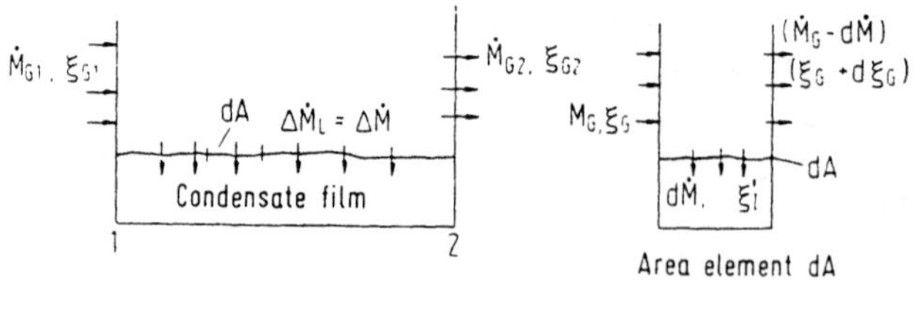

Fig. 6.12. Mass balance in a section of the condenser

the differentials by differences, then one obtains for a surface segment

$$\Delta A = \frac{-\dot{M}_G c_{pG}}{\alpha_G - |\dot{m}| c_{pG}} \frac{\Delta \vartheta_G}{\vartheta_G - \vartheta_I}. \tag{6.50}$$

As a further equation, one must employ the mass balance of the flowing vapor. For this, we shall consider a small surface segment dA lying between the cross sections 1 and 2, according to Fig. 6.12.

The mass balance for the more easily volatile component reads, when allowing for the total mass balance,

$$\dot{M}_G \xi_G = (\dot{M}_G - \mathrm{d}\dot{M})(\xi_G + \mathrm{d}\xi_G) + \mathrm{d}\dot{M}\xi'_I.$$

The mass fraction ξ_G of the lower boiling component is here, precisely as in Eq. (6.39), an integral mean value across a cross section of the vapor space. From the last equation, one obtains with $-\mathrm{d}\dot{M} = \mathrm{d}\dot{M}_G$ the so-called *Rayleigh equation*

$$\frac{-\mathrm{d}\dot{M}_G}{\dot{M}_G} = \frac{\mathrm{d}\xi_G}{\xi_G - \xi'_I}. \tag{6.51}$$

After integration between two cross sections 1 and 2, this equation yields the expression

$$\ln \frac{\dot{M}_{G2}}{\dot{M}_{G1}} = -\int_{\xi_{G1}}^{\xi_{G2}} \frac{\mathrm{d}\xi_G}{\xi_G - \xi'_I},$$

from which is calculated the condensate mass flow rate

$$\Delta \dot{M} = \dot{M}_{G2} - \dot{M}_{G1} = -\dot{M}_{G1}\left[1 - \exp \int_{\xi_{G1}}^{\xi_{G2}} \frac{-\mathrm{d}\xi_G}{\xi_G - \xi'_I}\right] \tag{6.52}$$

with $\Delta \dot{M} = \dot{m}\Delta A$.

The *practical calculation for binary mixtures* is now possible in various ways. The following procedure is recommended. Because the inlet and outlet temperatures of the vapor are generally given, one subdivides the difference of these two temperatures into equidistant segments $\Delta\vartheta_G$. If one would like to maintain definite concentration differences between the inlet and outlet cross sections, which occurs rather rarely, one can instead subdivide into equidistant segments $\Delta\xi_G$. One then comes to a calculation process, which proceeds analogously to that with the equal temperature segments $\Delta\vartheta_G$. This shall be discussed here generally. Still unknown is the surface segment ΔA, which is to be assigned to a temperature segment. In order to determine it, one proceeds as follows:

—One estimates the temperature ϑ_I at the condensate surface of a surface segment. Alternatively, of course, one can also estimate the heat flux $q_L = \alpha_L(\vartheta_I - \vartheta_w)$ through the condensate film. For a given ϑ_I, because of phase equilibrium, the mass fractions ξ'_I and ξ''_I at the condensate surface of the respective segment are fixed.

—One estimates further the change $\Delta\xi_G$ of the vapor composition in the respective segment and sets by approximation

$$\xi_G = (\xi_{G1} + \xi_{G2})/2 \quad \text{with} \quad \xi_{G2} = \xi_{G1} + \Delta\xi_G,$$

where ξ_{G1} is the vapor composition in the inlet cross section 1 and ξ_{G2} that in the outlet cross section 2 of the respective segment.

—With these estimated values, one can calculate the heat transfer coefficient α_G on the vapor side, as already discussed at the beginning of this section. In the same way, one obtains the heat transfer coefficient α_L of the condensate film, which is obtained for laminar film condensation from the Nusselt film condensation theory and for turbulent film condensation with the aid of Eq. (4.50) by Shah, for example. One obtains the mass transfer coefficient β_G in Eq. (6.39) as was explained at the beginning of this section. From Eq. (6.38) together with Eq. (6.39), one finds the temperature ϑ_I at the condensate surface of the respective segment. If this temperature does not agree with the estimated value, the calculation must be continued until Eqs. (6.38) and (6.39) are satisfied.

—If one has determined the temperature ϑ_I in this manner, then the requisite surface area ΔA of the respective segment is fixed, because of Eq. (6.50).

—In the next step one calculates with the aid of Eq. (6.52) the mass flow rate $\Delta\dot{M}$ of the condensate. If it agrees with that obtained from Eq. (6.39), then the initial estimate regarding the change $\Delta\xi_G$ of the vapor composition was correct. Otherwise, one must continue the calculation until the estimated and the calculated values agree sufficiently well.

—Only then can the requisite transfer surface area of the following segment be determined.

Table 6.2 summarizes the sequence of the calculation.

This calculation outline is valid even if one of the components is not condensable. One simply has to set the mass fraction of the condensing component in the liquid equal to unity, because this consists of only the one component.

If, by way of exception, the as yet ignored resistance for the mass exchange in the liquid plays a decisive role, one must modify the calculation procedure. The mass balance at the phase interface for the more easily volatile component with the mass fraction ξ then reads

$$j_{G,I} + \xi_I'' \frac{\Delta\dot{M}}{\Delta A} = j_{L,I} + \xi_I' \frac{\Delta\dot{M}}{\Delta A}, \tag{6.53}$$

from which is obtained the condensing mass flux

$$\frac{\Delta\dot{M}}{\Delta A} = \frac{j_{G,I} - j_{L,I}}{\xi_I' - \xi_I''}. \tag{6.54}$$

Table 6.2. Calculation of heat exchange area for condensation of binary mixtures

Step	Magnitude	Determined by
1	$\vartheta_G(i)$, $\Delta\vartheta_G$	Subdivision of temperatures between inlet and outlet cross section into equidistant segments. $\vartheta_G(i)$ is the temperature in section i
2	$\Delta\xi_G(i)$	estimation
3	$q_L(i)$ resp. $\vartheta_I(i)$	estimation
4	$\alpha_G, \alpha_L, \beta_G$	Eqs. for convective heat and mass transfer. Ackermann and Stefan correction factor to be considered, page 90.
5	ξ_I', ξ_I''	From conditions for phase equilibrium $\xi_I'(\vartheta_I(i))$, $\xi_I''(\vartheta_I(i))$ for given pressure
6	$\Delta\dot{M}/\Delta A$	Eq. (6.39)
7	ϑ_I	Eq. (6.38)
8	If $\vartheta_I \neq \vartheta_I(i)$ repeat calculation from Step 3 on. If $\vartheta_I = \vartheta_I(i)$ continue calculation	
9	ΔA	Eq. (6.50)
10	$\Delta\dot{M}$	Eq. (6.52)
11	If $\Delta\dot{M} \neq \Delta\dot{M}(i)$ repeat calculation from Step 2 on. If $\Delta\dot{M} = \Delta\dot{M}(i)$ continue calculation	
12	$i \Rightarrow i+1$	

With the assumptions for the mass transfer in the gas

$$j_{G,I} = \varrho_G \beta_G (\xi_I'' - \xi_G) \tag{6.55}$$

and in the liquid

$$j_{L,I} = \varrho_L \beta_L (\xi_L - \xi_I') \tag{6.56}$$

the following equation takes the place of Eq. (6.39):

$$\frac{\Delta\dot{M}}{\Delta A} = \frac{\varrho_G \beta_G (\xi_I'' - \xi_G) - \varrho_L \beta_L (\xi_L - \xi_I')}{\xi_I' - \xi_I''}. \tag{6.57}$$

In the case of a negligibly small mass exchange resistance in the liquid phase, because of $\xi_I' \rightarrow \xi_L$ the diffusional mass flux in the liquid phase also becomes diminishingly small, and one again obtains the previously introduced Eq. (6.39).

6.4.3 Binary Mixtures. Approximation According to Colburn and Hougen

A special case of the previously discussed procedure is that of Colburn and Hougen [6.17]. The Eq. (6.39) is here replaced by another one that is based upon simplifying assumptions of the film theory. This equation presupposes a one-dimensional, stationary mass flux directed vertically to the condensate surface. With this assumption, the mass balance for the considered component 1 reads

$$\frac{\partial \dot{M}_1}{\partial z} = 0,$$

where z designates the coordinate normal to the wall. Because of $\dot{M}_1 = M_1 \dot{N}_1$ with the molar mass M_1 and the molar flow rate $\dot{N}_1$ (SI-unit mole/s), one can also write for this equation

$$\frac{\partial \dot{N}_1}{\partial z} = 0. \tag{6.58}$$

The molar flow rate in the gas is composed of the diffusional flow rate $j_1 A$ (SI-unit of j_1:mole/m^{2}s) and the convective molar flow rate $y_1 \dot{N}$, if $\dot{N}$ is the entire molar flow rate:

$$\dot{N}_1 = j_1 A + y_1 \dot{N}. \tag{6.59}$$

In it

$$j_1 = -Dc \frac{\partial y_1}{\partial z} \tag{6.60}$$

according to Fick's law, with $c = n/V$. The coordinate z is again the coordinate normal to the condensate surface and is directed from the condensate surface into the vapor space. With Eq. (6.60), after division by $\dot{N}$, Eq. (6.59) can also be written as

$$\frac{\dot{N}}{A} = \frac{j_1}{\dfrac{\dot{N}_1}{\dot{N}} - y_1} = -Dc \frac{\partial y_1/\partial z}{\dfrac{\dot{N}_1}{\dot{N}} - y_1}. \tag{6.61}$$

Because the molar fluxes of each of the two components are said to be independent of the space coordinate z, the entire molar flux $\dot{N}/A$ and also the quotient $\dot{N}_1/\dot{N}$ are independent of z. One can, therefore, easily integrate Eq. (6.61). The integration is to extend from the condensate surface (Index I) into the vapor space (Index G). The thickness of the vapor boundary layer shall be δ. We assume here constant values of pressure and temperature. Assuming that the gas phase acts ideally, the diffusion coefficient and the molar volume concentration $c = n/V = p/(RT)$ are likewise independent of the space coordinate

z. The integration yields

$$\frac{\dot{N}}{A}\delta = Dc \ln \frac{\dfrac{\dot{N}_1}{\dot{N}} - y_{1,G}}{\dfrac{\dot{N}_1}{\dot{N}} - y_{1,I}}$$

and, after the introduction of a mass exchange coefficient $\beta_{G0} = D/\delta$,

$$\frac{\dot{N}}{A} = \beta_{G0} c \ln \frac{\dfrac{\dot{N}_1}{\dot{N}} - y_{1,G}}{\dfrac{\dot{N}_1}{\dot{N}} - y_{1,I}}. \tag{6.62}$$

In the procedure of Colburn and Hougen, this equation takes the place of the general Eq. (6.39). Equation (6.62) is valid assuming an isobaric-isothermal vapor flow and further incorporates the assumptions of the film theory, according to which the mass flow is stationary and is directed vertically to the condensate surface, which at sufficiently small condensation rates is approximately accurate.

In case one of the two components of the vapor is an inert gas, and, consequently, does not condense, the condensing molar flow rate of substance 1 is equal to the entire resulting condensate flow rate $\dot{N}_1 = \dot{N}$. Equation (6.62) is then simplified and one obtains

$$\frac{\dot{N}}{A} = \beta_{G0} c \ln \frac{1 - y_{1,G}}{1 - y_{1,I}}. \tag{6.63}$$

In this equation, y_1 is the molar fraction of the condensing component. Because the latter condenses at the phase interface, $y_{1,I} < y_{1,G}$ and thus the molar flux $\dot{N}/A$ according to Eq. (6.63) is negative. The flux of the condensing vapor is directed against the surface coordinate z chosen by us. Because only the absolute value of the molar flux is of interest,

$$\frac{|\dot{N}|}{A} = \beta_{G0} c \ln \frac{1 - y_{1,I}}{1 - y_{1,G}}, \tag{6.64}$$

from which, with $c = \varrho_G/M$ and $y_1 = p_1/p$, there is obtained Eq. (2.28) used already for the condensing mass flow rate $\dot{M} = |\dot{N}| M$:

$$\dot{N} = \varrho_G \beta_{G0} A \ln \frac{p - p_{1,I}}{p - p_{1,G}}. \tag{6.65}$$

For sufficiently small mass flow rates one can rearrange Eq. (6.64) in the following manner:

$$\frac{1 - y_{1,I}}{1 - y_{1,G}} = \exp \frac{|\dot{N}|}{A\beta_{G0} c} \approx 1 + \frac{|\dot{N}|}{A\beta_{G0} c}$$

or

$$\frac{|\dot{N}|}{A} = \beta_{\mathrm{G0}} c \frac{y_{1,\mathrm{G}} - y_{1,\mathrm{I}}}{1 - y_{1,\mathrm{G}}}. \tag{6.66}$$

In order to also apply this form of the equation to large mass fluxes, a corrected mass exchange coefficient

$$\beta_{\dot{\mathrm{G}}} = \beta_{\mathrm{G0}} \zeta$$

is inserted

$$\frac{|\dot{N}|}{A} = \beta_{\mathrm{G0}} \zeta c \frac{y_{1,\mathrm{G}} - y_{1,\mathrm{I}}}{1 - y_{1,\mathrm{G}}}. \tag{6.67}$$

The correction factor ζ is obtained by comparison with Eq. (6.64), whereupon

$$\frac{1 - y_{1,\mathrm{I}}}{1 - y_{1,\mathrm{G}}} = \exp \frac{|\dot{N}|}{A \beta_{\mathrm{G0}} c} \quad \text{or} \quad \frac{y_{1,\mathrm{G}} - y_{1,\mathrm{I}}}{1 - y_{1,\mathrm{G}}} = \exp \frac{|\dot{N}|}{A \beta_{\mathrm{G0}} c} - 1.$$

If one places this expression into Eq. (6.67), there results the correction factor

$$\zeta = \frac{\Phi}{\exp \Phi - 1} \quad \text{with} \quad \Phi = \frac{|\dot{N}|}{A \beta_{\mathrm{G0}} c}. \tag{6.68}$$

For very small molar flow rates, the correction factor approaches unity,

$$\lim_{\dot{N} \to 0} \zeta = 1.$$

and the corrected mass exchange coefficient β_{G} becomes identical with the mass exchange coefficient $\beta_{\dot{\mathrm{G}}0}$. This, in turn, can be calculated from the heat transfer coefficient according to Eq. (6.25).

6.4.4 Multi-component Mixtures

The energy balance, Eq. (6.38), established for binary mixtures remains valid for multi-component mixtures without any changes, so that the temperature ϑ_{I} at the phase interface can be calculated from it, at least in principle. Difficulties in the practical calculation of the temperature at the phase interface result from the fact that the accumulated condensate and the mass transfer coefficient in a multi-component mixture depend in a more complicated manner upon the concentration gradient and, therefore, cannot be determined in as simple a manner as in the case of binary mixtures.

In particular, the diffusional flow of a component depends not only upon its own concentration gradient but also upon those of all the other components. There are, therefore, K − 1 gradients, independent of one another in a mixture of K components. Because of the interaction among the individual components the diffusional flow of a component is influenced by that of all the others, and not only its own concentration gradient but also those of the other components

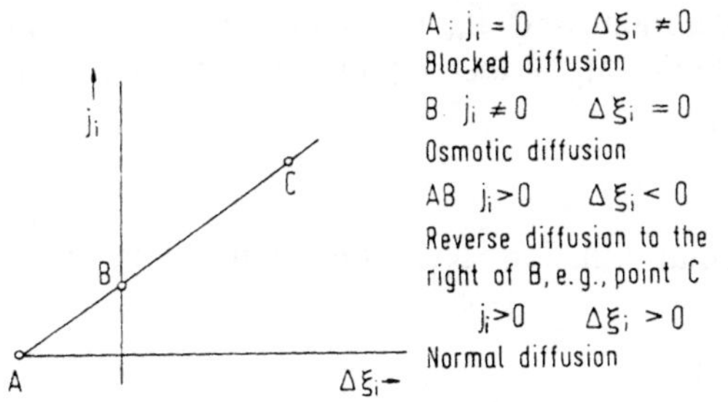

Fig. 6.13. Diffusion phenomena in a multi-component mixture. j_i is the diffusion flux and $\Delta\xi_i$ is the concentration gradient of substance i

act as a driving force. This can result in a component diffusing opposite to its own concentration gradient. This is known as *reverse diffusion*.

A component can also be transferred by diffusion, although its concentration gradient is zero. This phenomenon is called *osmotic diffusion*.

Finally, it is possible that no diffusion will result despite the finite concentration gradients of the components. One speaks of *blocked diffusion*.

Naturally the phenomena mentioned can be studied only if there is no convection so that no additional influences of the convective mass transfer can occur. Figure 6.13 illustrates the different effects of diffusion.

As yet, there has been no successful treatment of heat transfer during the condensation of multi-component mixtures. Instead, one has simply applied to the multi-component mixtures the assumptions of the film theory, as they were developed by Colburn-Hougen for binary mixtures. One, therefore, assumes that all mass flows are stationary and directed vertically to the condensate surface.

With this assumption, the heat and mass exchange during the condensation of vapor mixtures can be treated mathematically. Krishna and Standart [6.18] developed such a mathematical procedure. It presupposes that the boundary layers of all the participating substances in the gas space are equally large. Furthermore, the pressures are to be so moderate that the vapor phase can be regarded as an ideal gas.

In the following, the bases for this method are to be discussed. For the description of the diffusional flux $\boldsymbol{j}_i$[1] (SI unit mole/m^{2}s) of the component i in a mixture of K components, there are available Fick's law

$$\boldsymbol{j}_i = c \sum_{k=1, k \neq i}^{K} D_{ik} \boldsymbol{d}_k \tag{6.69}$$

and the Stefan/Maxwell equations [6.19]

$$\boldsymbol{d}_i = \sum_{k=1}^{K} \frac{y_i y_k}{c \mathfrak{D}_{ik}} \left(\frac{\boldsymbol{j}_k}{y_k} - \frac{\boldsymbol{j}_i}{y_i} \right), \quad i = 1, 2, \ldots, K.^2 \tag{6.70}$$

---

[1] Vectors and matrices are printed in bold-face italics.

[2] In [6.19], the Stefan/Maxwell equations are shown with the diffusional fluxes in the center-of-mass system. The equations are valid, however, for any other reference system, and, thus, also for the one used here, because the velocity of reference can be cancelled in the diffusional fluxes: The gradients d_i are invariant with regard to the reference system.

Here, D_{ik} designates the multi-component and $\mathfrak{D}_{ik}$ the binary diffusion coefficients, respectively, $c = n/V$ the molar density, y_i the molar fraction of component i, and $\boldsymbol{d}_i$ the gradient

$$\boldsymbol{d}_i = \text{grad } y_i,$$

if one disregards pressure and thermal diffusion. The diffusional flux is defined by

$$\boldsymbol{j}_i = c_i(\boldsymbol{w}_i - \boldsymbol{w}) \tag{6.71}$$

with the molar density $c_i = n_i/V$ of component i and the average molar velocity

$$c\boldsymbol{w} = \sum_{k=1}^{K} c_k \boldsymbol{w}_k \quad \text{or} \quad \boldsymbol{w} = \sum_{k=1}^{K} y_k \boldsymbol{w}_k. \tag{6.72}$$

For the diffusional fluxes, there thus results from Eq. (6.71) the continuity relationship

$$\sum_{k=1}^{K} \boldsymbol{j}_k = 0. \tag{6.73}$$

Furthermore, for the gradients, because of $\sum_{k=1}^{K} y_k = 1$ we have

$$\sum_{k=1}^{K} \boldsymbol{d}_k = 0, \tag{6.74}$$

Eq. (6.69) contains explicitly the diffusional fluxes. The disadvantage is, however, that the multi-component diffusion coefficients D_{ik} occur. These depend upon the binary diffusion coefficients $\mathfrak{D}_{ik}$ between the components of the mixture and on the molar fractions of all the components. By contrast, the binary diffusion coefficients $\mathfrak{D}_{ik}$ of gases at moderate pressures are, practically speaking, approximately independent of the gas composition and are constant for the assumed values of pressure and temperature.

The Stefan/Maxwell equations, which are linear in the diffusional fluxes, can be resolved according to these. For this purpose, we shall rearrange Eq. (6.70) somewhat. The Eq. (6.73) can also be written as

$$\boldsymbol{j}_K = -\sum_{k=1}^{K-1} \boldsymbol{j}_k = -\boldsymbol{j}_i - \sum_{k=1, k\neq i}^{K-1} \boldsymbol{j}_k.$$

We insert this expression into the Stefan/Maxwell equations, which we first multiply on both sides by the film thickness δ, and, then, using the abbreviations $\boldsymbol{d}_i^* = \boldsymbol{d}_i c\delta = c\delta \text{ grad } y_i$ and $\mathfrak{D}_{ij}/\delta = \beta_{ij}$, reformulate into

$$\boldsymbol{d}_i^* = \sum_{k=1, k\neq i}^{K-1} \frac{y_i \boldsymbol{j}_k}{\beta_{ik}} + \frac{y_i \boldsymbol{j}_K}{\beta_{iK}} - \boldsymbol{j}_i \sum_{k=1, k\neq i}^{K} \frac{y_k}{\beta_{ik}}.$$

Here the quantities β_{ij} are mass transfer coefficients of the binary mixture ij.

One obtains then

$$d_i^* = -\left[j_i \sum_{k=1,k\neq i}^{K} \left(\frac{y_i}{\beta_{iK}} + \frac{y_k}{\beta_{ik}}\right) - y_i \sum_{k=1,k\neq i}^{K} j_k\left(\frac{1}{\beta_{ik}} - \frac{1}{\beta_{iK}}\right)\right] \tag{6.75}$$

or in matrix notation

$$\begin{pmatrix} d_1^* \\ d_2^* \\ \vdots \\ d_K^* \end{pmatrix} = -\begin{pmatrix} a_{11}j_1 & a_{12}j_2 & \cdots & a_{1K}j_K \\ a_{21}j_1 & a_{22}j_2 & \cdots & a_{2K}j_K \\ \vdots & \vdots & & \vdots \\ a_{K1}j_1 & a_{K2}j_2 & \cdots & a_{KK}j_K \end{pmatrix}. \tag{6.76}$$

The elements a_{ii} of the main diagonal are given here by

$$a_{ii} = \frac{y_i}{\beta_{iK}} + \sum_{k=1,k\neq i}^{K} \frac{y_k}{\beta_{ik}}. \tag{6.77}$$

and the secondary diagonals a_{ij} with $i \neq j$ by

$$a_{ij} = -y_i\left(\frac{1}{\beta_{ij}} - \frac{1}{\beta_{iK}}\right). \tag{6.78}$$

The matrix notation in Eq. (6.76) is shortened to

$$d^* = -Aj, \tag{6.79}$$

where d^* is the vector d_i^*, A the matrix of the coefficients a_{ij}, and j the vector of the diffusional fluxes j_i with the indices i, j = 1, 2, ..., K. At the edge of the vapor boundary layer δ, we will characterize the values by the index b(= bulk); there is, therefore, the gradient

$$d_b^* = -A_b j_b, \tag{6.80}$$

or

$$j_b = -A_b^{-1} d_b^*, \tag{6.81}$$

in which, according to the rules of matrix calculation, there is $A_b^{-1} = A_{ad}/|A|$, if A_{ad} is the adjoint matrix of A and $|A|$ the determinant of the matrix A. In other words, by solving the set of equations that is linear in the diffusional fluxes j_i one obtains the diffusional fluxes j_i as a function of the gradients d_i. In order now to determine the gradients d_b^* in Eq. (6.81) we assume the validity of the film theory, according to which all molar flow rates $c_i w_i = \dot{n}_i$, $i = 1, 2, 3, \ldots, K$ are said to be vertical to the condensate surface and stationary. The coordinate z shall again be directed vertically from the condensate surface ($z = 0$) into the vapor space. Because the diffusional fluxes in the Stefan/Maxwell equations (6.70) are variable with the coordinate z, we replace the diffusional fluxes j_i by the molar fluxes, according to Eq. (6.71)

$$c_i w_i = \dot{n}_i = j_i + y_i \dot{n} \tag{6.82}$$

with

$$\dot{\boldsymbol{n}} = c\boldsymbol{w} = \sum_{k=1}^{k} c_k \boldsymbol{w}_k.$$

Again abbreviating, we simultaneously write $\boldsymbol{d}_i^* = c\mathrm{d}y_i/\mathrm{d}(z/\delta)$ and $\mathfrak{D}_{ij}/\delta = \beta_{ij}$. This then yields

$$\boldsymbol{d}_i^* = y_i \sum_{k=1,k\neq i}^{K} \frac{\dot{\boldsymbol{n}}_k}{\beta_{ik}} - \dot{\boldsymbol{n}}_i \sum_{k=1,k\neq i}^{K} \frac{y_k}{\beta_{ik}}. \tag{6.83}$$

Expanding the last term

$$\dot{\boldsymbol{n}}_i \sum_{k=1,k\neq i}^{K} \frac{y_k}{\beta_{ik}} = \dot{\boldsymbol{n}}_i \left(\sum_{k=1,k\neq i}^{K-1} \frac{y_k}{\beta_{ik}} + \frac{y_K}{\beta_{iK}} \right)$$

allowing for

$$y_K = 1 - y_i - \sum_{k=1,k\neq i}^{K-1} y_k$$

finally yields the following form of the Stefan/Maxwell equations:

$$\boldsymbol{d}_i^* = y_i \sum_{k=1,k\neq i}^{K} \left(\frac{\dot{\boldsymbol{n}}_i}{\beta_{iK}} + \frac{\dot{\boldsymbol{n}}_k}{\beta_{ik}} \right) - \dot{\boldsymbol{n}}_i \sum_{k=1,k\neq i}^{K} y_k \left(\frac{1}{\beta_{ik}} - \frac{1}{\beta_{iK}} \right) - \frac{\dot{\boldsymbol{n}}_i}{\beta_{iK}}. \tag{6.84}$$

In matrix notation, this equation reads

$$\begin{pmatrix} d_1^* \\ d_2^* \\ \vdots \\ d_K^* \end{pmatrix} = \begin{pmatrix} b_{11}y_1 & b_{12}y_2 & \cdots & b_{1K}y_K \\ b_{21}y_1 & b_{22}y_2 & \cdots & b_{2K}y_K \\ \vdots & \vdots & & \vdots \\ b_{K1}y_1 & b_{K2}y_2 & \cdots & b_{KK}y_K \end{pmatrix} + \begin{pmatrix} g_1 \\ g_2 \\ \vdots \\ g_K \end{pmatrix}. \tag{6.85}$$

The elements b_{ii} of the main diagonal are given by

$$b_{ii} = \frac{\dot{n}_i}{\beta_{iK}} + \sum_{k=1,k\neq i}^{K} \frac{\dot{n}_k}{\beta_{ik}} \tag{6.86}$$

and those of the secondary diagonals b_{ij} with $i \neq j$ by

$$b_{ij} = -\dot{n}_i \left(\frac{1}{\beta_{ij}} - \frac{1}{\beta_{iK}} \right), \tag{6.87}$$

while the vector $\boldsymbol{g}$ contains the elements

$$g_i = -\frac{\dot{n}_i}{\beta_{iK}}. \tag{6.88}$$

In shorthand notation, Eq. (6.85) can be written as

$$\boldsymbol{d}^* = \boldsymbol{B}\boldsymbol{y} + \boldsymbol{g}, \tag{6.89}$$

where $\boldsymbol{d}^* = c\mathrm{d}\boldsymbol{y}/\mathrm{d}\eta$ with $\eta = z/\delta$.

Equation (6.89) is thus a set of common linear differential equations for the unknown $\boldsymbol{y}$. The solution of this equation while allowing for the initial condition $\boldsymbol{y}_{\eta=1} = \boldsymbol{y}_b$ yields the concentration profile

$$\boldsymbol{y} - \boldsymbol{y}_b = (\exp(\boldsymbol{B}(\eta - 1)) - \boldsymbol{I})(\boldsymbol{y}_b + \boldsymbol{B}^{-1}\boldsymbol{g}). \tag{6.90}$$

In this, the unit matrix is designated by $\boldsymbol{I}$ whose elements are $\delta_{ij} = 0$ for $i \neq j$ and $\delta_{ii} = 1$ for $i = j$. One can easily confirm that this equation fulfills the differential Eq. (6.89) and the boundary condition $\boldsymbol{y}_{\eta=1} = \boldsymbol{y}_b$.

At the condensate surface (Index I), $\eta = 0$ and $\boldsymbol{y} = \boldsymbol{y}_I$. One then obtains from Eq. (6.90)

$$\boldsymbol{y}_I - \boldsymbol{y}_b = (\exp(-\boldsymbol{B}) - \boldsymbol{I})(\boldsymbol{y}_b + \boldsymbol{B}^{-1}\boldsymbol{g}). \tag{6.91}$$

Division of both equations yields

$$\frac{\boldsymbol{y} - \boldsymbol{y}_b}{\boldsymbol{y}_I - \boldsymbol{y}_b} = \frac{\exp(\boldsymbol{B}(\eta - 1)) - \boldsymbol{I}}{\exp(-\boldsymbol{B}) - \boldsymbol{I}}. \tag{6.92}$$

The previously unknown gradient $\boldsymbol{d}_b^*$ is obtained from this by differentiation with respect to η:

$$\boldsymbol{d}_b^* = c\left(\frac{\mathrm{d}\boldsymbol{y}}{\mathrm{d}\eta}\right)_{\eta=1} = c\frac{\boldsymbol{B}}{\exp(-\boldsymbol{B}) - \boldsymbol{I}}(\boldsymbol{y}_I - \boldsymbol{y}_b). \tag{6.93}$$

If one inserts this expression into Eq. (6.81) for the diffusional flux, one obtains

$$\boldsymbol{j}_b = \boldsymbol{A}_b^{-1} c\frac{\boldsymbol{B}}{\exp(-\boldsymbol{B}) - \boldsymbol{I}}(\boldsymbol{y}_b - \boldsymbol{y}_I). \tag{6.94}$$

We will now consider the limiting case of vanishingly small mass transfer $\dot{n}_1, \dot{n}_2, \ldots, \dot{n}_K \to 0$. By series expansion of the exponential function, one finds

$$\lim_{\dot{n}_1, \dot{n}_2, \ldots, \dot{n}_K \to 0} \frac{\boldsymbol{B}}{\exp(-\boldsymbol{B}) - \boldsymbol{I}} = 1. \tag{6.95}$$

The above equation thus converts into

$$\boldsymbol{j}_b = \boldsymbol{A}_b^{-1} c(\boldsymbol{y}_b - \boldsymbol{y}_I). \tag{6.96}$$

for $\dot{n}_1, \dot{n}_2, \ldots, \dot{n}_K \to 0$.

For vanishing mass flux at the phase interface, one defines as usual a mass transfer coefficient, which in this case represents a matrix

$$\boldsymbol{j}_b = \boldsymbol{k}_b c(\boldsymbol{y}_b - \boldsymbol{y}_I) \tag{6.97}$$

for $\dot{n}_1, \dot{n}_2, \ldots, \dot{n}_K \to 0$.

As the comparison with the previous equation shows, the mass transfer matrix is

$$\boldsymbol{k}_b = \boldsymbol{A}_b^{-1}. \tag{6.98}$$

It is given by inversion of the matrix with the elements a_{ii} and a_{ij}, Eqs. (6.77) and (6.78), and is thus traceable back to the binary mass transfer coefficients β_{ij} and the molar fractions y_{ib} in the vapor core. The inversion of a matrix can be easily done with the aid of a computer and standard programs. In order to keep the form of Eq. (6.97) for the finite mass transport to the phase interface, one follows the procedure for binary mixtures by introducing a mass transfer correction and writes

$$\boldsymbol{j}_b = \boldsymbol{k}_b \zeta c(\boldsymbol{y}_b - \boldsymbol{y}_I). \tag{6.99}$$

If one allows for $\boldsymbol{k}_b = \boldsymbol{A}_b^{-1}$, then the comparison with Eq. (6.94) gives the mass transfer correction as

$$\zeta = \frac{\boldsymbol{B}}{\exp(-\boldsymbol{B}) - \boldsymbol{I}}. \tag{6.100}$$

The elements b_{ii} and b_{ij} of the matrix $\boldsymbol{B}$ are, as a glance at Eqs. (6.86) and (6.87) indicates, dependent upon the molar flow rates $\dot{n}_i$ and the binary mass transfer coefficients. With the aid of Eqs. (6.99) and (6.100), one can calculate the diffusional fluxes of the individual components. For this, one must first calculate via Eqs. (6.77) and (6.78) the matrix $\boldsymbol{A}$, and then determine its inversion, by which according to Eq. (6.98) the mass transfer matrix $\boldsymbol{k}_b$ is fixed. Next, one calculates the matrix $\boldsymbol{B}$, whose elements are given by Eqs. (6.86) and (6.87). Thereby, the mass transfer correction ζ is fixed according to Eq. (6.100). In order to compute the diffusional fluxes, one needs only programs for the calculation and inversion of a matrix and for the formation of the product of two matrices.

Krishna and Standart [6.18] have recommended the following sequence of steps for *practical calculations*:

- One first calculates the matrix $\boldsymbol{A}$, whose elements are established by Eqs. (6.77) and (6.78), and from this by matrix inversion Eq. (6.98), the matrix of the mass transfer coefficients $\boldsymbol{k}_b$ for vanishing mass flow rates.
- One assumes the mass flow rates to be zero and thus the mass transfer correction ζ to be equal to unity.
- With these assumptions, one obtains then from Eq. (6.97) the diffusional fluxes for vanishingly small molar flow rates $\dot{n}_1, \dot{n}_2, \ldots, \dot{n}_K \to 0$.
- From these diffusional fluxes are obtained the associated molar flow rates, since because of Eq. (6.71) there is

  $$c_i \boldsymbol{w}_i = \dot{\boldsymbol{n}}_i = \boldsymbol{j}_i + y_i \dot{\boldsymbol{n}}.$$

  On the other hand, we neglected the diffusional resistance in the liquid phase; consequently, in the liquid, $\boldsymbol{j}_{1,L} = 0$ or $\boldsymbol{w}_{i,L} = \boldsymbol{w}_L$, and thus $(c_i \boldsymbol{w}_i)_L = (c_i \boldsymbol{w})_L = x_i (c\boldsymbol{w})_L$ or $\dot{\boldsymbol{n}}_{i,L} = x_i \dot{\boldsymbol{n}}_L$, from which, because of the balance at the phase interface $\dot{\boldsymbol{n}}_{i,L} = \dot{\boldsymbol{n}}_{i,G}$ and $\dot{\boldsymbol{n}}_L = \dot{\boldsymbol{n}}_G = \dot{\boldsymbol{n}}$ quite generally, as a result of $\dot{\boldsymbol{n}}_i = const$, there follows

  $$x_i = \dot{\boldsymbol{n}}_i / \dot{\boldsymbol{n}}.$$

Thus, for the point $\eta = 1$, one can write the above equation as

$$\dot{\boldsymbol{n}}_{ib} = \frac{\boldsymbol{j}_{ib}}{1 - y_{ib}/x_{ib}}$$

and thus calculate the associated molar flow rates from the diffusional fluxes.

—After the fluxes are known, there is also fixed the matrix $\boldsymbol{B}$, whose elements are given by Eqs. (6.86) and (6.87). The mass transfer correction, Eq. (6.100), can now be calculated.

—With the mass transfer correction ζ, one obtains new diffusional fluxes, and from these then new molar flow rates and a new mass transfer correction. The iteration is then repeated until convergence has been reached.

—If all molar flow rates are fixed, then the entire molar flow is known. With the aid of the energy balance at the phase interface, Eq. (6.38), one can now calculate the necessary exchange area, by employing the steps explained in Table 6.2.

7 Condensation of Vapors of Immiscible Liquids

Frequently, mixtures must be condensed that are immiscible in the liquid phase. Examples are mixtures of an organic compound as one component, such as refrigerant R-112 or R-113, and perchloric ethylene or p-xylol as the other component. Mixtures of water, benzene, phenol, butyl alcohol, etc. are also immiscible as liquids. The previous findings on the condensation of vapors, whose components are miscible in the liquid phase, cannot be extended to this case.

Experiments by Bernhardt et al. [7.1] on the condensation of vapors of water and different organic compounds, which are not miscible with water as liquids, have delivered several interesting discoveries about the condensate. Accordingly, as shown in Fig. 7.1, one sees in an extended liquid film I of the organic liquid small mobile water drops II of 0.03 to 0.04 mm diameter. There are also large stationary water drops III extending to the wall and having a diameter of 0.05 to 4 mm, and upon these tiny mobile drops IV of approximately 0.02 mm diameter of the organic liquid. Although this liquid has a higher boiling point than water and thus condenses first from the vapor, it surprisingly does not wet the entire surface. The small water drops moved rapidly with velocities between 2.5 and 127 mm/s, whereas the tiny drops of the organic liquid danced to and fro with speeds from 25 to 50 mm/s. Such forms of condensation are to be observed only if the wall temperature is sufficiently low; at a higher wall temperature, a homogeneous condensate film can also form.

In a condenser with the wall temperature decreasing along the path of flow, both kinds of condensate formation can occur one after the other: first the formation of a homogeneous condensate film, then the formation of both immiscible liquids. The occurrence of these different phenomena can be clarified easily with a phase diagram. Figure 7.2 shows the equilibrium of solutions with a miscibility gap for constant pressure. The points A and B state the composition of the two liquid phases; point C is the composition of the vapor in equilibrium with them. Point C is designated as the eutectic point and the temperature there as the eutectic temperature.

As long as both liquid phases and vapor are present and the pressure does not change, the temperature will remain constant during heat transfer until all the vapor is condensed. The vapor phase maintains its composition at point C and the liquid phases maintain their compositions at points A and B. This follows from the Gibbs phase rule, according to which in a binary mixture at

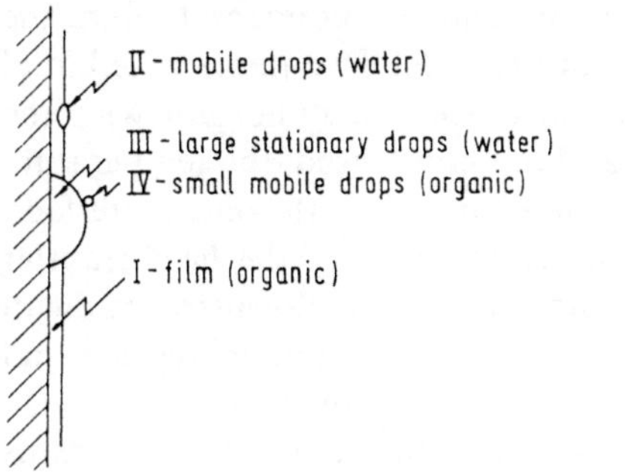

Fig. 7.1. Condensation of immiscible liquids

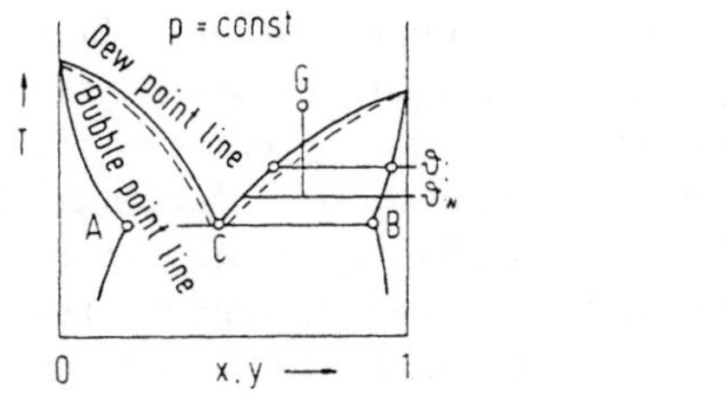

Fig. 7.2. Bubble and dew point lines of a mixture with miscibility gap

a given pressure, the two immiscible liquids and the vapor can coexist only at a definite temperature and with a definite composition. This temperature is lower than the boiling points of the pure components.

Figure 7.3 shows the equilibrium in the case of completely immiscible pure liquids, which practically occurs only approximately in the form shown, because each liquid contains both components.

If one has a vapor at the state *G* and if the wall temperature ϑ_w, as shown in Fig. 7.2, is above the temperature ϑ_C or is equal to ϑ_C, then the phase interface temperature ϑ_I is above the wall temperature and, thus, also above the eutectic temperature ϑ_C. There forms a homogeneous condensate film. Given the temperature ϑ_I, the liquid and vapor composition at the phase interface can be read from the equilibrium diagram. The calculation of the heat transfer proceeds according to the equations discussed in the previous chapter.

If the wall temperature is below the eutectic temperature, Fig. 7.3, then two cases are possible. In the presence of a sufficiently thick condensate film, the phase interface temperature can be higher than the eutectic temperature. Then

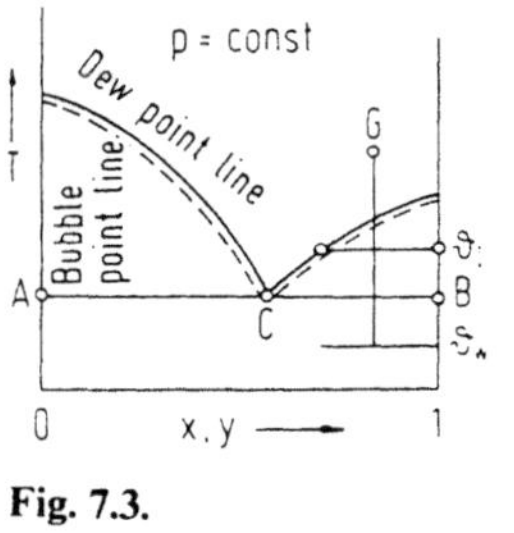

Fig. 7.3.

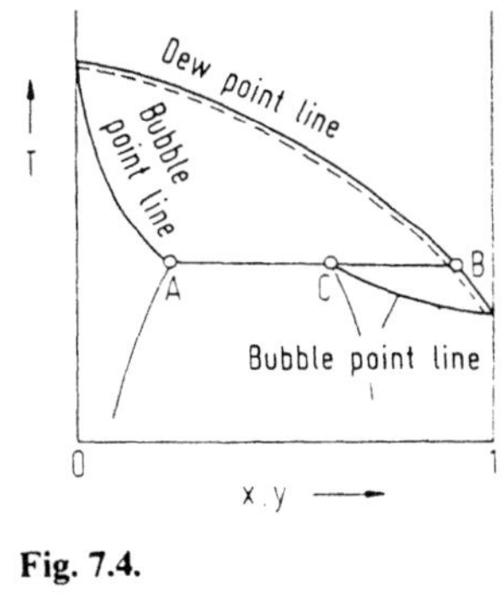

Fig. 7.4.

Fig. 7.3. Bubble and dew point lines for completely immiscible pure liquids

Fig. 7.4. Bubble and dew point line of a mixture with miscibility gap

a homogeneous condensate forms. In the case of a thin condensate film, the phase interface reaches the eutectic temperature $\vartheta_C = \vartheta_I$. Two immiscible liquid phases are formed, which at the phase interface are in equilibrium with the vapor C. As long as not all of the vapor has been condensed, the temperature and composition at the phase interface remain unchanged. At the phase interface, the vapor has the eutectic composition. The calculation of the heat transfer during condensation is simplified in so far as further downstream, with decreasing wall temperature, the temperature $\vartheta_I = \vartheta_C$ at the phase interface and the liquid composition remain unchanged, as long as vapor is present.

By no means are all the possibilities exhausted by the equilibrium diagrams as shown in Figs. 7.2 and 7.3. Equilibria of the type depicted in Fig. 7.4 have also been found. To the left of the eutectic point C the mixture behaves like a mixture according to Fig. 7.2 or 7.3. Depending upon the wall temperature or the phase interface temperature, a homogeneous or a heterogeneous condensate film can be formed. The same is true for vapor mixtures whose composition lies between C and B. Mixtures with a composition to the right of point B always form a homogeneous condensate film. Mixtures of the type in Fig. 7.4 are, for example, n-pentane/nitrobenzene, n-hexane/aniline, carbon dioxide/water, ammonia/toluene, benzene/sulfur, and water/nicotine.

As these considerations show, one must always first check, based upon equilibrium, whether homogeneous condensation is present—in which case the previously discussed calculation procedure can be used—or whether heterogeneous condensation is to be expected. In order to determine this in practical cases, there generally suffices a rough calculation of the phase interface temperature ϑ_I with the aid of the energy balance at the phase interface, Eq. (6.38).

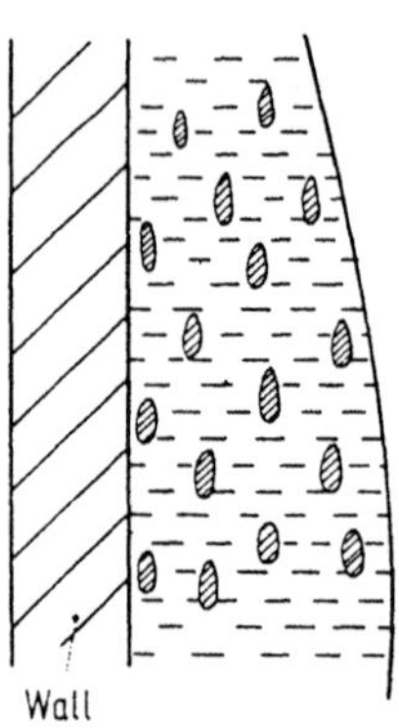

Fig. 7.5. Condensation forms. Homogeneously dispersed model

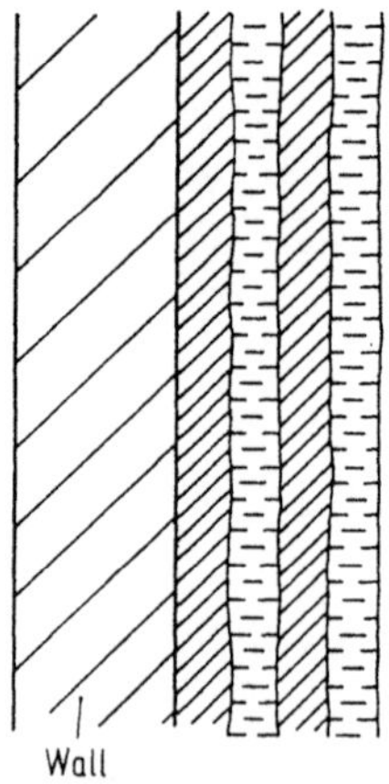

Fig. 7.6. Condensation forms. Heterogeneously dispersed model

There has been no lack of attempts to calculate the heat transfer during the condensation of immiscible liquids. They are based upon simplified model conceptions of the mechanism of heat transfer. These can be traced back essentially to two critical cases, which we shall call the homogeneously dispersed and the heterogeneously dispersed model.

In the homogeneously dispersed model, one imagines a phase completely uniformly distributed in the other, Fig. 7.5. The heterogeneously dispersed model is based upon the idealization that each of the two phases forms within itself cohesive layers, which are separated from those of the other phase, Fig. 7.6.

Equations for the calculation of the heat transfer generally presuppose a homogeneously dispersed model. According to a suggestion of Akers and Turner [7.2], one calculates for this the heat transfer coefficients from the Nusselt film condensation theory, Eq. (2.12a) for the vertical or Eq. (2.14a) for the horizontal tube, into which one must then insert the properties of the homogeneously dispersed mixture. Allowing for the different densities of liquid and vapor, for the *vertical tube* the earlier Eq. (2.12a) can be written in the following form:

$$\frac{\bar{\alpha}}{\lambda_L}\left(\frac{\eta_L^2}{\varrho_L(\varrho_L-\varrho_G)g}\right)^{1/3} = 0.925\left(\frac{\Gamma}{\eta_L}\right)^{-1/3}. \tag{7.1}$$

Correspondingly, one obtains from Eq. (2.14a) for the *horizontal tube*

$$\frac{\bar{\alpha}}{\lambda_L}\left(\frac{\eta_L^2}{\varrho_L(\varrho_L-\varrho_G)g}\right)^{1/3} = 0.959\left(\frac{\Gamma}{\eta_L}\right)^{-1/3}. \qquad (7.2)^1$$

For the thermal conductivity, one must use a mean value corresponding to the volumetric fractions of the two liquid phases. If one designates these volumetric fractions with $\varepsilon_A = \bar{V}_A/\bar{V}$ ($\bar{V}$ is the molar volume) and $\varepsilon_B = \bar{V}_B/\bar{V} = 1 - \varepsilon_A$, while the two phases of the phase diagrams in Figs. 7.2 to 7.4 are to be characterized by A and B, then

$$\lambda_L = \varepsilon_A\lambda_A + \varepsilon_B\lambda_B. \tag{7.3}$$

The volumes $\bar{V}_A$ and $\bar{V}_B$ are obtained according to the laws of mixed-phase thermodynamics [7.3] from

$$\bar{V}_A = x_A V_{01} + (1 - x_A)V_{02} + \Delta\bar{V}_A$$

$$\bar{V}_B = x_B V_{01} + (1 - x_B)V_{02} + \Delta\bar{V}_B.$$

The value

$$\varrho_L = \xi_A\varrho_A + \xi_B\varrho_B \tag{7.4}$$

is to be inserted into Eqs. (7.1) and (7.2) as the liquid density, and as the dynamic

[1] The numerical value 0.959 differs slightly from that in the original work [7.2], because Akers and Turner based their calculations upon the somewhat imprecise Nusselt value. Refer to the footnote on page 10.

viscosity that of the continuous phase, i.e., of the higher boiling point component, which forms the cohesive liquid film.

Bernhardt et al. [7.1] have stated a somewhat different procedure for calculation, which reproduces quite well the data known so far about the heat transfer of immiscible liquids. Accordingly, one forms the heat transfer coefficient of the homogeneously dispersed condensate film from the combination of the two phases A and B corresponding to their volumetric fractions

$$\alpha = \varepsilon_A \alpha_A + \varepsilon_B \alpha_B . \tag{7.5}$$

The heat transfer coefficients of the two liquid phases are calculated from one of the known heat transfer equations; for example, in the case of laminar film condensation, according to the Nusselt theory at the pressure and the temperature of the mixture. As temperature difference, one must insert into this equation the difference between the temperatures ϑ_I at the phase interface and ϑ_w at the wall; the phase interface temperature is, according to what has been said above, equal to the eutectic temperature ϑ_C of the mixture.

One can calculate both local and average heat transfer coefficients according to Eq. (7.5). This equation has been checked so far only on data for laminar film condensation and predicts their heat transfer coefficients with an error of up to $\pm 30\%$.

Tamir et al. [7.4, 7.5] have carried out several experiments on heterogeneously dispersed film condensation, by condensing vapor on a cold liquid film. The condensate formed was immiscible with the liquid film. The investigations showed that during this process the thermal resistance between the condensate surface and the vapor is no longer negligible. It can be of the same magnitude as the thermal resistance of the water film, as experiments on the condensation of n-pentane, methylchloride, R-113 or 1-1-dichloroethane on a water film have shown. The additional resistance at the phase interface can be explained by the fact that the condensation is possible only on tiny nuclei. According to the theory of nucleation, the rate of nucleation is given by $v \exp(-\Delta G/kT_\infty)$, where v is a frequency factor, ΔG the free enthalpy, and k the Boltzmann constant. Relying upon these findings, Tamir and Rachmilev [7.5] assumed that the mean heat transfer coefficient $\bar{\alpha}_I$ at the phase interface is proportional to the rate of nucleation. They represented their results by an empirical equation

$$\bar{\alpha}_I = 0.942 \cdot 10^{-6} \exp(6375/T_\infty) \tag{7.6}$$

where T_∞ is the saturation temperature of the vapor in **K** and $\bar{\alpha}_I$ is the heat transfer coefficient in W/m^{2}K. The total thermal resistance

$$\frac{1}{\bar{\alpha}} = \frac{1}{\bar{\alpha}_I} + \frac{1}{\bar{\alpha}_L}$$

is composed of the thermal resistance at the phase interface and the thermal resistance of the condensate film.

8 Enhancement of Heat Transfer During Condensation

8.1 Fundamentals

During film condensation of water vapor one achieves in practice mean heat transfer coefficients of between approximately 4000 W/m^2K and 10 000 W/m^2K. If vapors of organic substances are condensed, the mean heat transfer coefficients are seldom over 2500 W/m^2K. By contrast, heat transfer coefficients during drop condensation of organic vapors are ten to twenty times greater than those of film condensation. The heat transfer coefficients of the coolant lie between 4000 W/m^2K (achieved with, for example, cooling water of 35 °C, velocity 1.1 m/s, tube inner diameter 0.04 m, tube length 1.33 m), and 6000 W/m^2K (upon raising the water velocity to 1.8 m/s). Because the thermal resistance of the metallic wall can generally be ignored, the heat transfer coefficient on the condensate side controls the heat transfer, especially during the film condensation of organic vapors. It is, therefore, very worthwhile to seek ways to improve the heat transfer coefficient on the condensate side. In the literature, one finds a great number of suggestions, of which only a few have found their way into practice.

Measures for the improvement of the heat transfer differ clearly according to whether the decisive thermal resistance is that of the condensate film or that of the vapor.

During the condensation of pure vapors, the thermal resistance in the film is decisive; that of the vapor is negligible, as long as the shear stress at the condensate surface plays no role. Here, one will look for suitable ways to diminish the thermal resistance of the condensate film.

The relationships are different if the vapor contains inert gases or if a vapor mixture is being condensed. The condensing components then arrive by diffusion at the condensate surface, before they are condensed. Then, in general, the thermal resistance in the vapor is controlling, and in order to improve the heat transfer, one must reduce the thermal resistance on the vapor side.

In the case of the total condensation of mixtures, the diffusional resistance is, as we have seen, negligible on the vapor side. The decisive thermal resistance is then that of the condensate film, so that one must decrease it in order to improve the heat transfer.

8.2 Increasing the Vapor Velocity

During the condensation of vapors which contain inert gas, or during the partial condensation of mixtures, one will attempt to improve the heat transfer on the vapor side. This happens most simply by raising the vapor velocity. As one sees from Figs. 4.1 and 4.2 for the condensation of saturated and superheated vapors in tubes, by raising the vapor velocity in the inlet cross section from 40 to 60 m/s, improvements of 20 to 30% can be achieved in the heat transfer rate.

An impression of the influence of the vapor velocity on the heat transfer is also given in Fig. 8.1, in which local Nusselt numbers were calculated with the aid of different turbulence models, the mixing method model and the so-called k-ε model [8.1]. The local Nusselt number is plotted in this figure as a function of the Reynolds number of the condensate film, with the velocity in the inlet cross section as a parameter. The results are valid for a vertical condenser tube and assume that the vapor velocity, and, thus, also the friction along the tube, practically do not change. In real condenser tubes, however, the original high vapor velocity is rapidly reduced as the result of the condensation of the vapor, so that the improvement of the heat transfer caused by increasing the vapor velocity in the inlet cross section also rapidly decreases.

The influence of the vapor velocity on the heat transfer in horizontal condenser tubes with downward flowing vapor can be judged by using an empirical equation of Berman [8.2], which reproduces the experiments quite well. Accordingly, the heat transfer coefficient α related to the value α_0 of the stagnant vapor is

$$\frac{\alpha}{\alpha_0} = a + b \ln \pi \tag{8.1}$$

with

$$\pi = \frac{\lambda_L(\vartheta_I - \vartheta_w)}{\eta_L \Delta h_v} \frac{w_G^2}{dg}, \tag{8.1a}$$

$$0.24 \leqq \pi \leqq 12.1$$

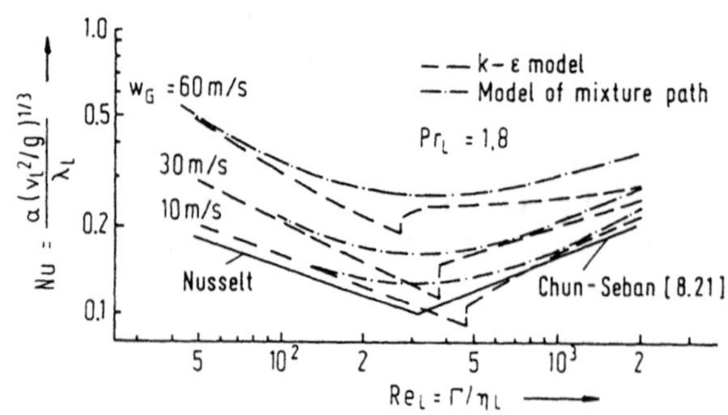

Fig. 8.1. Local Nusselt numbers at nearly constant vapor velocity

and $a = 1.45$, $b = 0.07$ for $\pi < 1$

$b = 0.12$ for $\pi > 1$.

The equation is valid within the stated area of the characteristic quantity π and, moreover, if

$$0.23 \leqq S = Pr \frac{\Delta h_v}{c_{pL}(\vartheta_I - \vartheta_w)} \left(\frac{\nu_G}{\nu_L}\right)^{1/2} \leqq 0.41.$$

The heat transfer coefficients $\alpha_2(\pi_2 = 12.1)$ and $\alpha_1(\pi_1 = 0.24)$ at the boundaries of the validity of Eq. (8.1) are in relation to one another as

$$\frac{\alpha_2}{\alpha_1} = 1.3.$$

With constant properties and temperatures, according to Eq. (8.1a), over this range there is an increase of the vapor velocity by the factor

$$\frac{w_{G2}}{w_{G1}} = \left(\frac{\pi_2}{\pi_1}\right)^{1/2} = 7.1.$$

The greater increases in the heat transfer calculated by other authors do not agree with the above values, because constant wall temperatures over the tube circumference were presupposed, and also because the separation of the vapor flow on the backside of the tube was not taken into account.

8.3 Roughened Tubes, Finned Tubes

Studies by Spencer and Ibele [8.3] of artificially roughened, vertical surfaces with a peak-to-valley-height of 12 μm showed that the heat transfer coefficients during film condensation in the region of very small Reynolds numbers of the condensate film of $Re = \Gamma/\eta_L < 35$ lay even below those of the smooth tube. Considerable quantities of condensate remain attached to the rough surface, causing the film to thicken during low rates of condensation. In this area of Reynolds numbers, the film thickness proved to be practically independent of the condensate quantity. The heat transfer coefficient was, consequently, approximately constant. The data for the mentioned roughnesses could be reproduced well by

$$\frac{\bar{\alpha}(\nu_L^2/g)^{1/3}}{\lambda_L} = 0.23 \quad \text{for} \quad Re = \Gamma/\eta_L < 35, \tag{8.2}$$

whereas according to the Nusselt theory (see Eq. (2.12a)),

$$\frac{\bar{\alpha}(\nu_L^2/g)^{1/3}}{\lambda_L} = 0.925\, Re^{-1/3} \tag{8.3}$$

in the region of Reynolds numbers $Re < 35$, much greater heat transfer co-

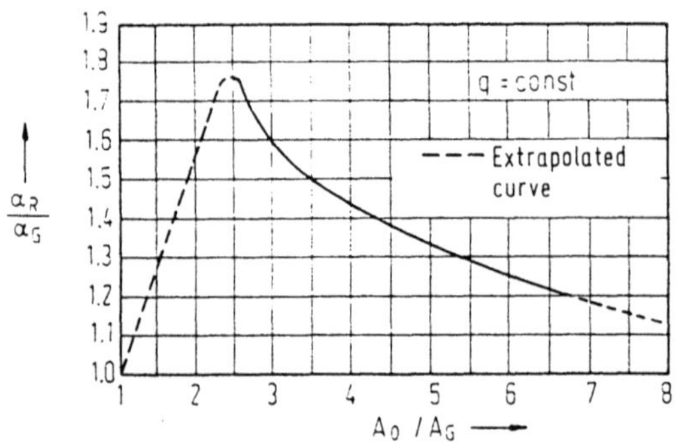

Fig. 8.2. Increase of the heat transfer coefficient on finned tubes in comparison to a smooth tube, according to [8.4]. Experiments with R-12 at 30 °C condensation temperature. A_0 is the surface of the finned tube and A_G is the surface of reference plain tube

efficients are calculated. Above a Reynolds number of 35, the heat transfer coefficients on a rough heating surface were still smaller than those on a smooth surface, in conformity with the Nusselt film condensation theory. With an increasing Reynolds number, however, the differences from the Nusselt theory became smaller, and, from a Reynolds number of 175 upward, the heat transfer on a rough condenser surface was greater than that on a smooth surface. Obviously, the increase of waviness and turbulence as a consequence of roughness has a greater effect in the region of larger Reynolds numbers.

The use of *finned tubes* in condensers gained in significance chiefly in refrigeration plants, because the properties of the chlorofluorocarbons used there as refrigerants produced very low heat transfer coefficients during condensation. As experiments of Henrici [8.4] on horizontal finned tubes showed, the average heat transfer coefficient relative to the entire surface of the finned tube is markedly greater than that for smooth tubes of the same diameter. This emerges clearly from Fig. 8.2, in which the average heat transfer coefficient α_R on a finned tube of copper relative to that of a smooth tube α_G, which has the same basic diameter, is plotted as a function of the surface ratio A_0/A_G of finned tube to smooth tube. The data relates to the film condensation of R-12 at 30 °C condensation temperature. The number of fins was between 400 and 500 per m tube; the fin thickness of 0.5 mm, and the increase of the finned surface was achieved mainly by greater fin heights of between 1.3 and 4.5 mm.

The heat transfer rate during condensation on finned tubes exceeds the surface increase compared to the smooth tube. In contrast to this, as is well known, the heat transfer increases less than proportional to the increase in surface during single-phase heat transfer. In order to understand this phenomenon, let us consider separately the condensation process on fins and on a smooth horizontal tube. A fin can be regarded as a vertical wall. If one compares the heat transfer coefficient on a vertical wall with that on a horizontal tube, Eqs. (2.12) and (2.14), one finds

$$\frac{\bar{\alpha}}{\bar{\alpha}_{hor}} = 1.3\left(\frac{d}{h}\right)^{1/4}. \tag{8.4}$$

When using fins of small height $h \ll d$, as is usually the case in industry, the heat transfer coefficient on a fin is significantly greater than that of a horizontal tube at the same surface temperature, because the thickness of the condensate film is small. Because of the greater length of flow and thus the greater average film thickness of the condensate, the average heat transfer coefficient decreases with the fin height. Given smaller fin intervals, the heat transfer coefficient increases because of the increase in the number of vertical short surfaces. If, however, the tube is too closely finned with more than 750 fins per meter, the condensate will remain attached between the fins as a result of surface forces, and the heat transfer will decrease. For fins of very small height, both the condensate flowing from the base tube as well as from the fins affect each other. This leads to a thicker condensate film around a fin, with the result that at a small fin height the heat transfer increases only little compared to a smooth tube. If the fins are higher than 4 to 5 mm, the surface temperature of the tube will increase appreciably in contrast to the experiments with tubes having low fins according to Fig. 8.2, which leads to deviations from the given curve. In practice, however, tubes with greater fin heights will not be considered for condenser tubes, either for reasons of manufacture or of economy.

The advantage in using finned condenser tubes lies above all in the fact that the heat transfer rate per unit of length is greater than when using smooth tubes. One can build, therefore, condensers of a prescribed output smaller and with fewer tubes, if one chooses finned tubes rather than smooth tubes. Because of the smaller number of tubes, one can select higher coolant velocities at the same pressure drop, which also leads to an increase in condenser output, and, in addition, lessens the danger of fouling on the coolant side.

For the *calculation of the heat transfer coefficient of finned tubes*, one imagines the surface area A_0 of the finned tube as divided into the area A_G of the horizontal base tube and of the vertical fins A_R. The total heat transferred is then

$$\Phi = \bar{\alpha}_G A_G (\vartheta_0 - \vartheta_a) + \bar{\alpha}_R A_R (\vartheta_0 - \vartheta_a), \tag{8.5}$$

if one makes the temperature ϑ_0 of the fin surface equal to that of the fin base, an exception that is very well fulfilled in the case of the low fins stipulated here for the condenser tubes. If one further assumes

$$\Phi = \bar{\alpha} A_0 (\vartheta_0 - \vartheta_a), \tag{8.6}$$

where A_0 is the entire outer surface area of the finned tube, then one obtains for the heat transfer coefficient of the finned tube

$$\bar{\alpha} = \bar{\alpha}_G \frac{A_G}{A_0} + \bar{\alpha}_R \frac{A_R}{A_0}. \tag{8.7}$$

Because the fins can be regarded as a vertical wall, there exists between the average heat transfer coefficient α_R of the fin and that α_G of the smooth tube a relationship that can be written according to Eq. (8.4) as

$$\frac{\bar{\alpha}_R}{\bar{\alpha}_G} = 1.3 \left(\frac{d}{H} \right)^{1/4}. \tag{8.4a}$$

Katz et al. [8.5] have suggested as the average height H for the vertical fin the value

$$H = \frac{\pi}{4} D \left(1 - \frac{d^2}{D^2} \right) \tag{8.8}$$

where d is the inner and D the outer diameter of the tube. The circular fin surface area is replaced here by a rectangle of equal area. If one enters this value for the fin height into Eq. (8.4a), one obtains values somewhat too large for the heat transfer coefficient, because the condensate collecting on the bottom of the tube flows off partially over the lower fin segment. Because of this, the condensate film is thicker there and the heat transfer coefficient is somewhat smaller than predicted by Eq. (8.4a). Henrici allowed approximately for this effect by replacing the constant 1.3 in Eq. (8.4a) with the somewhat smaller value 1.1. Thus, one obtains from Eq. (8.7) as the average heat transfer coefficient of a finned tube

$$\bar{\alpha} = \bar{\alpha}_G \left[1.1 \left(\frac{d}{H} \right)^{1/4} \frac{A_R}{A_0} + \frac{A_G}{A_0} \right] \tag{8.9}$$

with H according to Eq. (8.8) and the average heat transfer coefficient $\bar{\alpha}_G$ for the condensation on horizontal tubes, as given by the Nusselt film condensation theory, Eq. (2.14). According to Eq. (8.9), heat transfer coefficients could be reproduced very well on copper tubes with small fin height, i.e., on tubes such as are used in refrigeration plants.

8.4 Gregorig Tubes

Gregorig [8.6] was the first to prove that by using a finely undulated surface one can achieve a significant augmentation in heat transfer during film condensation. In Fig. 8.3 is depicted a profiled surface like the one suggested by him for vertical condenser tubes. As the figure shows, a thin condensate film forms on the peak and a thick one forms in the valley, which flows off downward. In the region of the peak, the heat transfer coefficient increases strongly because of the thin condensate film. Even if, therefore, one assumes that no heat is transferred in the runoff region in the wave valley, there results on the average a significant augmentation of the heat transfer coefficient in contrast to a smooth tube of the same condensation output.

The uneven film thickness can be explained by the fact that the pressure in the condensate film at the peak, according to the equation of Laplace, is larger by the fraction

$$\Delta p_1 = \sigma / r_1$$

while the pressure in the valley is smaller by the fraction

$$\Delta p_2 = \sigma / r_2$$

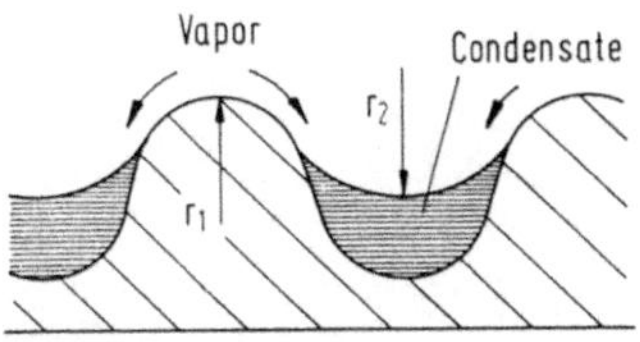

Fig. 8.3. Gregorig profile

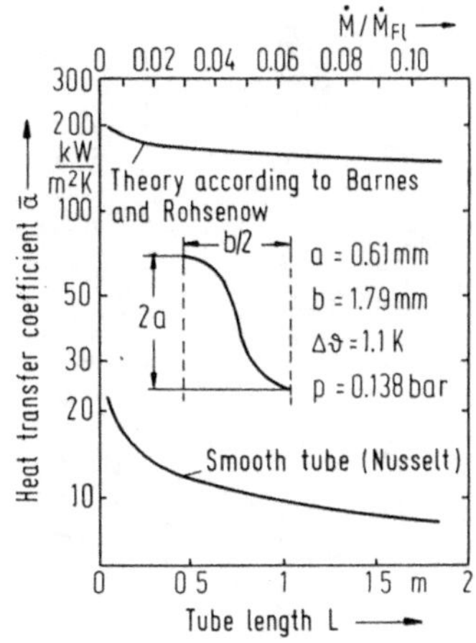

Fig. 8.4. Mean heat transfer coefficient for condensation of water vapor on a vertically profiled tube in comparison with a smooth tube, according to Barnes and Rohsenow [8.11]

than the vapor pressure. Here, r_1 and r_2 are the radii of curvature of the condenser surface on the peak and in the valley. In the condensate film, therefore, the pressure drops strongly from the peak to the valley. If the radii of curvature are sufficiently small, the condensate flows, even in vertical tubes, from the peak to the valley and then flows downward under the force of gravity. A very thin condensate film with high heat transfer is formed on the peak, whereas in the valley a thick film runs off with little heat transfer. Thus, the peak provides for good heat transfer, and the valley serves primarily for drainage of the condensate.

The ideas of Gregorig have been checked mathematically and experimentally by many different authors and have been developed further [among others, 8.7–8.10]. The original assertion of Gregorig was confirmed, namely, that, by a favorable formation of the profile, the heat transfer on the condensate side can be augmented by a factor of 10 in comparison with a smooth tube. As an example, Fig. 8.4 shows heat transfer coefficients according to Barnes and Rohsenow [8.11] as a function of tube length. Whereas the smooth tube shows the familiar length dependency of the heat transfer coefficient with the exponent $-1/4$, the heat transfer coefficient of the profiled tube is much less dependent on the length because of the condensate drainage in the valleys. The heat transfer decreases only if the profiled tubes are so long that the condensate fills the surface to such an extent that condensation is impeded on the profile flanks. A flooding can be prevented by use of condensate strippers. These should be placed where the condensate mass flow rate $\dot{M}$ is roughly 10% of the condensate mass flow rate $\dot{M}_{Fl}$, if the entire cross section were flooded by condensate [8.11]. The mass flow rate ratio $\dot{M}/\dot{M}_{Fl}$ can be read on the upper abscissa in Fig. 8.4.

If one designates with r_1 the radius of curvature of the condensate film on the peak, with $K = 1/r$ its curvature, and with s the arc length, then the pressure gradient dp/ds is the controlling factor for the flow of the condensate from the

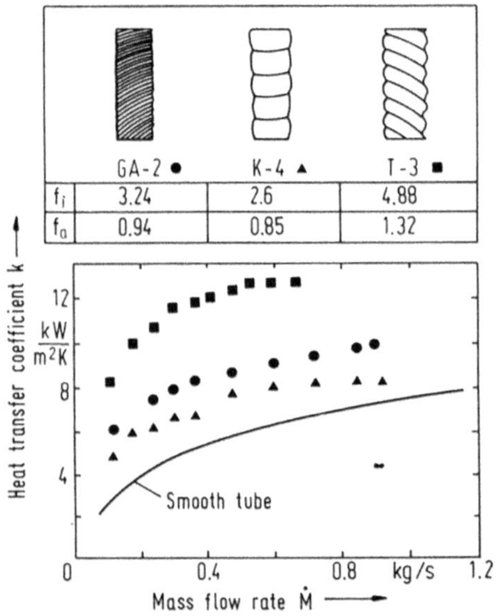

Fig. 8.5. Overall heat transfer coefficient and mean amplification factors f_i for heat transfer on the coolant side and f_a for that on the condensate side, according to Marto et al. [8.13]. $\dot{M}$ is the mass flow rate of cooling water

peak to the valley. From the equation of Laplace $\Delta p = \sigma / r_1$ follows

$$dp/ds = \sigma dK/ds \quad \text{with} \quad K = 1/r_1 . \tag{8.10}$$

A considerable augmentation of heat transfer can be achieved, only if this pressure gradient is large, i.e., if the curvature changes strongly from the peak to the valley. One must select, therefore, very finely profiled surfaces. As Adamek [8.12] showed on the basis of an optimization calculation, favorable profile heights are approximately 0.5 mm. Favorable intervals between two wave peaks are roughly just as large, which yields roughly 2000 profiles per meter. Because such a fine profiling is difficult to manufacture, and because it can be easily fouled in industrial processing, various attempts have been made to use more coarsely profiled tubes in condensers. The augmentation is, however, noticeably smaller than when the finely profiled Gregorig tubes are used.

Figure 8.5 shows as an example spiral-finned tubes and tubes with stamped, spiral-shaped grooves, that were used by Marto et al. [8.13] in experiments with condensing water vapor on horizontal tubes. The heat transfer coefficient is greater by a factor up to 5 than that for smooth tubes. This enhancement is, however, to be traced above all to the better convective heat transfer on the coolant side. The heat transfer coefficient for the water coolant was larger by the factor f_i given in the figure than the heat transfer coefficient for flow in a smooth tube. On the other hand, the heat transfer coefficient on the condenser side was different from that of the Nusselt film condensation theory only by the factor f_a. Therefore, condensation was scarcely influenced by the coarse profiles.

8.5 Condensation in Tubes

The heat transfer during condensation in tubes can be enhanced by fins on the inside or by the installation of static mixer inserts. A rough idea of the enhancement to be expected for the average heat transfer coefficient during condensation is given by Fig. 8.6, in which average heat transfer coefficients for condensation in four tubes with interior fins and in two with mixing inserts are plotted over the mass flux and are compared with the corresponding values for a smooth tube [8.14]. According to this, the heat transfer coefficient for an inner-finned tube is larger by a factor of 1.3 to 1.5 and that for a tube with mixing inserts is larger by roughly the factor 1.2 than that for a smooth tube.

Chiefly, the surface and thereby the heat transfer are augmented by interior finning. Additionally, the fins promote turbulent equilibrium between the phases and thus contribute to the increase in heat transfer [8.15–8.20]. However, this effect is partially nullified by the greater film thickness and the increase in temperature toward the fin peak [8.21].

Static mixer inserts made of twisted metal strips may be more economical than interior fins. The tangential component of velocity increases the residence time of the vapor, and the twisted strips increase the wetted circumference, without causing a noticeable change in the cross section of flow. All the effects together lead to the enhancement in heat transfer, as shown in Fig. 8.6.

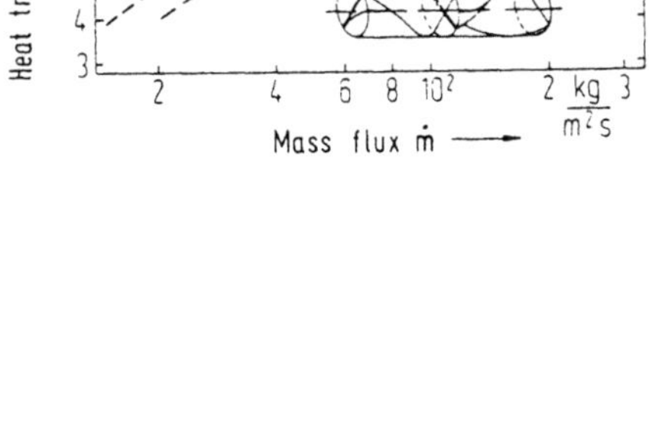

Fig. 8.6. Heat transfer coefficient during condensation in a smooth tube and in a tube with inner fins as well as in a tube with static mixer inserts, according to Azer and Said [8.14]

B Heat Transfer in Boiling

9 The Different Types of Heat Transfer During Boiling

Whereas heat transfer during convection can be described by physical properties, such as viscosity, density, thermal conductivity, thermal coefficient of expansion, as well as by geometric parameters, for boiling processes, there are of additional significance those variables that are connected with the phase transformation. To these belong: vaporization enthalpy, boiling temperature, density of vapor, and surface tension. The microstructure and the material of the heating surface also play a role. Because of this multiplicity of variables, it is more difficult to suggest equations for the calculation of heat transfer coefficients than it is in the case of other problems of heat transfer. Also, we are far from having worked out a complete theory, because the physical phenomena are too complicated and in no way have been sufficiently studied.

Causes for this are not only the many influential values, which play a role in boiling processes, but also the many different types of heat transfer which result depending upon the flow pattern and the magnitude of superheating.

9.1 Evaporation

If the liquid at a heated wall is only slightly superheated above the saturation temperature, then only a few or even no vapor bubbles are formed. In a vessel filled with liquid and heated from below, a temperature profile develops as depicted in Fig. 9.1. Above the heated base with the temperature ϑ_w there forms a boundary layer of the order of magnitude of 1 mm with a strong temperature drop, while in the core the liquid temperature is almost constant above the height z (mean value ϑ_L). At the free surface the temperature in a thin layer drops to the value ϑ_0, which is slightly above the saturation temperature ϑ_s. The difference $\vartheta_0 - \vartheta_s$ was measured for the first time by Prüger [9.1] for water at 1.01 bar as being 0.03 K, whereas for non-polar liquids such as carbon tetrachloride this value is approximately 0.001 K.

As important as this liquid superheating at the surface is for the kinetic consideration of vaporization, it can nevertheless be disregarded during technical calculations. In the following, therefore, the saturation temperature $\vartheta_0 = \vartheta_s$ is always ascribed to a vapor-forming surface.

In the thin layer near the wall, the temperature drops off steeply, as shown in Fig. 9.1. In this layer, conduction dominates. In the liquid below, the heat

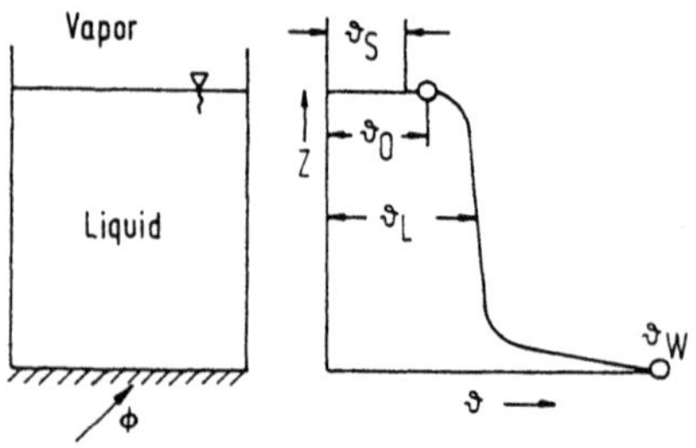

Fig. 9.1. Temperature profile in the liquid during surface evaporation

transport is provided by ascending and descending convective flows. They produce the uniform temperature in the core of the liquid. The two boundary layers above and below are differentiated from one another by the fact that the free surface is movable because of the vapor formation and that finite parallel velocities can also occur there, in contrast to the liquid at the wall.

The vaporization itself acts as a heat sink at the surface, which one could imagine as being replaced by another process, for example, by radiation. Because vaporization or evaporation results at the free surface, one speaks of "*quiet boiling*". This process belongs by its nature to the phenomena of free convection in closed spaces. Heat transfer coefficients from the heating surface to the liquid can be formed with the driving temperature difference $\vartheta_w - \vartheta_L$, where ϑ_w is the wall temperature of the heating surface and ϑ_L is the temperature of the liquid. Because the temperature of the liquid is not known in advance and, as explained above, deviates only slightly from the saturation temperature, it is practical to form the heat transfer coefficients with the temperature difference $\Delta\vartheta = \vartheta_w - \vartheta_s$. For evaporation, the laws of heat transfer for free convection apply. Thus, there is $\alpha = c_1 \Delta\vartheta^{1/4}$ during laminar flow and $\alpha = c_2 \Delta\vartheta^{1/3}$ during turbulent flow over a horizontal plate. Because the heat flux is given by $q = \alpha\Delta\vartheta$, there holds for laminar flow

$$\alpha = c_1 \Delta\vartheta^{1/4} \quad \text{or} \quad \alpha = c_1' q^{1/5} \tag{9.1}$$

or for turbulent flow

$$\alpha = c_2 \Delta\vartheta^{1/3} \quad \text{or} \quad \alpha = c_2' q^{1/4}. \tag{9.2}$$

9.2 Nucleate Boiling

If one raises the wall temperature by increasing the heat input, vapor bubbles form when a definite wall temperature is reached. As observation shows, these arise only at certain points of the heating surface. The number of vapor bubbles formed grows with the heat input. This type of heat transfer is characterized as *nucleate boiling*. Figure 9.2 shows a typical temperature pattern over a horizontal plate, according to the data of Jakob and colleagues [9.2–9.8], to whom we owe the first fundamental investigations of the process. In contrast

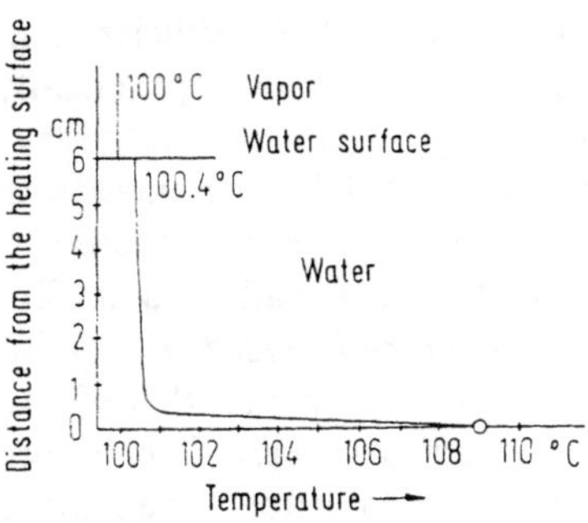

Fig. 9.2.

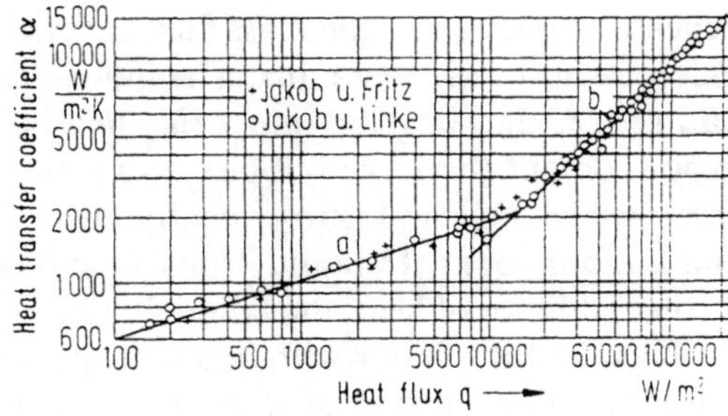

Fig. 9.3.

Fig. 9.2. Temperature profile above a heating surface during nucleate boiling, according to Jakob and Linke [9.5]. Heat flux $q = 22440\,\mathrm{W/m^2}$, temperature of heated surface $\vartheta_w = 109.1\,°\mathrm{C}$

Fig. 9.3. Heat transfer to boiling water at 100 °C from a horizontal heated surface, according to Jakob et al. [9.2, 9.4]. Curve *a*: region of free convection/evaporation; curve *b*: region of nucleate boiling

to Fig. 9.1, one notices that the temperature difference $\vartheta_w - \vartheta_L$ is substantially greater and that $\vartheta_L - \vartheta_s$ is much smaller. The bubble movement at the surface permits no exact measurement of the boundary layer. Once again, one forms the heat transfer coefficient with the temperature difference $\Delta\vartheta = \vartheta_w - \vartheta_s$ as for evaporation. The heat transfer is very much better than during evaporation and is approximately proportional to the third power of the temperature difference $\Delta\vartheta$. If one considers that the heat flux is given by $q = \alpha\Delta\vartheta$, then

$$\alpha = c_3 \Delta\vartheta^3 \quad \text{or} \quad \alpha = c_3' q^{3/4} \tag{9.3}$$

holds approximately in nucleate boiling. If one represents graphically $\alpha(q)$ from Eq. (9.1) or from Eqs. (9.2) and (9.3), there result two straight lines, if one chooses logarithmic scales for the ordinate and the abscissa. One gets two distinctly separate areas, one for evaporation and one for nucleate boiling, as shown by Fig. 9.3, which reproduces the data of Jakob and colleagues [9.2, 9.4].

9.3 Convective Boiling

Generally, vaporization occurs in technical equipment under forced convection. The flow conditions are determined to a great extent by the pressure difference along the heating surface. The vapor content increases along the path of flow up to the point of complete vaporization. Corresponding to the decreasing liquid supply, there result different boiling phenomena, upon whose heat transfer properties the local boiling temperature is dependent. In general, a liquid enters subcooled into a heated channel. Vapor bubbles formed at the wall condense again in the colder core of the liquid. If the liquid in the core is heated up to saturation temperature, then nucleate boiling results. The heat transfer coefficient is determined here chiefly by the heat flux. In a forced flow, the coefficient still depends weakly upon the mass flux, but with natural circulation it is practically independent of the mass flux. The individual bubbles grow together into large bubbles i.e., they develop into a *slug flow*, as shown in Fig. 9.4. With increasing vapor content the large bubbles also grow together, so that at first there is a *semi-annular flow* and, subsequently, there forms at the tube wall a liquid film and a vapor core with liquid drops. One speaks of *annular-dispersed flow*. With further addition of heat, the liquid film disappears downstream, and there flows through the tube vapor with liquid drops, a so-called *spray* or *mist flow*. Figure 9.4 shows these forms of flow in succession in a vertical tube. Even more complicated forms of flow, which occur in horizontal or inclined tubes, will be discussed later. Bubbly, slug, semi-annular, and annular flow represent different forms of *flow boiling* or *convective boiling*.

In technical applications, annular-dispersed flow occurs frequently or, if there are very small velocities, slug flow occurs. Bubbly flow occurs only with very small vapor content and high flow velocity. Increasing pressure, and, thus, a decreasing density difference between the vapor and the liquid, broadens the area of bubbly flow. In nucleate boiling, the heat transfer coefficient is chiefly dependent upon the heat flux and practically not at all upon the flow velocity, curve b in Fig. 9.3. On the contrary, in convective boiling, the heat transfer coefficient is primarily influenced by the velocity of flow or by the mass flux $\dot{m}$, but on the other hand is scarcely influenced by the heat flux. This is shown in

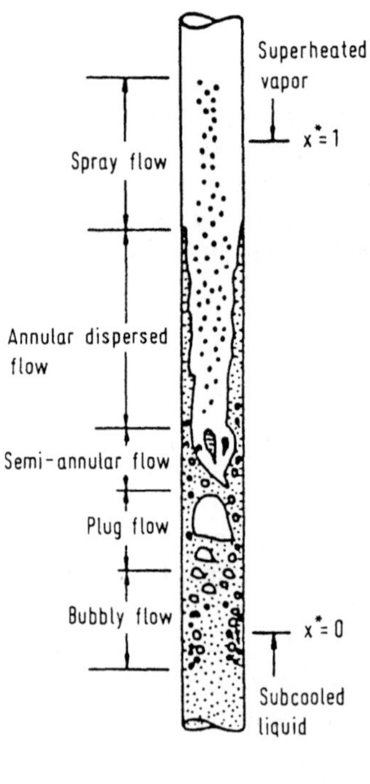

Fig. 9.4. Flow patterns in a vertical heated tube

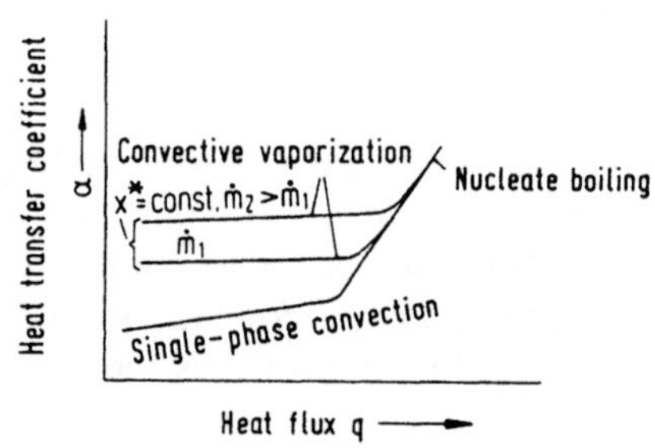

Fig. 9.5. Heat transfer coefficient during nucleate and convective boiling (qualitative)

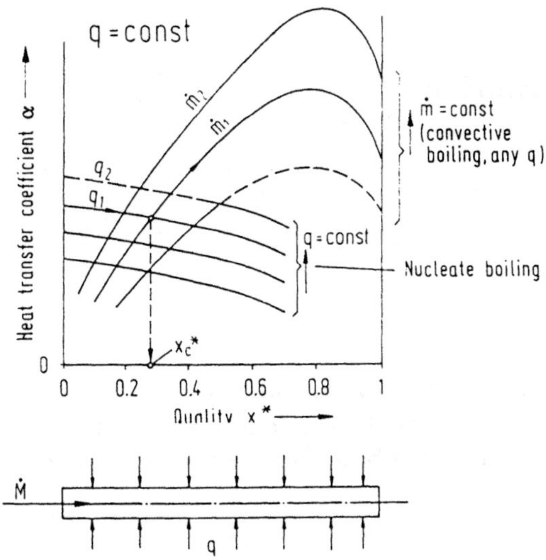

Fig. 9.6. Trends of the heat transfer coefficient α for a horizontal evaporator tube

Fig. 9.5, in which the areas of nucleate boiling and of convective boiling are clearly separated from one another.

An additional independent variable is the quality x^*. By this is meant the ratio of the vapor mass flow rate $\dot{M}_G$ to the total mass flow rate $\dot{M}$

$$x^* = \dot{M}_G/\dot{M}.$$

With increasing quality, the curves for the convective boiling in Fig. 9.5 are shifted to higher heat transfer coefficients α.

The fundamental dependence of the heat transfer coefficient on the quality is presented in Fig. 9.6. There is nucleate boiling in the area of low quality x^*, and the heat transfer coefficient depends chiefly upon the heat flux. Downstream, the quality increases. The average velocity thus increases also. The supplied heat is essentially transferred by convection from the tube wall to the vapor-liquid flow. The nucleate boiling converts into convective boiling, as is indicated in Fig. 9.6 by the arrows at the curves q_1 and $\dot{m}_1$. In the area of convective boiling, the local heat transfer coefficient is practically independent of the heat flux q and depends strongly upon the mass flux and the quality. At large qualities, the wall is essentially dry and the heat transfer coefficient again decreases because of the lower thermal conductivity of the vapor in comparison with the liquid.

Useful for the calculation of the heat transfer coefficient are equations in the form

$$\alpha = c q^n \dot{m}^s f(x^*),$$

where c is dependent on the fluid properties. In the area of convective boiling, $n \approx 0$, whereas s lies between 0.6 and 0.8. In the area of nucleate boiling, n is approximately 3/4 and s around 0.1 to 0.3.

10 Physical Fundamentals of Vapor Bubble Formation

The heat transfer during vaporization is more easily understood if one knows how vapor bubbles form at heating surfaces. In the following, we shall first discuss the formation and growth of vapor bubbles before we treat the different types of heat transfer in detail.

10.1 Bubble Formation and Liquid Superheating

The following considerations hold for the equilibrium of a vapor bubble, assumed to be spherical, with the liquid surrounding it, as shown in Fig. 10.1. Between the bubble (gas = index G) and the surrounding liquid (liquid = index L) there is *thermal equilibrium*

$$\vartheta_G = \vartheta_L = \vartheta. \tag{10.1}$$

If, as depicted in Fig. 10.1, one cuts out of the vapor bubble a surface element of a spherical shell, whose edge lengths are $r\mathrm{d}\varphi$, then the forces $\sigma r\mathrm{d}\varphi$ exerted by the surface tensions σ (σ is a force per unit of length) act upon the edges. The resultant F_R of these forces is given by

$$\mathrm{d}^2F_R = 2\sigma r\mathrm{d}\varphi^2.$$

Also of influence are the forces resulting from the gas and liquid pressures

$$p_L(r\mathrm{d}\varphi)^2 + \mathrm{d}^2F_R = p_G(r\mathrm{d}\varphi)^2.$$

From this follows the condition for *mechanical equilibrium*

$$p_G = p_L + 2\sigma/r. \tag{10.2}$$

If one has a vapor bubble with a randomly curved surface with the main curvature radii r_1 and r_2 of a surface element, then there follows from Fig. 10.2, as resultant of the surface forces:

$$\mathrm{d}^2F_R = \sigma r_2\mathrm{d}\varphi\mathrm{d}\psi + \sigma r_1\mathrm{d}\psi\mathrm{d}\varphi.$$

These surface forces are in equilibrium with the forces exerted by the gas and liquid pressures

$$p_L r_1\mathrm{d}\psi r_2\mathrm{d}\varphi + \mathrm{d}^2F_R = p_G r_1\mathrm{d}\psi r_2\mathrm{d}\varphi.$$

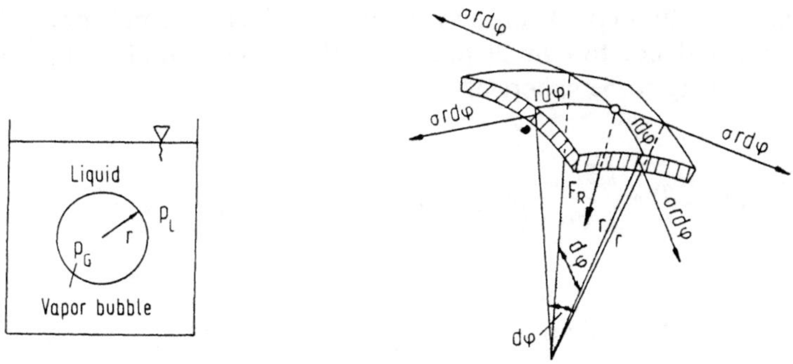

Fig. 10.1. Mechanical equilibrium between a spherical vapor bubble and the surrounding liquid

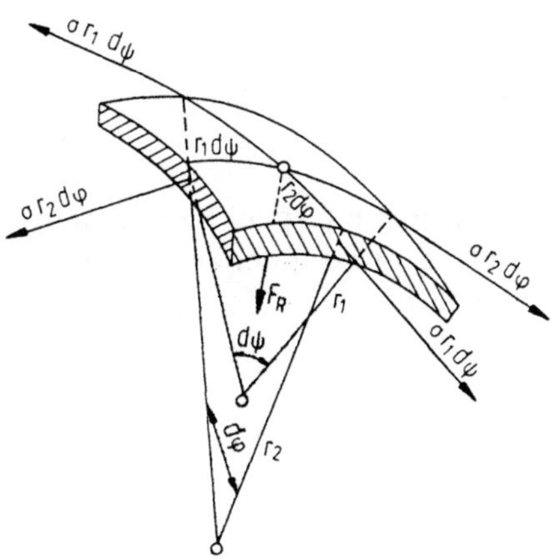

Fig. 10.2. Mechanical equilibrium between a vapor bubble with a surface of random curvature and the surrounding liquid

One obtains as the condition for mechanical equilibrium

$$p_G = p_L + \sigma\left(\frac{1}{r_1} + \frac{1}{r_2}\right), \tag{10.2a}$$

which as a special case of the spherical bubble ($r_1 = r_2 = r$) converts into Eq. (10.2) derived above.

Finally, there also applies the condition for the *material equilibrium*

$$\mu_G(p_G, \vartheta) = \mu_L(p_L, \vartheta), \tag{10.3}$$

according to which the chemical potentials[1] (SI unit J/kg or J/mole, here J/kg) of the two phases are equal. The equilibrium is completely described by the

[1] For pure substances like the ones treated here, the chemical potential is equal to the specific free enthalpy $g = h - Ts$, which, in material equilibrium between gaseous and liquid phases, is $g_G = g_L$. The reader who is not acquainted with the concept of chemical potential can pass over the following derivations up to the Thomson equations (10.7) and (10.8).

above conditions. Equation (10.3) can be further simplified for small pressure differences $p_G - p_L$, and not too small radii r of the vapor bubble, which is assumed in the following to be spherical.

There is

$$\mu_G(p_G, \vartheta) = \mu_G(p_L, \vartheta) + \left(\frac{\partial \mu_G}{\partial p}\right)_\vartheta (p_G - p_L) + \cdots$$

or

$$\mu_G(p_G, \vartheta) = \mu_G(p_L, \vartheta) + v_G(p_G - p_L) + \cdots. \tag{10.4}$$

Here, $v_G(p_L, \vartheta)$ is the specific volume. Considering Eq. (10.2), there results

$$\mu_G(p_G, \vartheta) = \mu_G(p_L, \vartheta) + v_G \frac{2\sigma}{r}, \tag{10.5}$$

for which, with Eq. (10.3), one can also write

$$\mu_G(p_G, \vartheta) = \mu_L(p_L, \vartheta) = \mu_G(p_L, \vartheta) + v_G \frac{2\sigma}{r}. \tag{10.6}$$

As stipulated, this equation does not hold for extremely small bubble radii. It includes, however, the limiting case of the plane phase interface $r \to \infty$. The equation is then $\mu_G(p_G, \vartheta) = \mu_G(p_L, \vartheta)$ and, therefore,

$$p_G = p_L = p_0,$$

so that Eq. (10.6) can be rearranged into

$$\mu_L(p_L, \vartheta) - \mu_L(p_0, \vartheta) = \mu_G(p_L, \vartheta) - \mu_G(p_0, \vartheta) + v_G \frac{2\sigma}{r}.$$

In case the bubble radii are not small, then $p_L - p_0$ is small. Thus, one can develop the chemical potential in a Taylor's series and truncate the series after the first term when allowing $(\partial \mu / \partial p)_\vartheta = v$. If, as usual, one abbreviates specific volumes in the saturated state with $v_L(p_0, \vartheta) = v'$ and $v_G(p_0, \vartheta) = v''$, there results

$$v'(p_L - p_0) = v''(p_L - p_0) + v_G \frac{2\sigma}{r}.$$

In addition, because $p_L - p_0$ is small, we put by approximation $v_G(p_L, \vartheta) \approx v_G(p_0, \vartheta) = v''$. With $\varrho = 1/v$ there results from this

$$p_L = p_0 - \frac{\varrho'}{\varrho' - \varrho''} \frac{2\sigma}{r} \tag{10.7}$$

or with Eq. (10.2)

$$p_G = p_0 - \frac{\varrho''}{\varrho' - \varrho''} \frac{2\sigma}{r}. \tag{10.8}$$

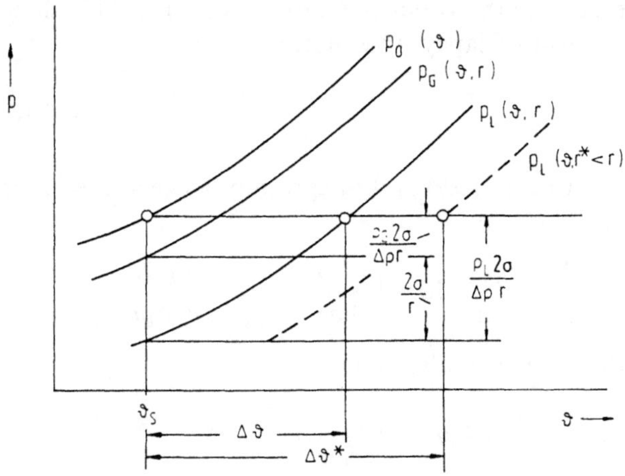

Fig. 10.3. Vapor and liquid pressures at equilibrium between liquid and a spherical vapor bubble

Equation (10.7) or, as the case may be, Eq. (10.8) is called the *Thomson equation.* It produces a connection between the vapor pressure $p_0(\vartheta)$ at a plane phase interface, the liquid pressure $p_L(\vartheta, r)$, and the vapor pressure $p_G(\vartheta, r)$ at the surface of a vapor bubble of radius r. The interrelationships are clearly depicted in Fig. 10.3. Given the temperature ϑ, then the vapor pressure p_G corresponding to Eq. (10.8) is smaller by

$$p_0 - p_G = \Delta p_G = \frac{\varrho''}{\varrho' - \varrho''} \frac{2\sigma}{r}$$

than the vapor pressure p_0 at the plane phase interface. Because the surface tension σ is temperature dependent, the curves for the vapor pressure p_G and the liquid pressure p_L do not run parallel to the vapor pressure curve p_0 for the plane phase interface.

If one is not given the temperature ϑ of the liquid-vapor system but rather its pressure p_0, then the liquid must be superheated by $\Delta\vartheta$ in contrast to the system with a plane phase interface, so that a vapor bubble of radius r is in equilibrium with the liquid, as shown in Fig. 10.3. One recognizes also that the necessary superheating $\Delta\vartheta$ is the greater the smaller the radius r of the vapor bubble, because, for smaller radii $r^* < r$, the curves of the vapor pressure $p_G(\vartheta, r^*)$ and of the liquid pressure $p_L(\vartheta, r^*)$ are shifted further to the right in Fig. 10.3. Conversely, at a given liquid superheat $\Delta\vartheta$, a vapor bubble of definite radius r is in equilibrium with the superheated liquid.

For the calculation of the necessary superheating, we differentiate Eq. (10.7) with respect to temperature

$$\frac{dp_L}{d\vartheta} = \frac{dp_0}{d\vartheta} - \frac{d}{d\vartheta}\left(\frac{\varrho'}{\varrho' - \varrho''} \frac{2\sigma}{r}\right). \tag{10.9}$$

The derivative $\mathrm{d}p_0/\mathrm{d}\vartheta$ is the slope of the vapor pressure curve $p_0(\vartheta)$. This slope can be calculated from the Clausius-Clapeyron equation

$$\frac{\mathrm{d}p_0}{\mathrm{d}\vartheta}=\frac{\Delta h_v \varrho'' \varrho'}{T_s(\varrho'-\varrho'')}. \tag{10.10}$$

Substitution of this equation into Eq. (10.9) and integration between the limits ϑ_s and $\vartheta_s+\Delta\vartheta$ yields, with $\Delta\varrho=\varrho'-\varrho''$,

$$p_L(\vartheta_s+\Delta\vartheta)-p_L(\vartheta_s)=\int_{\vartheta_s}^{\vartheta_s+\Delta\vartheta}\frac{\Delta h_v\varrho''\varrho'}{T_s\Delta\varrho}\,\mathrm{d}\vartheta-\left(\frac{\varrho' 2\sigma}{\Delta\varrho r}\right)_{\vartheta_s+\Delta\vartheta}+\left(\frac{\varrho' 2\sigma}{\Delta\varrho r}\right)_{\vartheta_s},$$

for which, because of Eq. (10.7), one can also write

$$p_L(\vartheta_s+\Delta\vartheta)=p_0(\vartheta_s)+\int_{\vartheta_s}^{\vartheta_s+\Delta\vartheta}\frac{\Delta h_v\varrho''\varrho'}{T_s\Delta\varrho}\,\mathrm{d}\vartheta-\left(\frac{\varrho' 2\sigma}{\Delta\varrho r}\right)_{\vartheta_s+\Delta\vartheta}.$$

Furthermore, as seen from Fig. 10.3,

$$p_L(\vartheta_s+\Delta\vartheta)=p_0(\vartheta_s),$$

and we obtain, therefore, for the relation between bubble radius r and liquid superheating $\Delta\vartheta$

$$\left(\frac{\varrho' 2\sigma}{\Delta\varrho r}\right)_{\vartheta_s+\Delta\vartheta}=\int_{\vartheta_s}^{\vartheta_s+\Delta\vartheta}\frac{\Delta h_v\varrho''\varrho'}{T_s\Delta\varrho}\,\mathrm{d}\vartheta. \tag{10.11}$$

As one recognizes, the bubble radius becomes infinitely large, $r\to\infty$, if the liquid superheat is reduced to zero, $\Delta\vartheta\to 0$, whereas, conversely, a small vapor bubble can exist in equilibrium only if the liquid is sufficiently superheated.

Because the liquid superheating is generally very small, the equation can be further simplified by developing the expressions on both sides into a Taylor's series and truncating the series after the terms that are linear in $\Delta\vartheta$. If one then solves the equation for the bubble radius r, one finds

$$r=\frac{2\sigma T_s}{\varrho''\Delta h_v\Delta\vartheta}\left(1+\frac{\Delta\vartheta}{T_s}\omega\right) \tag{10.12}$$

with

$$\omega=\frac{\Delta\varrho T_s}{\varrho'\sigma}\frac{\mathrm{d}}{\mathrm{d}\vartheta}\left(\frac{\varrho'\sigma}{\Delta\varrho}\right).$$

According to an investigation of Mitrović and Stephan [10.1], numerical values of the factor ω for water and also for low-boiling liquids lie between 0 and 1. Therefore, for small temperature differences $\Delta\vartheta$ and not extremely low boiling temperatures T_s, the supplementary term $(\Delta\vartheta/T_s)|\omega|\ll 1$ so that it can be disregarded in Eq. (10.12).

According to the above considerations, the liquid in equilibrium is superheated by $\Delta\vartheta$, and associated with that superheat of the liquid is a quite definite bubble radius r, at which the bubble is in equilibrium with the liquid. Such bubbles, whose radii are $r^* < r$, are in equilibrium with the liquid only if the superheating is $\Delta\vartheta^* > \Delta\vartheta$, as shown in Fig. 10.3. A liquid superheated only by $\Delta\vartheta$ is too cold. Therefore, bubbles that are too small condense again. Bubbles with the radius $r^* > r$, on the other hand, are in a liquid that is superheated, and they can continue to grow. Actually, the lifetimes of the bubbles, especially in the proximity of the wall are so short that no equilibrium can occur, and the actual liquid superheat is many times larger than $\Delta\vartheta$. A definite critical bubble radius also belongs to this actual superheat. With boiling water at 1 bar, according to Eq. (10.12) the bubble diameter is $2r \approx 0.155\,\text{mm}$, based upon a superheat in the core of the liquid of 0.4 K. Only a bubble of this size is "viable" and can continue to grow.

Such a bubble contains approximately $3 \cdot 10^{20}$ water molecules. So many molecules with the well above average energy of vapor molecules can scarcely collect at a definite point in the interior of the liquid by chance, form a vapor bubble there, and then continue to grow. Thus, there arises the question how bubbles form at all.

As observation teaches, no bubbles can form in the interior of completely pure, carefully degassed liquids, unless one were to superheat the liquid extremely or to send ionizing rays through it, for example. Observation shows further that, over a rather long period of time, the bubbles form again and again at the same points on the surface of the heating element, with a frequency variable with time and distributed approximately according to an error function. Obviously, it is a question of highly active centers that catalyze the transition of the unstable superheated liquid into the stable vapor. Such centers are gas or vapor remnants in the depressions in the surface, which are not completely displaced by the liquid, because the liquid cannot completely fill up the fine depressions in the surface, even if the liquid is highly wetting. During heat input, the gas or vapor remnants expand until a definite critical size is reached, which corresponds to the size of a viable bubble. Then, based upon the prevailing liquid superheat, the bubble can grow until finally the forces of adhesion have become smaller than the buoyant and dynamic forces, so that the bubble frees itself from the heating surface. After the bubble breaks off, there remains locked in the depression a gas or vapor remnant, which is now cooled by the cold liquid flowing from the liquid interior to the wall and which is then heated again by heat input from the wall and thus grows into a new "nucleus" for a vapor bubble. From these observations it becomes understandable why the surface structure is an important limiting quantity for the heat transfer.

Vapor bubbles are formed almost always at especially favorable points of solid surfaces or on suspended particles. One is dealing almost always with *heterogeneous nucleus formation*. The *homogeneous nucleus formation*, by which bubbles arise "by themselves" on the basis of natural fluctuating motions of the molecules, plays a subordinate role. After vapor bubbles break off, new ones can form by homogeneous nucleus formation only in special cases, as will be shown in Sect. 10.2.

10.2 Bubble Contour and Limiting Volume

The vapor bubble formed at the wall grows in size by means of vaporization until it reaches a limiting volume, breaks off, and rises. In order to calculate the limiting volume, one must know the contour of a vapor bubble. This is determined during slow growth by buoyant and surface forces and by the adhesion conditions at the wall.

The differential equation for the bubble contour is obtained from a balance of forces. Assuming that the forces of buoyancy and gravity are equal, there holds for an element of volume in the vapor bubble

$$\frac{\mathrm{d}p_G}{\mathrm{d}z} = \varrho_G g,$$

where p_G is the pressure in the vapor bubble and z is the coordinate in the direction of the gravitational force, Fig. 10.4. The corresponding balance of forces for the surrounding liquid is

$$\frac{\mathrm{d}p_L}{\mathrm{d}z} = \varrho_L g.$$

Subtracting these two equations, there follows

$$\mathrm{d}(p_G - p_L) = (\varrho_G - \varrho_L) g \mathrm{d}z.$$

This equation is valid for the infinitesimally narrow area, which in Fig. 10.4 is bounded by the dashed lines on both sides of the bubble surface and which coincides with the solid wall at the base of the bubble. Integration of this equation from the top of the bubble ($z = 0$) to any random point z yields

$$(p_G - p_L) - (p_G - p_L)_{z=0} = (\varrho_G - \varrho_L) g z,$$

or allowing for Eq. (10.2a)

$$\sigma\left(\frac{1}{r_1} + \frac{1}{r_2}\right) - \sigma\left(\frac{1}{r_1} + \frac{1}{r_2}\right)_{z=0} = (\varrho_G - \varrho_L) g z.$$

We now assume that the bubble is axially symmetrical and designate the radius of curvature at the tip as $d, d = r_1 = r_2$ for $z = 0$. Thus, one obtains

$$\sigma\left(\frac{1}{r_1} + \frac{1}{r_2}\right) - \frac{2\sigma}{d} = (\varrho_G - \varrho_L) g z. \qquad (10.13)$$

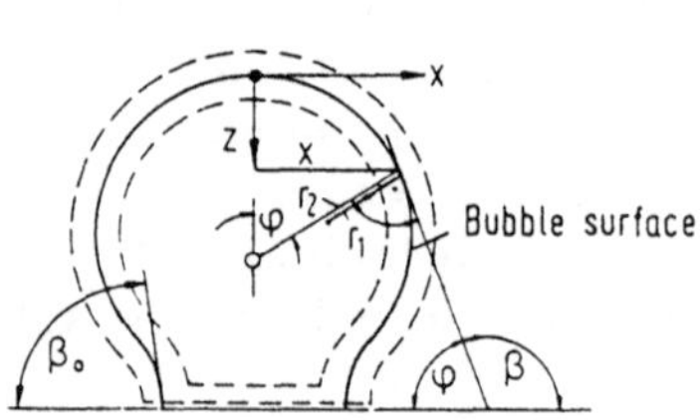

Fig. 10.4. Definition of parameters for the differential equation for the bubble contour. r_1 is the radius of principal curvature in the plane of the stretch and r_2 is the radius during rotation of the axis z

According to the rules of differential geometry, the two main radii of curvature r_1 and r_2, Fig. 10.4, are connected with the contour $z(x)$ in accordance with

$$\frac{1}{r_1} = \frac{z''}{(1+z'^2)^{3/2}} \quad \text{and} \quad \frac{1}{r_2} = \frac{\sin\varphi}{x} = \frac{1}{x}\frac{z'}{(1+z'^2)^{1/2}},$$

where $z' = \mathrm{d}z/\mathrm{d}x$ and $z'' = \mathrm{d}^2z/\mathrm{d}x^2$. Thus, the *differential equation of the bubble contour* is

$$\sigma\left(\frac{z''}{(1+z'^2)^{3/2}} + \frac{1}{x}\frac{z'}{(1+z'^2)^{1/2}}\right) - \frac{2\sigma}{d} = (\varrho_G - \varrho_L)gz \tag{10.14}$$

with the initial conditions $x = 0$: $z = 0$ and $z' = x/d$.

This equation was first solved numerically by Bashfort and Adams [10.2]. Their charts contain the volume associated with every angle $\beta = \pi - \varphi$ of the bubble contour. Starting from this solution, Fritz [10.3] later showed that there is a maximum volume V_A of a vapor bubble, which can be represented in the form

$$\left(\frac{V_A}{b^3}\right)^{1/3} = f(\beta_0) \tag{10.15}$$

where β_0 is the contact angle between bubble and heating wall, Fig. 10.4. The parameter b is the so-called Laplace constant

$$b = \sqrt{\frac{2\sigma}{g(\varrho_L - \varrho_G)}}. \tag{10.16}$$

If this is calculated with the saturation values ϱ' and ϱ'' of the densities, one obtains, for example, for water at 100 °C $b = 3.538$ mm and for the refrigerant R-11 at 25 °C $b = 162$ mm. Table 10.1 contains several other values for air bubbles in a liquid.

Equation (10.15) was confirmed by Fritz and Ende [10.4] and by Kabanow and Frumkin [10.15] by measurements from photos, Fig. 10.5. The data in this diagram can be well approximated [10.6] by

$$\left(\frac{V_A}{b^3}\right)^{1/3} = 0.686\beta_0. \tag{10.17}$$

Table 10.1. Laplace constant for air bubbles in liquids at 20 °C

	Liquid		
	Water	Ethanol	Mercury
b in mm	3.82	2.26	2.69

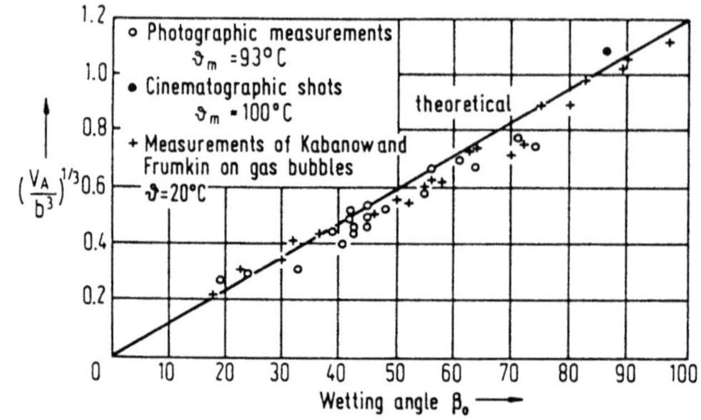

Fig. 10.5. Departure volume V_A of vapor and gas bubbles above a horizontal heating surface, according to Fritz and Ende [10.4]. b is the Laplace constant

From this one obtains the break-off diameter of the bubble, regarded as spherical $d_A = (6V_A/\pi)^{1/3}$, as

$$d_A = 0.851\beta_0 \sqrt{\frac{2\sigma}{g(\varrho_L - \varrho_G)}}, \qquad (10.17a)^1$$

with the contact angle β_0 in radians. Actual break-off diameters deviate somewhat from this and are also dependent upon the shape of the cavity [10.7] and upon the wall superheating [10.8]. For this, see Sect. 10.3.

The bubble shape before the break off is influenced additionally by intensive vaporization directly at the heating surface. Thus, there results a bubble form as shown in Fig. 10.6a. A "bubble neck" is formed between the heating surface and the part of the bubble away from the surface. As Mitrović [10.9] has shown, the bubble breaks off when the bubble neck is completely constricted, Fig. 10.6b. The entire vapor mass then separates into two parts of different size. A small vapor remnant remains stuck to the heating surface. The greater part of the vapor mass is in the broken off bubble. Because of the small radii of curvature at its lower peak, there is a high capillary pressure, which seeks to equalize itself and which causes the bubble to oscillate, if it is not extremely small. A capillary pressure acts in a similar manner on the vapor residue on the heating surface. It can be of the magnitude of 4 bar, as Mitrović [10.9] has proven with an example. Because of this excess pressure, the vapor residue is partially or even completely liquefied. Especially during slight superheating of the heating wall, the vapor residue will condense to a great extent. It takes therefore a long

[1] The constants of Eqs. (10.17) and (10.17a) resulted from a numerical solution of the differential Eq. (10.14) calculated by the author. By graphic evaluation of the previous solution of Bashfort and Adams [10.2], Fritz [10.3] had calculated the only slightly less precise constants 0.647 instead of 0.686, and 0.837 instead of 0.851.

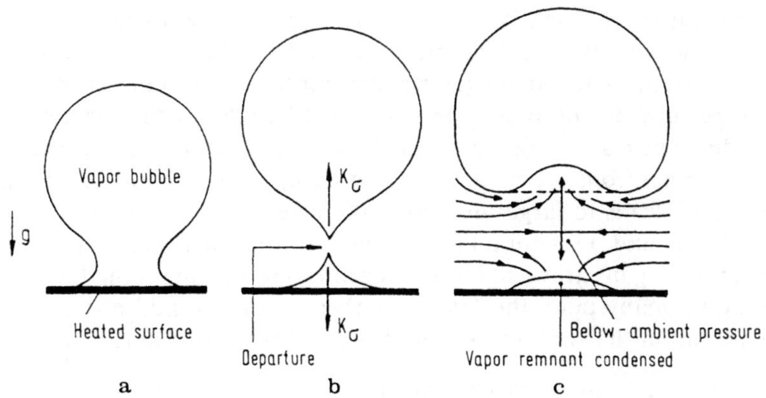

Fig. 10.6 a–c. Departure process, according to Mitrović [10.9]. **a** form of the adhering bubble; **b** constriction and break off of the bubble; **c** condensation of the vapor residue and liquid flow

time before the next bubble forms, whereas, with sufficiently large superheating, the next bubble forms from the vapor residue without delay. By removal of the capillary forces, the peaks in Fig. 10.6b disappear, and there results the bubble form shown in Fig. 10.6c with the diverging flow lines below the bubble. In the liquid beneath the bubble, a below-ambient pressure develops which keeps the departing bubble close to the wall for a short time. Because of the low pressure, the liquid is in a metastable condition with high superheating. The superheated liquid itself can serve as the nucleus for new bubbles, so that, in addition to the heterogeneous bubble formation in the depressions of the wall, a homogeneous bubble formation close to the wall is possible, as the observations of Mitrović [10.9] have shown. This, as well as the flow of the liquid, quickly re-establishes ambient pressure in the liquid.

As these observations show, the convection is an additional influence on the growth and departure of the bubbles. Previous findings on the influence of convection are still inconsistent, as emerges from summarizing presentations [among others 10.10]. In a theoretical work, Tokuda [10.11] has shown that heat conduction determines further growth only immediately after bubble formation, and that, subsequently, with increasing bubble size, the radial convection is decisive. After the departure of the bubble, heat conduction, radial convection, and axial convection are influential.

10.3 Application Equations for Departure Diameter and Frequency

During the slow period of bubble growth, the departure diameter is determined by the equilibrium of buoyancy and surface forces, and from the adhesive conditions on the wall. The departure diameter follows then from Eq. (10.17a).

If the wall temperature is much larger than the saturation temperature, then the bubble grows so rapidly that dynamic forces such as those of inertia, flow resistance, and pressure can no longer be disregarded. Substantial dynamic forces occur, especially during boiling in subcooled liquids, so that departure diameters are different from those predicted by Eq. (10.17a). Forces of inertia and flow resistance act from the surrounding liquid upon the bubble and keep it at the wall, which leads to larger departure diameters.

As Kutateladze and Gogonin [10.12] have established, the departure diameter during boiling in saturated liquids depends not only upon the Laplace constant, but additionally upon the Jakob number *Ja*, the Prandtl number *Pr* of the liquid and upon the Archimedes number *Ar*. These are defined by

$$Ja = \frac{c_{pL}\Delta\vartheta\varrho'}{\varrho''\Delta h_v}, \quad Pr = \frac{\nu_L}{a_L} = \frac{\nu_L c_{pL}\varrho'}{\lambda_L} = \frac{\eta_L c_{pL}}{\lambda_L} \quad \text{and} \quad Ar = \frac{g}{\nu_L^2}\left(\frac{\sigma}{\varrho' g}\right)^{3/2}.$$

Measurements of many authors for different substances can be described by the empirical equation

$$d_A = 0.25\sqrt{\frac{\sigma}{g(\varrho'-\varrho'')}}\left[1+\left(\frac{Ja}{Pr}\right)^2\frac{1}{Ar}\right]^{1/2} \tag{10.18}$$

which, according to Fig. 10.7, represents quite precisely data in the region

$$5\cdot 10^{-7} \leqq \left(\frac{Ja}{Pr}\right)^2\frac{1}{Ar} \leqq 10^{-1}.$$

The departure volume calculated from this equation agrees well with the values calculated according to Eq. (10.17) for vapor bubbles of water and of refrigerants at atmospheric pressure, if one uses the contact angle given by Fritz of 40° to 45° for water and approximately 35° for refrigerants.

In subcooled liquids, the formed vapor bubbles condense again. Vapor formation and condensation are present simultaneously. While the bubble is still growing at the base, condensation is already occurring at the top. As Fig. 10.8 shows, maximum bubble radius and condensation time decrease with increasing subcooling $\Delta\vartheta$. Also, an increasing flow velocity impedes bubble growth. The previous equations hold, therefore, only during vanishingly small subcooling and free convection. With large subcooling in forced flow, the condensation takes place much more rapidly than the growth of bubbles. The maximum volume and life span of the bubbles are, consequently, very small.

Equations for the calculation of the frequencies of vapor bubbles f started originally from the fact that one can set by approximation $f d_A = \text{const}$, where the constant for water and carbon tetrachloride was found to be 100 mm/s [10.23]. The constant was then later expressed in terms of the physical properties of the boiling liquid as

$$f d_A = 0.59\left(\frac{g\sigma(\varrho'-\varrho'')}{\varrho'^2}\right)^{1/4},$$

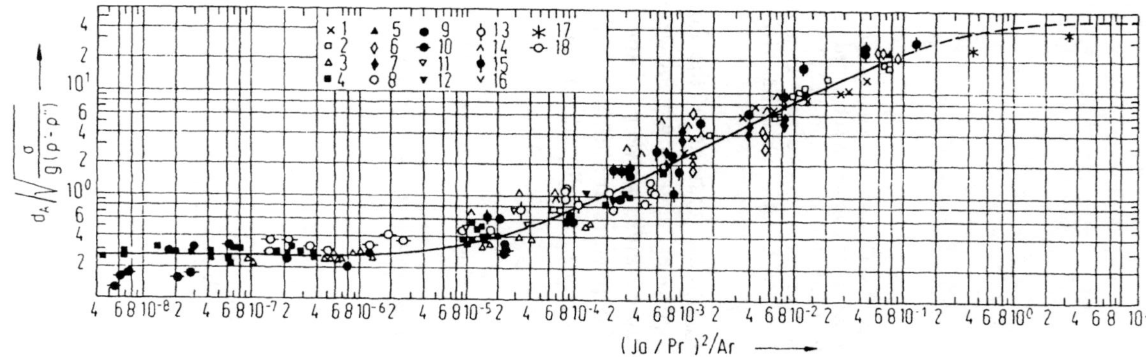

Fig. 10.7. Break-off diameter of vapor bubbles. *1* water [10.13]; *2* ethanol [10.13]; *3* benzene [10.14]; *4* ethanol [10.14]; *5* ethanol [10.15]; *6* water [10.16]; *7* methanol [10.16]; *8* pentane [10.16]; *9* water, freshly worked heating surface [10.17]; *10* water, aged heating surface [10.17]; *11* alcohol [10.18]; *12* water [10.18]; *13* nitrogen [10.18]; *14* ethanol [10.19]; *15* water [10.19]; *16* water [10.20]; *17* water [10.15]; *18* R-12 [10.21]

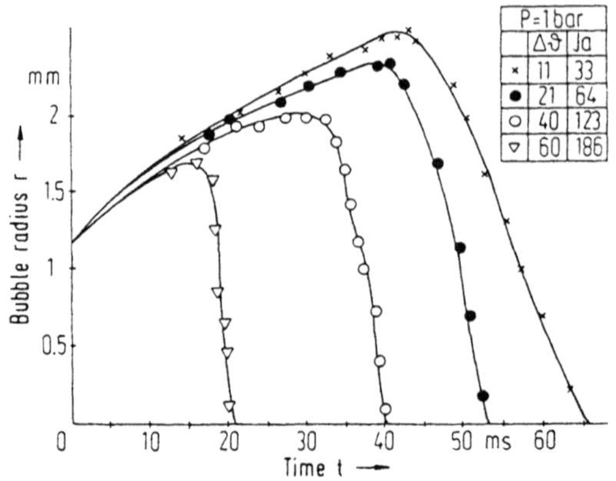

Fig. 10.8. Growth and condensation of water vapor bubbles in subcooled liquid at a pressure of 1 bar [10.22]. The bubbles were formed on a nozzle; therefore, the initial bubble radius is finite

whereas other authors [10.24, 10.25, 10.16] suggested equations of the form $fd_A^n = \text{const}$ with $n = 1/2$. More precise investigations have shown, however, that the exponent n is not constant, but rather assumes values between 0.5 and 2.

Assuming that the bubbles at the moment of departure are not accelerated and that their velocity is equal to the ascent velocity in the liquid, Malenkov [10.27] found for the frequency the expression

$$f = \frac{1}{d_A \pi} \left[\frac{d_A g(\varrho' - \varrho'')}{2(\varrho' + \varrho'')} + \frac{2\sigma}{d_A(\varrho' + \varrho'')} \right]^{1/2}, \tag{10.19}$$

which can be further simplified for the region of pressures well below the critical pressure, $\varrho' \gg \varrho''$, to

$$f = \frac{g^{1/2}}{d_A^{1/2} \pi \sqrt{2}} \left(1 + \frac{4\sigma}{d_A^2 \varrho' g} \right)^{1/2}. \tag{10.20}$$

These equations hold only if the heat flux is so low that the bubbles do not noticeably influence one another. Otherwise, the frequency is raised by a factor dependent upon the heat flux

$$1 + \frac{q}{\varrho'' \Delta h_v w}$$

with the velocity

$$w = \left[\frac{d_A g(\varrho' - \varrho'')}{2(\varrho' + \varrho'')} + \frac{2\sigma}{d_A(\varrho' + \varrho'')} \right]^{1/2}.$$

Then, one obtains the frequency f as

$$f = \frac{1}{d_A \pi} \left(1 + \frac{q}{\varrho'' \Delta h_v w} \right) w, \tag{10.21}$$

an equation, that can be simplified as before, if $\varrho' \gg \varrho''$. In the limiting case of sufficiently small heat fluxes, this equation converts into the above equation found for single bubbles. Malenkov tested these equations, using numerous data of different authors, and found the equations and the data to be in good agreement.

As one recognizes from Eq. (10.21), for large departure diameters, the second term in the brackets is small. Then $fd_A^{1/2} = \text{const}$. If, conversely, the departure diameter is sufficiently small, the second term in the brackets becomes large and $fd_A^{3/2} = \text{const}$. If both of the terms are of the magnitude of unity, then $fd_A = \text{const}$. The equations thus contain the relations between frequency and departure diameter established by various authors.

According to Eq. (10.21), the bubble frequency increases with the heat flux. To be sure, the size of the nucleation site is also important. Nucleation sites with a rather small aperture diameter emit bubbles with higher frequency than do those with a larger aperture diameter. The above equations do not allow for this effect and reproduce only the mean values for nucleation sites of varying size. Nor is the influence of subcooling considered. With increasing subcooling, the frequency decreases, as shown by Fig. 10.8, because the bubble growth is impeded by condensation.

11 Heat Transfer During Boiling of Pure Substances in Free Convection

11.1 Fundamentals

We shall next deal with boiling in free convection. The dimensions of the container in which the vapor is formed should be large in comparison to the vapor bubbles. A flow should result only by means of the buoyancy of the produced bubbles and by the differences in density. This is referred to simply as *pool boiling*. The two phase flow in a narrow evaporator tube, even if it is caused in a vertical tube only by the buoyancy of the vapor, will not be investigated. As already explained at the beginning of Chap. 9, there is evaporation in the region of small superheating of the heating wall. Heat is transferred only by free convection, curve a in Fig. 9.3. Nucleate boiling sets in only at a definite superheating of the wall. We will consider here the common case of a liquid at saturation temperature. The driving temperature difference for the heat transfer is then the difference $\vartheta_w - \vartheta_s = \Delta\vartheta$ between the wall and the saturation temperature. As Jakob and Linke [11.1] first discovered, the heat flux in the region of the completely developed nucleate boiling, curve b in Fig. 9.3, can be described by simple empirical equations of the form

$$q = c'\Delta\vartheta^{m}. \tag{11.1}$$

If one defines a heat transfer coefficient by

$$\alpha = q/\Delta\vartheta, \tag{11.2}$$

then one can also write Eq. (11.1) as

$$\alpha = cq^{n}, \tag{11.3}$$

with $c = (c')^{1/m}$ and $n = (m - 1)/m$. In this equation, as shown by much data, the values of n, and thus also of m, depend to some extent upon the type of boiling liquid, the material and shape of the heating surface, and the prevailing pressure. In general,

$$0.6 < n < 0.8.$$

Smaller values $n \approx 0.5$ have been found only for low-boiling substances such as helium. The value c depends strongly upon the properties of the boiling liquid and on the surface structure of the heating surface. If, for example, one knows c for one liquid, that boils at the prescribed pressure on smoothly polished steel tubes, one can no longer use heat transfer coefficients calculated with this value if the same liquid is boiling on rough steel tubes or on copper tubes.

Equation (11.1) or, as the case may be, Eq. (11.3) is valid only in the region of intensive nucleate boiling. As shown first by Nukijama [11.2], there are additional regions of vaporization. In Fig. 11.1, the individual regions are depicted that occur during boiling of water under atmospheric pressure and during free convection. The heat flux q is plotted as a function of the wall superheat $\Delta\vartheta = \vartheta_w - \vartheta_s$.

The left ascending branch of the curve designates the region of developed nucleate boiling. The heat flux increases with the wall temperature (or wall superheat). After reaching a maximum, the transferred heat flux decreases in spite of a rising wall temperature. We designate this falling region of the "boiling curve" as partial film boiling, because the heating surface is covered partially by vapor. After surpassing a minimum heat flux, the heat flux increases again with the wall temperature. We characterize the right ascending region of the boiling curve as the region of film boiling, because the heating surface there is completely covered by a vapor film.

This N-shaped curve, which at first seems rather strange and which can be observed in many other physical and chemical processes, is physically plausible. With increasing wall temperature, more and more vapor bubbles are formed, which set the liquid close to the wall into strong motion. As a result, the bubble formation promotes the heat transfer from the wall to the liquid.

Because the vapor isolates the heating surface more and more from the liquid, the heat transfer is impeded all the more strongly as more vapor is formed. One can easily imagine that, with increasing wall temperature, the isolating effect of the vapor finally predominates and the heat flux decreases again in spite of a rising wall temperature. The heat flow increases again only from sufficiently high wall temperatures on, because the thickness of the vapor film then increases only slightly with the wall temperature. The maximum of the nucleate boiling characteristic is frequently characterized as the *critical* or *peak heat flux*. Useful, but less accurately chosen designations, are burnout or DNB (= departure from nucleate boiling).

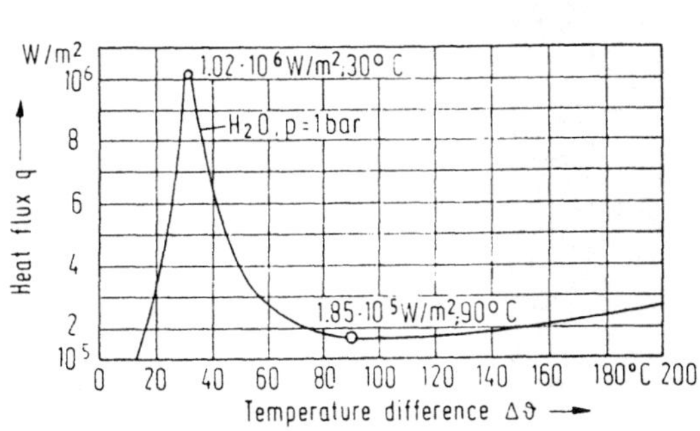

Fig. 11.1. Pool boiling curve for water

11.2 Stability During Boiling in Free Convection

With controlled heat input, if one raises the temperature of a heating surface beyond the value associated with the peak heat flux, one frequently notices that the surface is suddenly covered by a vapor film. The temperature of the wall increases sharply within fractions of a second and can even reach the melting temperature of the material (burnout). An example is depicted in Figs. 11.2 and 11.3. Figure 11.2 shows nucleate boiling of the refrigerant R-11 on a horizontal tube, which was electrically heated with a heat flux of approximately $2 \cdot 10^5$ W/m^2. This heat flux is only slightly smaller than the peak heat flux. After a very small increase in the heat flux, the temperature will increase so much that the tube melts through. In Fig. 11.3, one recognizes clearly that this has already happened in part of the tube.

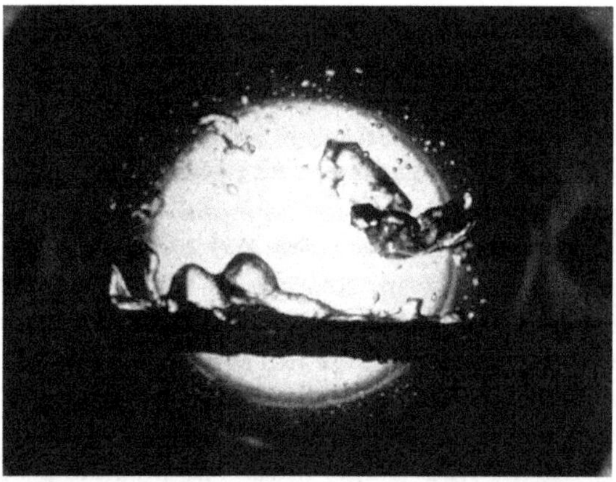

Fig. 11.2. Electrically heated tube immediately before reaching the peak heat flux. $q_{max} = 2.2 \cdot 10^5$ W/m^2, boiling R-11, $p = 1$ bar

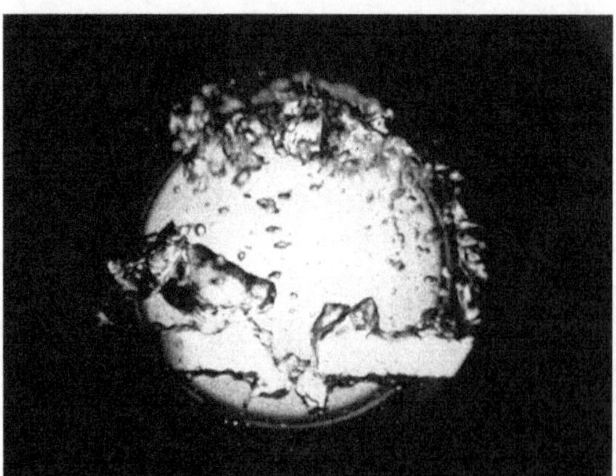

Fig. 11.3. Electrically heated tube of Fig. 11.2 at the moment of burnout

In order to avoid this phenomenon, which is especially to be feared in reactor technology, one must maintain a sufficient "safety margin" from the peak heat flux. Obviously, not all of the conditions depicted in Fig. 11.1 are stable. Under certain circumstances, instabilities can occur.

In the following, such instability phenomena are to be made understandable. All value pairs $(q, \Delta\vartheta)$, which are possible during the steady operation of the evaporator, lie on a boiling characteristic, as shown in Fig. 11.1. Of these value pairs, only quite specific ones come into question as working points on the characteristic curve of the evaporator. Just as the working point of a compressor is only determined by the compressor characteristic and the "piping characteristic", so also the working point of a boiler is determined by the boiling characteristic and the characteristic of the evaporator. The latter, assuming temperature-independent properties of the heating surface, is a straight line in the $(q, \Delta\vartheta)$-diagram, which can be clarified most simply with the example of a plane wall, Fig. 11.4.

With the notations of Fig. 11.4, the heat flux transferred by the wall is

$$q = \frac{1}{R_w}(\vartheta_L - \vartheta_w) \tag{11.4}$$

with the thermal resistance

$$R_w = \frac{1}{\alpha_i} + \frac{s}{\lambda}, \tag{11.5}$$

where α_i is the heat transfer coefficient between the heating agent and wall, s is the wall thickness, and λ is the thermal conductivity of the wall. For Eq. (11.4), one can also write

$$q = \frac{-1}{R_w}(\vartheta_w - \vartheta_s) + \frac{1}{R_w}(\vartheta_L - \vartheta_s). \tag{11.6}$$

If now additional energy is generated in the heating wall, as for example, by nuclear fission or by electrical dissipation, then the heat flux is raised by q_0, and

$$q = \frac{-1}{R_w}(\vartheta_w - \vartheta_s) + \frac{1}{R_w}(\vartheta_L - \vartheta_s) + q_0. \tag{11.7}$$

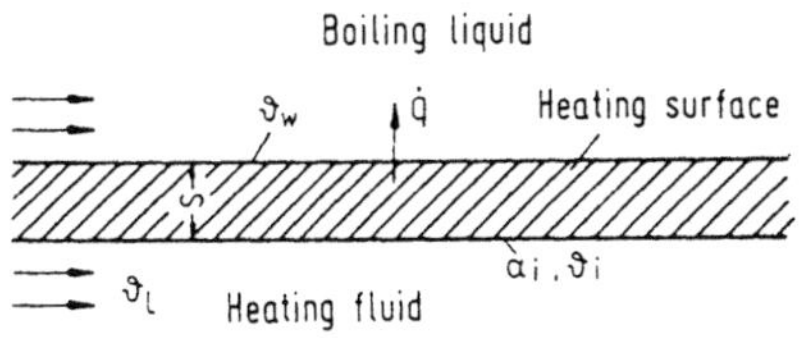

Fig. 11.4. Heat transfer to a boiling liquid

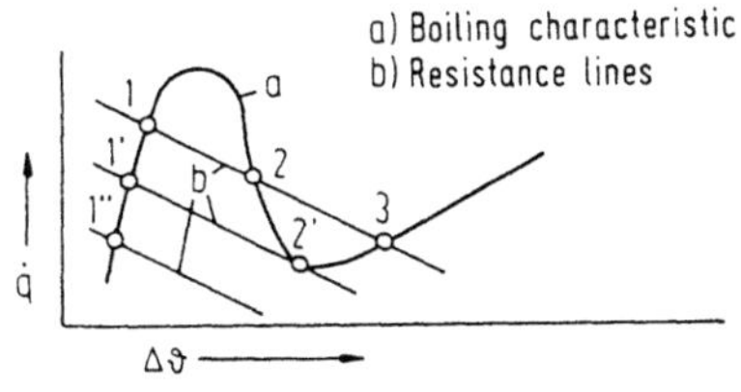

Fig. 11.5. Operating points on the characteristic curve of the evaporator

The required characteristic of the evaporator is thus fixed. It is a straight line in the $(q, \Delta\vartheta)$-diagram. All the possible operating points during the steady operation of the evaporator must lie on it. Because Eq. (11.7) corresponds in its structure to Ohms law, we designate the characteristic as a resistance line. Because of the negative slope, the heat flux from the heating surface decreases with increasing wall temperature ϑ_w. This initially surprising result becomes understandable, if one considers that with increasing wall temperature ϑ_w, the difference between wall temperature ϑ_w and fluid temperature ϑ_L diminishes, whereby the transferred heat flux decreases.

Working points of the evaporator are now the points of intersection of the boiling curve and of the resistance line. As Fig. 11.5 shows, both curves can have in common one point of intersection, one point of intersection and one point of contact, or three points of intersection. These are the fundamentally possible steady points of operation of the evaporator. The analysis of stability [11.3], which we will not go into here, shows that a working point is only stable if the slope of the boiling curve at the operating point is larger than the slope of the resistance line. For the usual N-shaped boiling curve, we can say the same thing in different words as follows:

Given three points of intersection, the two outer ones are stable, whereas the middle one is characterized by an unstable condition. If only one point of intersection is present, then it is always stable.

With these explanations, the understanding of the so-called burnout phenomenon becomes easy. If one plots into the $(q, \Delta\vartheta)$-diagram the resistance line for the case of nuclear energy production or for electrical resistance heating, then in the vicinity of the peak heat flux the boiling characteristic and the resistance line have three points of intersection, Fig. 11.6, of which points *1* and *2* are in close proximity to one another. A slight disturbance is sufficient to convert the system from the stable condition *1* to the unstable condition *2*. Because the heat input is fixed, the wall superheat increases until at point 3 a new stable condition is achieved. In most systems, there is associated with this new steady condition a wall temperature that is larger than the melting temperature of the heating surface, so that it melts before the steady condition is reached. In order to avoid this undesirable event, one must choose, as curve a in Fig. 11.6 shows, the resistance line so steep that it does not intersect the boiling characteristic a second time in the vicinity of the maximum. A steep resistance line results

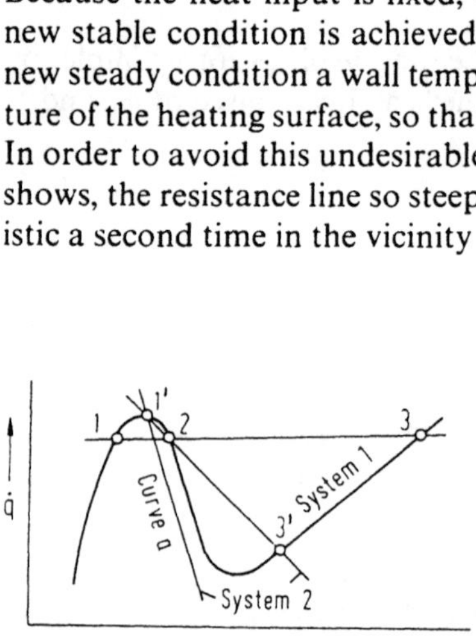

Fig. 11.6. Stability at peak heat flux

with Eq. (11.6), if one chooses in Eq. (11.5) α_i large and the thermal resistance of the wall small. Then, one can also achieve a stable operation at the peak heat flux.

11.3 Models of Heat Transfer

Almost all models for the calculation of heat transfer start from the fact that vapor bubbles are formed from nuclei, that are gas or vapor residues contained in the depressions caused by the roughness of the heating surface. There are differences of opinion concerning how the energy is primarily transmitted from the heating surface.

One of the first models, derived by Jakob and Linke [11.4], assumes that by the formation, growth, and departure of vapor bubbles, strong local flows close to the wall are produced, whose effect on the heat transfer can be described by means of dimensionless numbers with the equation

$$\frac{\alpha b}{\lambda_L} = 42.4\left(\frac{q}{\varrho''\Delta h_v w}\right)^{0.8}. \tag{11.8}$$

In it b is the Laplace constant, Eq. (10.16), a characteristic length, which is related to the departure volume, and thereby with the departure diameter d_A of the vapor bubble. The value ϱ'' is the density of the saturated vapor, Δh_v is the vaporization enthalpy, and λ_L is the thermal conductivity of the liquid; w is an empirical constant having the dimension of a velocity. According to Jakob and Linke, w is approximately equal to the product fd_A of frequency and departure diameter; they used the value 0.0778 m/s determined by measurements on water and carbon tetrachloride. The significance of this equation lies above all in the fact that it was the first equation that had a crude model as a basis. It satisfactorily reproduced the data for water, carbon tetrachloride, and several other liquids at atmospheric pressure, but failed, however, for other pressures.

One of the first equations to better determine the influence of pressure on heat transfer, was that of Rohsenow [11.5]. He agreed with Jakob and Linke in assuming that the good heat transfer during nucleate boiling is traceable principally to the convective heat exchange between the heating surface and the adjacent liquid. The turbulent flow in the superheated liquid layer close to the wall necessary for this is produced, according to this conception, by the vapor bubbles being formed at the heating surface. Because the convective heat exchange during turbulent flow can be represented by $Nu = F(Re, Pr)$, Rohsenow also assumed such an equation for nucleate boiling and used the power law equation

$$Nu = C Re^m Pr^n.$$

The characteristic length in the Nusselt and Reynolds numbers is the bubble departure diameter according to Eq. (10.17a). The characteristic velocity is found from the heat flux according to $w = q/\varrho''\Delta h_v$.

Forster and Zuber [11.6] also assume that heat is transferred principally by convection during nucleate boiling. Like Rohsenow, they used the power law equation $Nu = CRe^m Pr^n$. As characteristic length and velocity, they used expressions for bubble radius and rate of bubble growth, which they derived from their theory on bubble growth [11.7].

Han and Griffith [11.8] subdivided the heating surface into two regions, as shown in Fig. 11.7. In the region *1*, called the region of "bulk-convection", there is formed, by unsteady heat conduction from the heating surface to the adjacent liquid, a superheated boundary layer, which, in turn, induces bubble formation at nucleation sites of the heating surface. In this region, which corresponds, according to Han and Griffith, to twice the bubble diameter, a bubble displaces the superheated boundary layer from the heating surface. After reaching the departure diameter, the bubble frees itself from the heating surface and carries with it the previously dislodged warmer boundary layer liquid into the colder liquid core and mixes itself with this. Simultaneously colder liquid flows to the heating surface and into the space behind the departed bubble. A thermal boundary layer is formed once again by intensive heat exchange as the result of unsteady heat conduction.

In region 2, outside the range of influence of the bubbles, heat is given off by free convection from the heated wall to the liquid. Based upon this model, Han and Griffith developed an equation for the calculation of the heat flux, into which the nucleation site density and the dynamic contact angle are to be inserted as experimentally determined parameters.

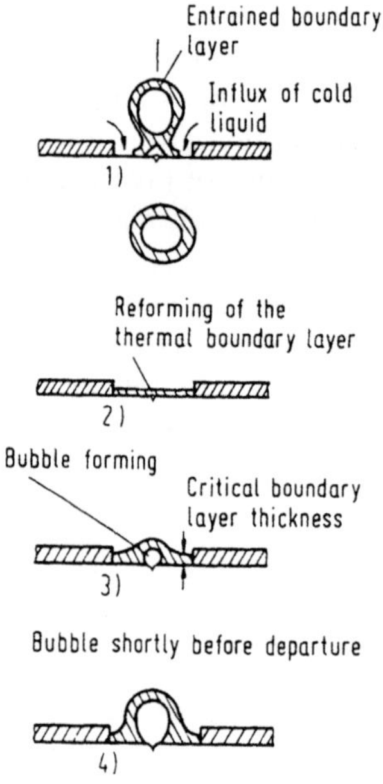

Fig. 11.7. Model for heat transfer during nucleate boiling, according to Han and Griffith [11.8]

Beer [11.9] extended the model of Han and Griffith and proved that behind the rising bubble there develops a drift current that exerts a suction effect on the boundary layer. Thus, a boundary layer volume of half the size of the bubble volume is removed and intermixed with colder liquid at some distance from the heating surface. Because the drift current deforms the temperature profile of the boundary layer, an additional heat transport takes place by means of convection and unsteady heat conduction, Figs. 11.8 and 11.9. According to Beer, the so-called Marangoni effect also has a marked influence on the heat transfer between the vapor bubble and the surrounding liquid. By this is meant the following. Because the vapor bubble growing in the superheated boundary layer is subjected to a temperature gradient, there occurs at its phase interface a gradient of the temperature-dependent surface tension. The imbalance in surface tension forces results in a flow along the phase interface, the velocity of which Beer estimated to be 9 m/s.

Moore and Mesler [11.10] could show with very sensitive methods of testing that the surface of a heated wall exhibits strong temporary fluctuations of temperature during nucleate boiling. For water at atmospheric pressure, a temperature drop of more than 16 K was determined within a time span of approximately 2 ms. Corresponding to this strong temperature drop in the heating wall during this short time, is a very large heat flux, which is larger by approximately a factor of six than the average heat flux and which cannot be explained alone by heat conduction and convective heat exchange to the boiling liquid. This experimental result is in harmony with the hypothesis that, during the bubble growth, a thin, liquid layer adjacent to the wall (microlayer) is evaporated into the bubble, Fig. 11.10. The microlayer evaporation results in very effective cooling of the heated surface.

Building upon this and upon his own data, van Stralen [11.11] developed the so-called "Microlayer-Theory". According to it, a part of the bubble is surrounded by superheated liquid, which supplies the vaporization enthalpy to the bubble and by so doing is cooled itself, Fig. 11.11. After the departure of the

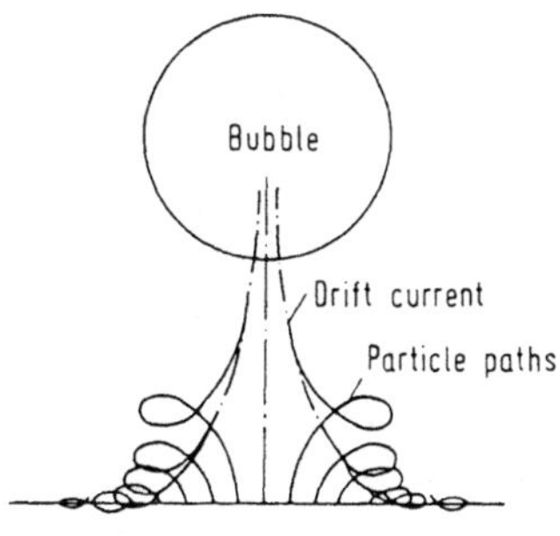

Fig. 11.8. Drift flow behind a rising bubble

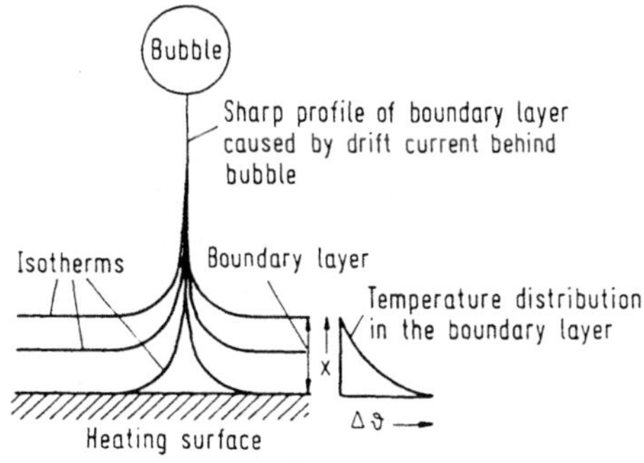

Fig. 11.9. Boundary layer and temperature profiles resulting from drift flow

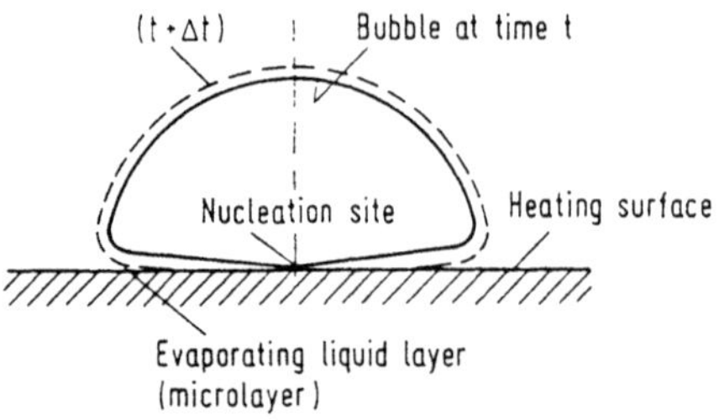

Fig. 11.10. Description of bubble formation within a thin wall adjacent liquid film (microlayer-theory)

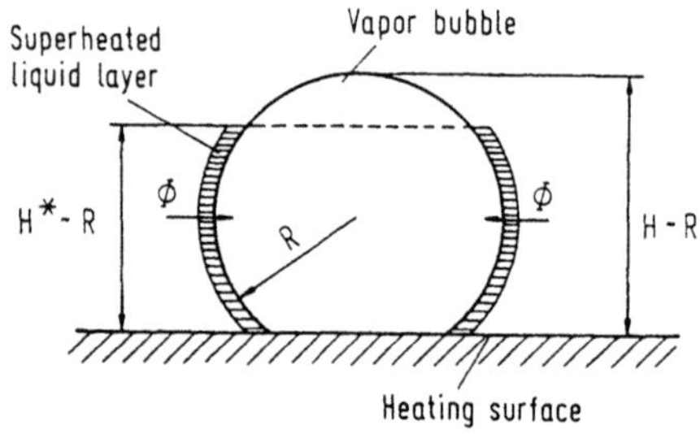

Fig. 11.11. Bubble growth and superheated liquid layer

bubble, the boundary layer at the wall is heated once again, until the next bubble forms at the nucleation site, and the process begins again. The attempt to quantitatively describe these processes by inclusion of the time dependency of the heat exchange, leads, after a series of simplifying assumptions, to expressions for the velocity of bubble growth, adhesion time, waiting time, bubble frequency, the departure radius, and the peak heat flux for both pure substances and binary mixtures. In these equations, the thickness of the superheated liquid layer appears as a parameter, which is usually unknown, so that the practical use of the equations for the calculation of heat transfer coefficients is difficult.

Frost and Kippenhan [11.12] suggested for heat transfer a model, in which, in addition to the convective heat transfer, the mass transfer is also of decisive influence. Accordingly, a bubble growing at the wall in a superheated boundary layer displaces the superheated liquid, which is then replaced on the upper side of the bubble by colder liquid from the surrounding area (bulk convection). A temperature gradient is thus formed, which makes possible the evaporation from superheated liquid at the wall into the bubble and also the condensation of vapor on the upper side of the bubble. This combined heat and mass transfer is suggested as being responsible for a large part of the entire heat transfer, particularly during forced flow. Moreover, the process should be augmented by the addition of substances, that lower the surface tension. Also, Styrikovich and colleagues [11.13] pointed to the fact that the mass transfer between the core flow and the boundary layer is responsible for the good heat transfer during nucleate boiling and defined as the parameter for the mass transfer the ratio of the quantity of liquid, that flows to the wall and the quantity of vapor produced at the heating surface.

11.3.1 Evaluation of the Models

Although each of the models to this point in time can describe individual parts of processes, there is as yet no complete theory. Obviously the heat transfer during nucleate boiling is so complicated that several mechanisms of exchange are active together. One or the other mechanism will predominate, depending upon the prevailing conditions, such as the magnitude of the heat flux, pressure, wettability, forced convection, subcooling, etc. Presumably, this also explains why the previous theories agree well in certain ranges of conditions and for certain substances, but fail in other areas and for other substances.

It can be stated fundamentally that previous models contain one or several of the following mechanisms for heat exchange:

1. Microconvection in the wall boundary layer as a result of rapid growth of vapor bubbles and also of their collapse in subcooled liquids.
2. Displacement of hot liquid from the wall by the growing and departing bubbles and return flow of colder liquid from the interior of the liquid to the wall.
3. Unsteady heat conduction to the liquid carried along in the wake of the bubble.
4. Unsteady heat conduction in the liquid near the wall during the growth of a bubble in its vicinity.
5. Vapor formation from the thin, superheated liquid layer under the growing bubble. If the core liquid is subcooled, vapor simultaneously condenses on the bubble tip.

It is not rare for heat transfer coefficients calculated according to different equations to deviate by a factor of as much as two compared with values obtained experimentally. Such deviations are traceable not only to the deficiencies in the theory, but also can be explained partially by errors in measurement and imprecise knowledge of properties. Moreover, the departure diameter of the bubbles, the wetting angle, and the bubble frequency during fixed external conditions are not constant, as often assumed, but rather are subjected to strong statistical variations, and it is not yet explained how one has to consider these fluctuations in the theory.

A disadvantage in all but a few of the given equations is the fact that the composition of the heating surface has not been taken into consideration, although it has been known for a long time that the heat transfer during nucleate boiling on rough surfaces is noticeably larger than on smooth surfaces. For the first time, the surface roughness was dealt with approximately by using the depth of surface smoothness R_p (according to DIN 4762, DIN stands for "Deutsche Industrie Norm"-German Industry Standard) [11.14]. The dependency found for close to atmospheric pressure $\alpha \sim R_p^{0.133}$ has been confirmed frequently in the meantime [11.15–11.19].

11.4 Empirical Correlations and Application Equations

11.4.1 Heat Transfer in the Vicinity of Ambient Pressure

In order to find an equation with the broadest range of application, it is expedient to state the properties decisive for heat transfer as nondimensional quantities. For the general law of heat transfer it will be advisable to choose an exponential equation in these quantities, because such equations have generally proven themselves for the representation of heat transfer.

Regression analysis represents an effective tool for establishing a relation between the Nusselt number and the other dimensionless variables. With its aid can be decided which characteristic quantities are important for heat transfer and which are of lesser importance.

Stephan and Abdelsalam [11.20] have examined critically approximately 5000 known data for heat transfer in nucleate boiling and have established empirical correlations. As the regression analysis showed, certain nondimensional quantities proved to be important for some substances but unimportant for others. The chosen characteristic data could, therefore, be best reproduced, if one divided the measured substances into four groups (hydrocarbons, cryogenic fluids, refrigerants, and water) and used a different equation of characteristic quantities for each of these groups. In addition, an equation valid for all substances could be stated, the accuracy of which is, to be sure, somewhat less than the equations for the mentioned groups of substances.

Because most measurements were conducted close to atmospheric pressure, which corresponds for organic liquids to approximately a normalized boiling pressure of $p/p_{cr} = 0.03$, one can increase the accuracy of such correlations, if one limits their validity to pressures in the vicinity of the ambient pressure. Such an equation has been established by Stephan and Preusser [11.21]. It reads:

$$Nu = 0.0871\left(\frac{qd_A}{\lambda' T_s}\right)^{0.674}\left(\frac{\varrho''}{\varrho'}\right)^{0.156}\left(\frac{\Delta h_v d_A^2}{a'^2}\right)^{0.371} \cdot \left(\frac{a'^2 \varrho'}{\sigma d_A}\right)^{0.350}(Pr')^{-0.162}. \tag{11.9}$$

Quantities designated by "relate to the boiling liquid and those designated by" relate to the saturated vapor. The Nusselt number is defined by $Nu = \alpha d_A/\lambda'$. The departure diameter d_A is given by Eq. (10.17a), in which $\varrho_L = \varrho'$ and $\varrho_G = \varrho''$ are substituted. For the wetting angle β_0, there is to be inserted for water $\pi/4\,\text{rad} = 45°$, for cryogenic liquids $0.01745\,\text{rad} = 1°$, and for other liquids $0.611\,\text{rad} = 35°$.

Measurements on *water* in the range of fully developed nucleate boiling for heat fluxes between

$$10^4\ \text{W/m}^2 < q < 10^6\ \text{W/m}^2$$

and saturation pressures between

$$0.5\,\mathrm{bar} < p < 20\,\mathrm{bar}$$

can be reproduced well, according to the investigations of Fritz [11.22], by the simple empirical equation

$$\alpha = 1.95 q^{0.72} p^{0.24} \tag{11.10}$$

where α is in $\mathrm{W/m^2K}$, q in $\mathrm{W/m^2}$, and p in bar.

11.4.2 Influence of Boiling Pressure and Heat Flux

In the empirical Eq. (11.3)

$$\alpha = c q^{n}$$

the value c for a given boiling liquid and heated surface is strongly dependent upon the properties of the boiling liquid and, thus, upon its saturation temperature or saturation pressure $c = c(p)$. We set

$$c = c_0 F(p),$$

where c_0 expresses the properties of liquid and heating wall at a definite reference pressure and $F(p)$ expresses the pressure influence itself. Let $F(p_0) = 1$. Then, the above equation reads

$$\alpha = c_0 F(p) q^{n}. \tag{11.11}$$

For a random set of values α_0 and q_0 at reference pressure p_0 there holds

$$\alpha_0 = c_0 q_0^{n}. \tag{11.12}$$

From both equations follows

$$\frac{\alpha}{\alpha_0} = F(p)\left(\frac{q}{q_0}\right)^{n}. \tag{11.13}$$

The pressure function occurring in this equation must be dimensionless, and thus independent of the system of units with which one measures pressures. In the theory of similitude, such functions are said to be homogeneous in their dimensions. In such functions, one can measure each value with the other values of the same dimension. One can thus measure the pressure via linkage with another pressure; it is expedient to choose the critical pressure for this purpose. We set, therefore, $p^* = p/p_{cr}$. The pressure function in Eq. (11.13) can thus be represented by $F(p^*)$, so that Eq. (11.13) converts into

$$\frac{\alpha}{\alpha_0} = F(p^*)\left(\frac{q}{q_0}\right)^{n}, \tag{11.14}$$

with $F(p_0^* = p_0/p_{cr}) = 1$. If one, therefore, knows α_0, and q_0 at reference pressure p_0, then one can calculate from this the complete development $\alpha(q)$, if only the exponent n of the heat flux is known. Heat transfer coefficients at pressures other than the reference pressure p_0 are obtained, if one additionally knows the function $F(p^*)$. The parameter α_0, q_0 at reference pressure p_0 can be determined by measurements or can be calculated from Eq. (11.9).

An empirical equation for the *pressure function* $F(p^*)$ was first established by Danilowa [11.24] for several refrigerants up to $p^* = 0.5$; it was then expanded by Haffner [11.25] and later by Bier and colleagues [11.26], so that measurements at even higher normalized pressures can be reproduced well. A modification of these equations, which allowed for numerous previously known data, was made by Gorenflo [11.23]. Accordingly, there holds for organic fluids, sulfurhexafluoride, and ammonia

$$F(p^*) = 2.1p^{*0.27} + \left(4.4 + \frac{1.8}{1 - p^*}\right)p^*, \tag{11.15}$$

and for water and cryogenic liquids

$$F(p^*) = 2.55p^{*0.27} + \left(9 + \frac{1}{1 - p^{*2}}\right)p^{*2}. \tag{11.16}$$

Both equations were established for a heat flux $q_0 = 20\,000\ \mathrm{W/m^2}$. The reference pressure p_0 is here so chosen that for the normalized pressure $p_0^* = 0.03$, the pressure function becomes $F(p_0^*) = 1$. Correspondingly, α_0 and q_0 in Eq. (11.14) are values of the heat transfer coefficient and heat flux at the normalized pressure $p_0^* = 0.03$, respectively.

Figures 11.12 and 11.13 show the course of the function $F(p^*)$ according to Gorenflo [11.23]. In this representation, α_0 is the arbitrary heat transfer coefficient at a heat flux of $q_0 = 20\,000\ \mathrm{W/m^2}$ and the normalized pressure $p_0^* = 0.03$. As is recognized, at close to the critical point, the heat transfer coefficient rises sharply.

Numerous data have shown that the exponent n of the heat flux in Eq. (11.14) is not constant, but rather decreases with increasing saturation pressure. According to Gorenflo [11.23], for organic liquids, sulfurhexafluoride, and ammonia

$$n = 0.9 - 0.3p^{*0.3}. \tag{11.17}$$

Thus $n = 0.8$ is for $p_0^* = 0.03$ and $n = 0.62$ is for $p_1^* = 0.8$. The pressure dependency for water and cryogenic liquids is somewhat weaker

$$n = 0.9 - 0.3p^{*0.15}. \tag{11.18}$$

For water at 1 bar, there follows from this $n = 0.77$, for $p_0^* = 0.03$ is $n = 0.72$ and for $p_1^* = 0.8$ is $n = 0.61$.

Thus, one can reproduce *measurements on water* by the following empirical equation obtained from Eq. (11.14) with $\alpha_0 = 3\,800\ \mathrm{W/m^2K}$, $q_0 = 20\,000\ \mathrm{W/m^2K}$ and $p_0^* = 0.03$, as

$$\frac{\alpha}{3800\,\dfrac{\mathrm{W}}{\mathrm{m^2K}}} = F(p^*)\left(\frac{q}{20000\,\dfrac{\mathrm{W}}{m^2}}\right)^{0.9-0.3p^{*0.15}} \tag{11.19}$$

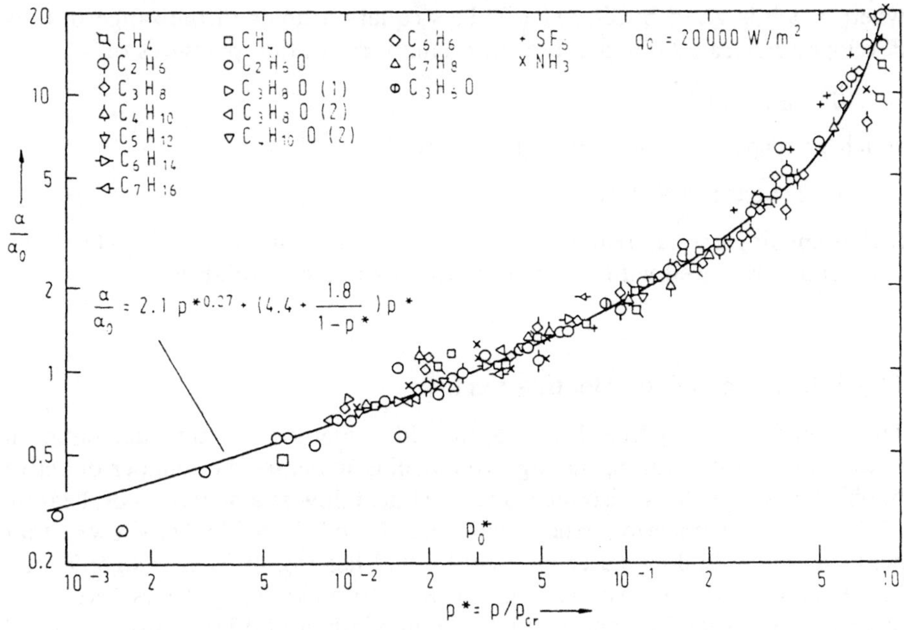

Fig. 11.12. Pressure dependence of the heat transfer coefficient during nucleate boiling for organic liquids, sulfur hexafluoride, and ammonia, according to [11.23]

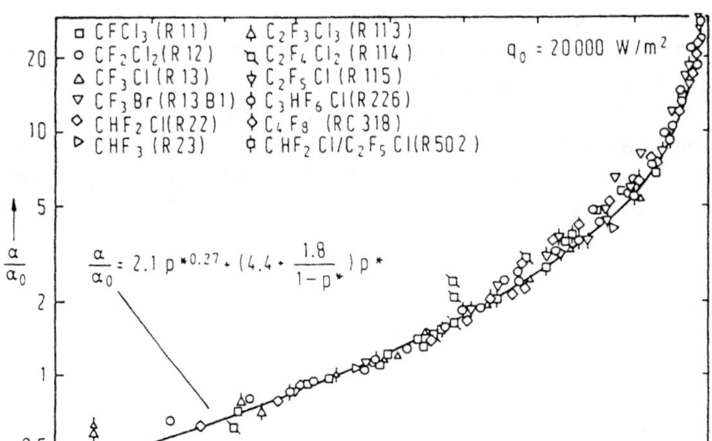

Fig. 11.13. Pressure dependence of the heat transfer coefficient during nucleate boiling with halogen hydrocarbons (refrigerants), according to [11.23]

where $F(p^*)$ is given by Eq. (11.16). This equation holds in the range of fully developed nucleate boiling and at normalized saturation pressures of

$$10^{-4} \leqq p^* \leqq 0.9,$$

which corresponds to saturation pressures of

$$0.0221\,\text{bar} \leqq p \leqq 199\,\text{bar}.$$

It thus encompasses a greater range of pressure than the simpler Eq. (11.10) by Fritz and agrees well with the latter within its range of validity.

11.4.3 Influence of the Heating Surface

The roughening of surfaces is one of the oldest methods for the augmentation of the heat transfer during boiling. Roughening increases the number of active bubble nuclei and, thus, also the transferred heat flow at a definite superheating of the wall. In pioneering experiments, Jakob and Fritz [11.27] showed that heat transfer of boiling water on sand-blasted heating surfaces is significantly better than on polished surfaces. If one were to make fine grooves vertical to one another and of 0.15 mm depth, 0.17 mm width, and 0.3 mm distance, there would again result an increase of the heat transfer by more than double that of the roughened surface, at least initially.

The influence of the construction material and of the structure of the heating surface upon the heat transfer during boiling is for the most part unclear. Especially over time, surface structures of heating walls can be changed by precipitates or chemical reactions with the fluid, which also has as a consequence a temporal change in heat transfer coefficients. Experimenting with heating surfaces that had been machine processed by turning or drawing, Stephan [11.28] found during measurements on refrigerants and water at close to atmospheric pressure that for such surfaces the heat transfer coefficient is dependent upon the depth of surface smoothness R_p (defined by DIN 4762) as follows:

$$\alpha \sim R_p^{0.133}. \tag{11.20}$$

The depth of surface smoothness is generally measured in μm. Allowing for the influence of roughness on the heat transfer, Eq. (11.14) then reads

$$\frac{\alpha}{\alpha_0} = F(p^*)\left(\frac{q}{q_0}\right)^{n}\left(\frac{R_p}{R_{p0}}\right)^{m} \tag{11.21}$$

with $m = 0.133$ for pressures close to atmospheric pressure, $R_{p0} = 1\,\mu\text{m}$ and $0.1\,\mu\text{m} \leqq R_p \leqq 10\,\mu\text{m}$. Consequently, one must multiply the heat transfer coefficients according to Eq. (11.14) or, for boiling water, according to Eq. (11.10) or Eq. (11.19), by the factor

$$\left(\frac{R_p}{R_{p0}}\right)^{0.133}.$$

Danilowa and Belskij [11.17] also arrived at a similar equation on the basis of measurements using the refrigerants R-12 and R-113. Instead of the depth of surface smoothness, they used the maximum peak-to-valley height R_z of the heating surface and recommended multiplying the heat transfer coefficient by the factor

$$\left(\frac{R_z}{R_{z0}}\right)^{0.2},$$

with $R_{z0} = 1\,\mu\text{m}$ in order to include the influence of roughness. According to Nishikawa and colleagues [11.29], the exponent of the depth of surface smoothness in Eq. (11.21) is variable with the pressure and for refrigerants at atmospheric pressure is somewhat larger than 0.133. By measurements with boiling R-11, R-21, R-113 and R-114, they found for the exponent m in Eq. (11.21)

$$m = 0.2(1 - p/p_{cr}) \tag{11.22}$$

valid for the pressure range

$$0.076 \leqq p/p_{cr} \leqq 0.9$$

and for depths of surface smoothness

$$0.022\,\mu\text{m} \leqq R_p \leqq 4.31\,\mu\text{m}.$$

According to this equation, using $m = 0.185$, one would calculate for a depth of surface smoothness of 4.31 μm and a normalized pressure of 0.076—which, for example, in the case of R-11 corresponds to a boiling pressure of 3.35 bar—the factor

$$\left(\frac{R_p}{R_{p0}}\right)^{0.185} = 1.31$$

whereas with $m = 0.133$, one would obtain the factor

$$\left(\frac{R_p}{R_{p0}}\right)^{0.133} = 1.21.$$

The difference is less than the error in the measurements used for the determination of the exponents m.

11.5 Heat Transfer on Finned Tubes

Data on the heat transfer to boiling refrigerants on finned tubes [11.30] showed that the heat transfer coefficients relative to the entire outer surface at equal pressure and equal heat flux are clearly larger than on smooth tubes. Especially in the range of low heat flux during beginning nucleate boiling, the heat transfer is substantially augmented when compared to a smooth tube. The heat transfer coefficient is formed, as is usual in the case of finned tubes, with the mean temperature at the base of the fin and not with the true, locally variable surface temperature of the finned tube.

The heat transfer coefficient increases more than in proportion to the increase of area. For example, for R-113 at a boiling pressure of 1 bar and a temperature difference of 10 K, the heat flux achievable with a finned tube with a surface ratio[1] $\varphi = 3.3$ is five times that of the value for a smooth tube. The heat transferred per tube length is then $5 \cdot 3.3 = 16.5$ times greater than with a smooth tube whose outer diameter is equal to the core diameter of the finned tube. This can be explained by an additional convective heat transfer, which is caused by the vapor bubbles rising on the fin sides.

As Gorenflo [11.30] found, the pressure dependence of the heat transfer coefficient can also be reproduced by Eq. (11.14), if one adds the area ratio φ as a further parameter.

The heat transfer from finned tubes increases less with pressure than in smooth tubes. One can allow approximately for the weaker pressure influence by introducing $p^*/\sqrt{\varphi}$ as the argument in the pressure function $F(p^*)$ in the place of the normalized pressure. One can then further calculate heat transfer coefficients with the aid of Eq. (11.14); one need only replace the normalized pressure p^* by $p^*/\sqrt{\varphi}$ in the pressure functions, Eqs. (11.15) and (11.16). According to Gorenflo [11.23, 11.30], these pressure functions should be used for finned tubes only within the range

$$0.03 \leqq p^* \leqq 0.3$$

and

$$1 \leqq \varphi \leqq 4.9.$$

Because, up to the present, measurements have been made only on copper tubes, experimental testing is advisable for finned tubes of other materials. For $p^*/\sqrt{\varphi} = 0.03$, the pressure function of finned tubes assumes the value $F(0.03) = 1$. The reference values α_0 and q_0 in Eq. (11.14) for finned tubes are obtained either through measurement or by calculation of the reference value α_0 from Eq. (11.9) for a large heat flux q_0 of 10^5 W/m^2. Because heat transfer coefficients for finned tubes increase more weakly with the heat flux than for smooth tubes, the two heat transfer coefficients for a heat flux above 10^5 W/m^2 can not be distinguished. Therefore, the exponent n of the heat flux in Eq. (11.14) is somewhat smaller than for a smooth tube:

$$n_R = n - 0.1 H/t_1, \tag{11.23}$$

where n_R is to be inserted into Eq. (11.14) as the exponent of the heat flux of the finned tube. Here, the value n is the exponent of the heat flux for a smooth

[1] The surface ratio is defined as the surface of the finned tube relative to the surface of a smooth tube with the base diameter of the finned tube.

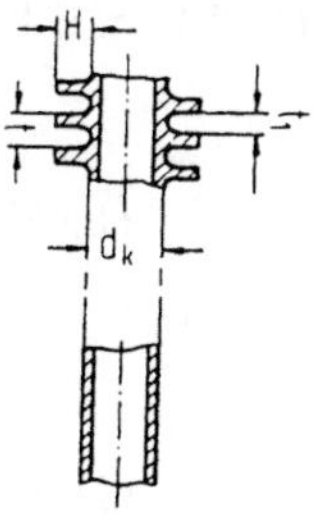

Fig. 11.14. Dimensions of a finned tube

tube, Eqs. (11.17) and (11.18), H is the fin height, and t_1 the interior fin interval, Fig. 11.14.

A simple approximation formula for the calculation of the heat transfer coefficient during boiling from finned tubes is obtained in the following manner. Let α_0 be the heat transfer coefficient in a smooth tube of equal core diameter as the finned tube, which is measured at a heat flux $q_0 = 20\,000\ \mathrm{W/m^2}$ and the normalized pressure $p_0^* = 0.03$. These values, α_0 and q_0, which belong to one another, can also be calculated from Eq. (11.9). Now the heat transfer coefficient α_1 for the smooth tube agrees with that for the finned tube α_{1R} for a heat flux $q_1 = 10^5\ \mathrm{W/m^2}$ and a definite pressure, for example, $p_0^* = 0.03$. Then, because $F(p_0^* = 0.03) = 1$, Eq. (11.14) can be written in the following form:

$$\frac{\alpha_1}{\alpha_0} = \left(\frac{q_1}{q_0}\right)^n \tag{11.24}$$

with $n = n(p^*)$ according to Eq. (11.17) or, as the case may be, Eq. (11.18). For the finned tube one, therefore, obtains

$$\frac{\alpha_{1R}}{\alpha_{0R}} = F(p_0^*/\sqrt{\varphi})\left(\frac{q_1}{q_0}\right)^{n_R}, \tag{11.25}$$

where α_{0R} is the heat transfer coefficient for the heat flux $q_0 (q_0 = 20\,000\ \mathrm{W/m^2})$ and for the pressure $p_0^*/\sqrt{\varphi} = 0.03$. In conformity with Eq. (11.23), the exponent n_R is to be formed with the pressure $p_0^* = 0.03$. From Eqs. (11.24) and (11.25), one obtains by division

$$\frac{\alpha_{0R}}{\alpha_0} = \frac{1}{F(p_0^*/\sqrt{\varphi})}\left(\frac{q_1}{q_0}\right)^{n-n_R}$$

or, noting that $q_1 = 10^5\ \mathrm{W/m^2}$ and $q_0 = 2 \cdot 10^4\ \mathrm{W/m^2}$ and because of equation (11.23) $n - n_R = 0.1H/t_1$, then

$$\frac{\alpha_{0R}}{\alpha_0} = \frac{1}{F(p_0^*/\sqrt{\varphi})}\,5^{0.1H/t_1}. \tag{11.26}$$

The heat transfer coefficient α_{0R} for the finned tube can be expressed, therefore, by that of the smooth tube of equal core diameter.

In general, the heat transfer coefficient for finned tubes, Eq. (11.14), can be written in the following form:

$$\frac{\alpha}{\alpha_{OR}} = F(p^*/\sqrt{\varphi})\left(\frac{q}{q_0}\right)^{n_R}, \tag{11.27}$$

from which together with Eq. (11.26) there follows:

$$\frac{\alpha}{\alpha_0} = \frac{F(p^*/\sqrt{\varphi})}{F(p_0^*/\sqrt{\varphi})} 5^{0.1H/t_1}\left(\frac{q}{q_0}\right)^{n_R}. \tag{11.28}$$

In order to calculate approximately from this the heat transfer coefficient during boiling for finned tubes, one must first determine by measurement the heat transfer coefficient α_0 for a smooth tube of equal core diameter or calculate from Eq. (11.14) the heat transfer coefficient for the smooth tube at a heat flux $q_0(q_0 = 20\,000\ \mathrm{W/m^2})$ and a normalized pressure $p_0^*(p_0^* = 0.03)$. The function $F(p^*/\sqrt{\varphi})$ is given by Eq. (11.15) or Eq. (11.16), in which the value $p^*/\sqrt{\varphi}$ is to be inserted in the place of the pressure p^*. The exponent n_R follows from Eq. (11.23).

11.6 Heat Transfer on Horizontal Smooth or Finned Tube Bundles

Heat transfer coefficients during boiling on the periphery of smooth or finned tubes in a bundle are greater than for the single tube, as shown in Fig. 11.15, which was taken from the work of Bier and colleagues [11.31]. For the lowest

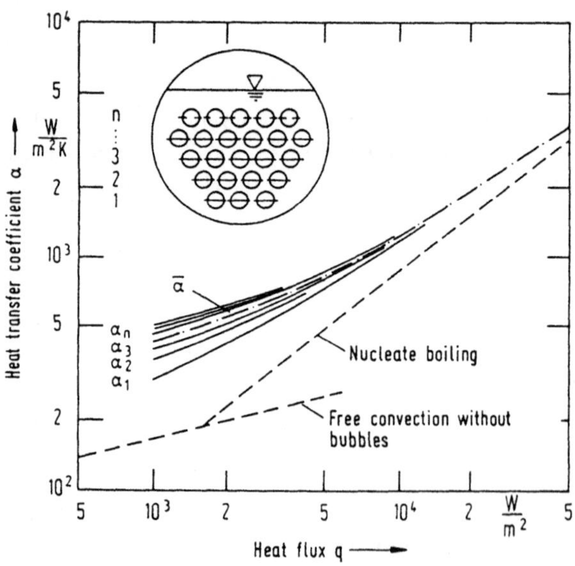

Fig. 11.15. Heat transfer coefficient α as a function of the heat flux q during boiling of R-113 on a single tube and on a smooth tube bundle, according to [11.31]. Saturation pressure $p = 1$ bar; mean depth of surface smoothness $R_p = 1\ \mu m$. Dashed curve: heat transfer coefficient for single tube (data); dash-dotted curve: mean heat transfer coefficient for the bundle; solid curves: heat transfer coefficient for the individual tube rows of the bundle

tube row, the heat transfer is augmented by the increased oncoming flow; in the bundle itself, the rising vapor bubbles provide for a better convection. The influence of the convection on the entire heat transfer is clearly noticeable, especially when the heat transfer rates are not all too high. Consequently, the mean heat transfer coefficient $\bar{\alpha}$ of the bundle is markedly greater than that for the single tube. With increasing heat flux, the influence of nucleate boiling becomes greater. The heat transfer coefficients of the individual tubes approach one another and become similar to that of the single isolated tube. For the calculation of heat transfer coefficients one must, therefore, consider both the oncoming flow of the lowest tube row and also the flow in the bundle. By evaluation of the data of various authors, Bier and colleagues [11.31] found that the improvement of the mean heat transfer coefficient in the bundle is to a great extent independent of the saturation pressure, that at $q = 2000\ \mathrm{W/m^2}$ it is larger approximately by a factor of 1.25 than that for the lowest tube, but that at heat fluxes above $10^4\ \mathrm{W/m^2}$ the improvement becomes insignificant. Approximately, there resulted [11.23]

$$\frac{\bar{\alpha}}{\alpha_1} = 1 + \frac{1}{2 + q\varphi/(1000\ \mathrm{W/m^2})}, \tag{11.29}$$

where α_1 is the heat transfer coefficient of the lowest tube. The equation holds for both smooth and finned tube bundles in the range

$$1000\ \mathrm{W/m^2} \leqq q\varphi \leqq 20\,000\ \mathrm{W/m^2}$$

and for pressures close to the ambient pressure. The heat flux q is related to the entire outer tube surface. The value φ is the surface ratio of the finned tube, see p. 156; for a smooth tube, $\varphi = 1$. The heat transfer coefficient α_1 at the lowest tube row is obtained following a suggestion of Slipčević [11.32] from the heat transfer coefficient α_B for nucleate boiling on a single tube and from the heat transfer coefficient α_K for free convection without nucleate boiling, also for a single tube, according to

$$\alpha_1 = \alpha_B + f\alpha_K, \tag{11.30}$$

where f is a factor which is dependent on the magnitude of the inlet flow velocity. For small bundles, f is around 0.5; for large ones, $f = 1$.

Unfortunately, the previous data do not permit as yet any more precise statements concerning the factor f. Equations (11.29) and (11.30) assume constant heat flux throughout the bundle. If the assumption is not valid, for example in the case of liquid-heated bundles, one must subdivide the bundle into segments of approximately equal heat transfer rate and calculate each of these segments according to the previous equations.

It should be pointed out that the equations hold only for fully developed nucleate boiling. In the range of very small heat fluxes or during an operation with increasing heat flux, it can very easily occur that there are practically no vapor bubbles present and heat is transferred only by free convection. One tests best whether free convection or fully developed nucleate boiling is present by plotting the heat transfer coefficient for both cases as a function of the heat flux. If the supplied heat flux lies in the vicinity of the intersection of both curves, then it is best to assume that the heat was transferred solely by free convection and, therefore, to be on the safe side, to design for the lower heat transfer coefficient.

11.7 Peak Nucleate Boiling Heat Flux

With increasing pressure and at constant wall superheating, more vapor bubbles are formed per surface unit. The heat transfer increases, therefore, with the pressure. By increasing the number of bubbles per surface unit, there is formed at sufficiently high pressure a more or less complete vapor film, which, in turn, leads to a reduction in the heat transfer. Accordingly, the peak heat flux q_{max} also increases with the pressure, reaches a relative maximum, and then decreases again. As an example, Fig. 11.16 shows the peak heat flux during nucleate boiling of water, which derives from Kutateladze, Eq. (11.34). One also finds a similar curve for other substances. The dependence of the peak heat flux upon pressure can be depicted for all investigated substances by a similar curve, if one plots $q_{max}/q_{max,0}$ with respect to the normalized pressure $p^* = p/p_{cr}$. Here, the heat flux $q_{max,0}$ is the peak heat flux at a reference pressure p_0^*, for which we again choose, as in the previous sections, the pressure $p_0^* = 0.03$, which for most organic substances lies close to the ambient pressure. Figure 11.17 shows such a plot, which was formed on the basis of an evaluation of data of many authors by Nikolaev and Skripov [11.33][1]. The data can be reproduced well by

$$\frac{q_{max}}{q_{max,0}} = 3.51 p^{*0.35}(1 - p^*)^{0.9}. \tag{11.31}$$

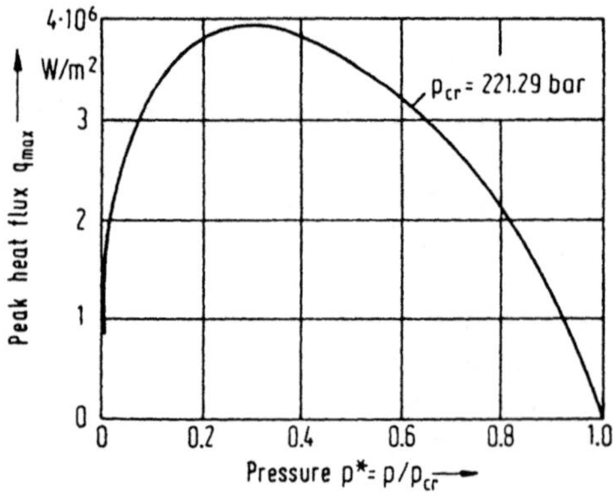

Fig. 11.16. Peak heat flux during nucleate boiling of water, calculated according to Eq. (11.34) with $K = 0.13$

[1] Nikolaev and Skripov chose the reference heat flux $q_{max,0}$ at the pressure $p_1^* = 0.31$. Their values were recalculated in Fig. 11.17 for the reference heat flux $q_{max,0}$ at the pressure $p_0^* = 0.03$.

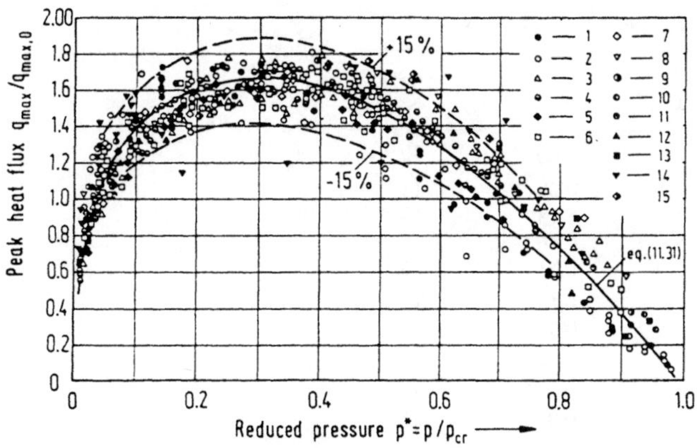

Fig. 11.17. Dependence of peak heat flux upon pressure, according to [11.33], $q_{max,0}$ at pressure $p_0^* = 0.03$. *1* and *2* n-pentane, n-hexane (horizontal tube); *3* and *4* ethyl and methyl alcohol (horizontal plate); *5* methyl alcohol (Ni–Cr–wire); *6–8* ethyl alcohol (flat plate, different positions, water); *9–13* ethyl alcohol, n-pentane, n-hexane (fouled heating surface), n-heptane benene (chromium-plated Cu-plate), *14* water (horizontal plate); *15* benene (horizontal tube)

Almost all the data lie within an error range of $\pm 15\%$. The maximum in the heat flux q_{max} lies at a pressure $p^* = 0.28$ and reaches there the value $q_{max}/q_{max,0} = 1.67$. By evaluating the data, empirical correlations for the difference $\Delta\vartheta_{max}$ between wall and saturation temperature at the peak heat flux could be established. They read

$$\frac{\Delta\vartheta_{max}}{\Delta\vartheta_{max,0}} = 0.264 p^{*-0.38} \quad \text{for} \quad 0.03 \leqq p^* \leqq 0.3 \tag{11.32}$$

and

$$\frac{\Delta\vartheta_{max}}{\Delta\vartheta_{max,0}} = 0.59(1 - p^*) \quad \text{for} \quad 0.3 \leqq p^* \leqq 1. \tag{11.33}$$

The temperature difference $\Delta\vartheta_{max,0}$ is to be taken at the reference pressure $p_0^* = 0.03$. In order to calculate from these equations the peak heat flux or the temperature difference during peak heat flux, which is generally of less interest, one needs a single value of these quantities.

For the calculation of the peak (or critical) heat flux one can find numerous empirical equations in the literature. To be sure, their accuracy is limited, because many limiting quantities have not yet been researched adequately. For example, it is known from experiments, that one can strongly alter interfacial tensions at the heating surface by annealing or corrosion and thereby raise the peak heat flux by as much as 30%. According to experiments by Hesse [11.35], the peak heat fluxes during heating by a hot fluid are approximately 20% greater than during electrical heating. Apparently this results from the fact that, during electrical heating, the wall superheating necessary for the formation of vapor bubbles is achieved at a lower heat flux and that the bubble frequency thus

Table 11.1. Wall superheat $\Delta\vartheta_{max}$ at peak heat flux

Substance	Saturation pressure, bar	p^*	$\Delta\vartheta_{max}$ K
Water	1	0.0045	30
R-114	3	0.09	15
	6	0.18	13
	15	0.46	8
	20	0.61	5.5
R-12	7	0.17	13
	14	0.34	11
	30	0.73	4
R-113	0.5	0.015	25
	1	0.03	21

becomes higher than during heating by a fluid. The peak heat flux is achieved, therefore, earlier with electrical heating.

Measurements of the peak heat flux are usually taken on electrically heated tubes. These data can be reproduced quite well by an equation, that was obtained by Kutateladze [11.36] and also by Zuber [11.37] in almost the same form,

$$q_{max} = K\Delta h_v \sqrt{\varrho''}[\sigma(\varrho' - \varrho'')g]^{1/4}, \tag{11.34}$$

where K is a constant, that is to be determined by experiments, and which lies between 0.13 and 0.16. A mean value is $K = 0.145$.

Data on the wall superheats $\Delta\vartheta_{max}$ belonging to the peak heat flux are very sparse, because usually only the peak heat fluxes are measured. Reference values are given in Table 11.1.

11.8 Transition Boiling and Film Boiling

If one increases the wall temperature of the heating surface, a substantial part of the heat flow is transferred from the wall to the liquid by convection. With progressive elevating of the wall temperature, more vapor is formed, so that the flow of liquid to the wall is hindered and a more and more complete film is formed. As observations show, the phase interface between vapor and liquid is initially irregularly and agitated intensively. The more the vapor film grows with the wall temperature, however, the smoother and calmer the phase interface becomes. The transferred heat flux decreases with increasing wall temperature, passes through a minimum value and then increases again during further increase of the wall temperature. Heat is transferred by conduction, convection, and radiation by the vapor film.

As experiments by Hesse [11.35] showed, in contrast to the assumption of several authors, there is formed, even before the attainment of the minimum heat flux a complete vapor film, whose phase interface is strongly agitated toward the liquid. The point $q, \Delta\vartheta$ of minimum heat flux, occasionally called the *Leidenfrost point*, is accordingly not that point at which film boiling begins, if the wall temperature is continuously raised from the range of nucleate boiling.

As Nikolaev and Skripov [11.33] found by the evaluation of data, the minimum heat flux is also dependent upon the pressure. Figure 11.18 shows the course of the minimum heat flux q_{min} relative to its value $q_{min,0}$ at the pressure $p_0^* = 0.03$ as a function of the normalized pressure p^*. The reference heat flux $q_{min,0}$ was chosen at the pressure $p_1^* = 0.31$ by Nikolaev and Skripov. The minimum heat fluxes published by them were converted into the reference heat flux $q_{min,0}$ at the pressure $p_0^* = 0.03$. The data can be reproduced by the compensating curve

$$\frac{q_{min}}{q_{min,0}} = 2.36 p^{*0.24} (1 - p^*)^{0.61}. \tag{11.35}$$

In order to derive from this an equation for the calculation of the minimum heat flux q_{min}, if the heat flux $q_{min,0}$ is unknown, we form with Eq. (11.31) the expression

$$\frac{q_{max}}{q_{min}} \frac{q_{min,0}}{q_{max,0}} = 1.49 p^{*0.11} (1 - p^*)^{0.29}. \tag{11.36}$$

If one assumes that the quotient q_{max}/q_{min} can be described as function of the normalized pressure p^*, then the ratio $q_{max,0}/q_{min,0}$, which is to be formed

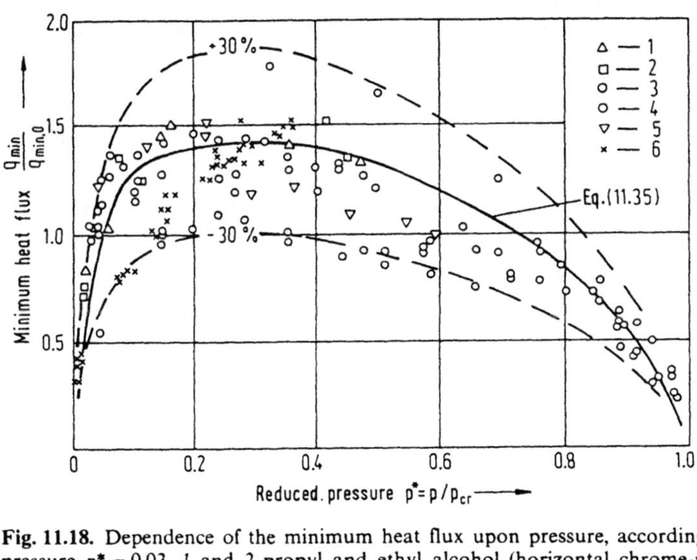

Fig. 11.18. Dependence of the minimum heat flux upon pressure, according to [11.33], $q_{min,0}$ at pressure $p_0^* = 0.03$. *1* and *2* propyl and ethyl alcohol (horizontal chrome-nickel plate); *3* and *4* n-pentane and n-hexane (horizontal tube); *5* iso-octane (horizontal cylindrical fin); *6* water (chrome-nickel wire)

at the pressure $p_0^* = 0.03$, is a fixed value independent of properties. This ratio, however, can be calculated, because for boiling water at $p_1 = 1$ bar or $p_1^* = 4.52 \cdot 10^{-3}$, there is $q_{max} = 1.02 \cdot 10^6$ W/m^2, see also Fig. 11.1. With this, $q_{max,0} = 1.92 \cdot 10^6$ W/m^2 is obtained from Eq. (11.31). The minimum heat flux of water at an equal pressure equals $q_{min} = 1.85 \cdot 10^5$ W/m^2. Then the reference value $q_{min,0} = 2.86 \cdot 10^5$ W/m^2 is obtained from Eq. (11.35). With these values for $q_{max,0}$ and $q_{min,0}$, Eq. (11.36) converts into

$$\frac{q_{max}}{q_{min}} = 10 p^{*0.11} (1 - p^*)^{0.29}. \tag{11.37}$$

From this, one can determine the minimum heat flux, if the peak heat flux is known. The latter is given by Eqs. (11.31) and (11.34), as was shown in the previous section.

The ratio q_{max}/q_{min} reaches, according to Eq. (11.37), a maximum value of 7.9 at the pressure $p^* = 0.275$. In the range of the normalized pressures $0.2 \leqq p^* \leqq 0.5$, the dependence upon pressure is weak, which is also confirmed by data of Hesse [11.35]. A mean value in this pressure range equals $q_{max}/q_{min} \approx 7.8$.

For the stable range of *film boiling*, Bromley [11.38] has developed a theory that is based upon assumptions similar to those of the Nusselt film condensation theory. It is assumed that the heating surface is covered by a continuous, laminar film of superheated vapor, that separates wall and liquid. The wall temperature ϑ_w is assumed to be constant, and the phase interface between vapor and liquid is assumed to be at saturation temperature. Without allowing for radiation, thus assuming pure heat conduction, one then finds by means of a mass and momentum balance, analogous to the Nusselt film condensation theory, the thickness of the vapor film on a vertical flat plate

$$\delta = a_0 \left[\frac{\eta_G \dot{M}}{\varrho_G (\varrho_L - \varrho_G) g b} \right]^{1/3}. \tag{11.38}$$

Here, b is the width of the flat plate and $\dot{M}$ is the vapor mass flow rate. If the liquid flows upward with the same velocity as the vapor at the phase interface, then the constant a_0 is identical with that of the Nusselt film condensation theory, Eq. (2.8), $a_0 = \sqrt[3]{3} = 1.44$. If it is assumed that the liquid is not agitated, then the constant a_0 is raised to $\sqrt[3]{12} = 2.29$. The properties of the vapor are to be inserted at the mean temperature $\bar{\vartheta} = (\vartheta_w + \vartheta_s)/2$. The density of the liquid in the state of boiling is $\varrho_L = \varrho'$.

The influence of the thermal radiation is allowed for by a heat transfer coefficient α_r, which is defined by

$$\alpha_r = \frac{\sigma_r}{\dfrac{1}{\varepsilon_w} + \dfrac{1}{\varepsilon_L} - 1} \, \frac{T_w^4 - T_S^4}{\vartheta_w - \vartheta_S}. \tag{11.39}$$

Here, σ_r is the radiation coefficient of a black body, ε_w the emissivity of the wall, and ε_L that of the liquid surface. The processes of heat conduction and heat radiation are not independent of one another. Vapor is also produced by radiation, the mass flow rate of the vapor in Eq. (11.38) is thereby increased, and the heat flow transferred by conduction is diminished. The local heat transfer coefficient α is composed of the local heat transfer coefficients for conduction and radiation together. If, analogous to the Nusselt film condensation theory, it is assumed that the main thermal resistance to conduction is determined by the thickness of the vapor film, then, assuming a linear temperature distribution, the local heat transfer coefficient for conduction is given by λ_G/δ, and is

$$\alpha = \frac{\lambda_G}{\delta} + \alpha_r . \tag{11.40}$$

In the equation for the film thickness δ, Eq. (11.38), the mass flow rate of the vapor is still unknown. In order to determine it, we use the energy balance for the vapor formed along the height dz

$$\Delta h \mathrm{d}\dot{M} = \alpha \Delta \vartheta b \mathrm{d}z . \tag{11.41}$$

The value Δh is the difference of the enthalpy of superheated vapor (at the temperature $\bar{\vartheta} = (\vartheta_w + \vartheta_s)/2$ and the enthalpy of saturated liquid. With Eq. (11.40) and the film thickness $\delta = a_1 \dot{M}^{1/3}$ according to Eq. (11.38), one obtains by integration between the plate leading edge and a point z the following equation:

$$\frac{\Delta \vartheta b z}{\Delta h} = \int_{\dot{M}=0}^{\dot{M}} \frac{\mathrm{d}\dot{M}}{\dfrac{\lambda_G}{a_1 \dot{M}^{1/3}} + \alpha_r} . \tag{11.42}$$

If the substitution $u = \dot{M}^{-1/3}$ is introduced, then the right side can be integrated according to a decomposition to partial fractions. With the abbreviation $b_1 = \alpha_r a_1/\lambda_G$, one obtains

$$\frac{\Delta \vartheta b z}{\Delta h} = \frac{3a_1}{\lambda_G b_1^4}\left[\tfrac{1}{3} b_1^3 \dot{M} - \tfrac{1}{2} b_1^2 \dot{M}^{2/3} + b_1 \dot{M}^{1/3} - \ln(1 + b_1 \dot{M}^{1/3})\right] . \tag{11.43}$$

In the limiting case of negligible thermal radiation $\alpha_S \to 0$, the integration of Eq. (11.42) or the substitution of $b_1 \to 0$ in Eq. (11.43) yields

$$\frac{\Delta \vartheta b z}{\Delta h} = \frac{3a_1}{4\lambda_G} \dot{M}^{4/3} , \tag{11.44}$$

where $\dot{M}$ is the mass flow rate of the vapor produced at a random location z, if heat is transferred only by conduction. Thereby, one obtains the local heat transfer coefficient α by forming the differential quotient d$\dot{M}$/dz and inserting it into Eq. (11.41). Allowing for the definition of a_1 for stagnant liquid

$$a_1 = \left[\frac{12\eta_G}{\varrho_G(\varrho_L - \varrho_G) g b}\right]^{1/3} , \tag{11.45}$$

one finds for the local heat transfer coefficient in the vapor film

$$\alpha_G = \left[\frac{\Delta h \varrho_G(\varrho_L - \varrho_G) g \lambda_G^3}{16\eta_G \Delta \vartheta} \frac{1}{z}\right]^{1/4}. \tag{11.46}$$

From this is obtained the mean heat transfer coefficient

$$\bar{\alpha}_G = \frac{1}{z_1} \int_0^{z_1} \alpha_G \mathrm{d}z$$

as

$$\bar{\alpha}_G = \tfrac{4}{3}\alpha_G. \tag{11.47}$$

With this and the heat transfer coefficient α_r for radiation, one can further simplify Eq. (11.43). As follows from the energy balance, Eq. (11.41), by integration

$$\Delta h \dot{M} = \bar{\alpha}\Delta\vartheta b z_1$$

and

$$\Delta h \dot{M}_G = \bar{\alpha}_G \Delta\vartheta b z_1,$$

if the mean heat transfer coefficient is designated by $\bar{\alpha}$. The resulting vapor mass flow rates $\dot{M}$ with allowance for radiation and $\dot{M}_G$ during pure heat conduction stand, therefore, in the same relation as the corresponding mean heat transfer coefficients

$$\frac{\dot{M}}{\dot{M}_G} = \frac{\bar{\alpha}}{\bar{\alpha}_G}. \tag{11.48}$$

Then, the expression $b_1 \dot{M}^{1/3} = \alpha_S a_1 \dot{M}^{1/3}/\lambda_G$ in Eq. (11.43) can be converted into

$$b_1 \dot{M}^{1/3} = \frac{\alpha_r a_1 \dot{M}_G^{1/3}}{\lambda_G}\left(\frac{\bar{\alpha}}{\bar{\alpha}_G}\right)^{1/3}.$$

Here $a_1 \dot{M}_G^{1/3} = \delta_G$ is the thickness of the vapor film, if heat is transferred only by conduction, and $\alpha_G = \lambda_G/\delta_G$ is the accompanying heat transfer coefficient. This yields

$$b_1 \dot{M}^{1/3} = \frac{\alpha_r}{\alpha_G}\left(\frac{\bar{\alpha}}{\bar{\alpha}_G}\right)^{1/3} = \frac{4}{3}\frac{\alpha_r}{\bar{\alpha}_G}\left(\frac{\bar{\alpha}}{\bar{\alpha}_G}\right)^{1/3}.$$

In order to abbreviate, we set

$$\frac{\alpha_r}{\bar{\alpha}_G} = x \quad \text{and} \quad \frac{\bar{\alpha}}{\bar{\alpha}_G} = y, \tag{11.49}$$

and can, therefore, write

$$b_1 \dot{M}^{1/3} = \tfrac{4}{3} x y^{1/3}, \tag{11.50}$$

where the quantities x and y are defined by Eq. (11.49). In Eq. (11.43), there is also the factor

$$\frac{\Delta\vartheta bz}{\Delta h}\frac{b_1^4\lambda_G}{3a_1}$$

for which one can write, when allowing for $a_1 = \delta_G/\dot{M}_G^{1/3}$,

$$\frac{\Delta\vartheta bz}{\Delta h}\frac{b_1^4\lambda_G}{3a_1} = \frac{\dot{M}_G}{\bar{\alpha}_G}\left(\frac{\alpha_r a_1}{\lambda_G}\right)^4 \frac{\lambda_G}{3a_1} = \frac{\alpha_r^4}{3\bar{\alpha}_G}\left(\frac{\delta_G}{\lambda_G}\right)^3 = \frac{\alpha_r^4}{3\bar{\alpha}_G\alpha_G^3}$$

or, after introduction of the mean heat transfer coefficients, Eq. (11.47), and, allowing for the definition of the quantity x, Eq. (11.49),

$$\frac{\Delta\vartheta bz}{\Delta h}\frac{b_1^4\lambda_G}{3a_1} = \frac{64}{81}\left(\frac{\alpha_r}{\bar{\alpha}_G}\right)^4 = \frac{64}{81}x^4.$$

Then, Eq. (11.43) can be written as

$$\tfrac{64}{81}x^4 = \tfrac{64}{81}x^3 y - \tfrac{8}{9}x^2 y^{2/3} + \tfrac{4}{3}xy^3 - \ln\left(1 + \tfrac{4}{3}xy^3\right). \tag{11.51}$$

This is a transcendental equation from which one can obtain the unknown $y = \bar{\alpha}/\bar{\alpha}_G$ for a prescribed value of $x = \alpha_r/\bar{\alpha}_G$, and from this can be obtained the mean heat transfer coefficient $\bar{\alpha}$ for film boiling. The equation was first published in this form by Roetzel [11.39]. As Roetzel showed, the solution of the equation can be well approximated by

$$y = 1 + \frac{1}{5}x\left(4 + \frac{1}{1 + 3/x}\right). \tag{11.52}$$

The greatest error in $y = \bar{\alpha}/\bar{\alpha}_G$ in the entire range $0 \leqq x = \alpha_r/\bar{\alpha}_G < \infty$ is smaller than 1% compared to the exact solution according to Eq. (11.51). Equation (11.52) is similar to an approximate equation stated earlier by Bromley [11.38], but it is more exact.

12 Heat Transfer in Falling Film Evaporators

Falling film evaporators consist of vertically arranged heating surfaces. Tubes are usually used. From these, a liquid flows off in the form of a thin film and is evaporated by the addition of heat. Frequently, in such apparatus, a solvent is to be extracted by evaporation from a solution consisting of a solvent (water or organic solvent) and the substances dissolved in it (salts or other substances). Examples are the concentrating of fruit juices, the recovering of solvents during the production of lacquer, or the extraction of fresh water from sea water.

This type of evaporator construction exhibits a series of advantages. The heat transfer coefficients are very large and, depending upon the properties of the solution to be evaporated, lie between 700 and 4000 W/m^2K; also, the residence time of the solution and the filling capacities are very small. In contrast to flooded evaporators, practically no pressure differences occur as a result of the static head of liquid, and the frictional pressure drop is usually small, so that the evaporation temperature over rather large ranges of an apparatus is changed very little. The vapor formed is usually drawn off downward. Vapor and liquid, therefore, flow parallel to one another, and the liquid cannot be blocked as in the case of countercurrent flow. As a result of parallel flow and the large volume of vapor, substantial vapor velocities can develop downstream, especially with evaporation inside tubes. These velocities exert a large shear stress on the liquid film and produce a thin liquid film with a correspondingly good heat transfer. In most cases, the falling film evaporator works in the range of surface evaporation. Vapor is formed only on the film surface, without any vapor bubbles being formed on the wall. Because of the slight superheating, there is less danger of thermal decomposition with this type of evaporation, and a fouling deposit on the heating surface does not occur as easily as during nucleate boiling.

In order to guarantee that vapor is formed only on the film surface, the superheating of the wall must not be too large in comparison with the saturation temperature and, depending upon the properties of the substances to be evaporated, should not exceed 3 to 10 K. In general, one must determine in prior experiments the admissible superheating at which the bubble formation sets in. A criterion for the determination of this admissible superheating is the fact that, in the range of surface evaporation, the heat transfer coefficient is independent of the superheating, but, in the range of nucleate boiling, is strongly dependent upon the superheating.

Whereas during film condensation on vertical tubes the film becomes thicker toward the bottom, as the initially laminar film converts into a turbulent film, during surface evaporation, the liquid film is usually so thick initially that the flow in the film is turbulent. With increasing concentration by evaporation, the film thickness decreases, so that the film flow can be laminar at the lower end of a falling film evaporator. When one concentrates solutions, the viscosity also increases at the lower end of the tube, and, therefore, the Reynolds number decreases.

12.1 Heat Transfer Coefficients for Vertical Tubes

During surface evaporation, the Nusselt number is dependent only on the Reynolds and Prandtl numbers. Equations and a method of computation were derived in Sect. 2.4.1 for the film condensation of stagnant vapors and in Sect. 4.2.3 for the film condensation of flowing vapors. The equations for surface evaporation on a falling film are of a similar nature, because, just as in the case of film condensation, the Nusselt number is dependent only upon the Reynolds and Prandtl numbers formed with the film thickness. In contrast to film condensation, the film thickness decreases with the length of flow.

12.1.1 Heat Transfer with Negligible Vapor Shear Stress

Because falling film evaporators are usually run with small differences between wall and saturation temperatures, the rate of evaporation and the vapor velocity are frequently so small that the influence of vapor shear stress on the liquid film can be neglected.

The liquid fed in at the upper tube end often runs downward as a turbulent film. The decisive factor for this is the Reynolds number $\mathrm{Re} = \Gamma/\eta_{\mathrm{L}}$, where Γ is the mass flow rate $\dot{M}$ of the added liquid relative to the circumference b. As was explained earlier, and as was shown by Fig. 2.7, during condensation, the region of turbulent flow joins the region of the laminar film after a transition region. According to investigations of Chun and Seban [12.1] the flow in a falling film is completely turbulent if

$$Re = \Gamma/\eta_{\mathrm{L}} \geqq 1460\, Pr^{-1.05} \tag{12.1}$$

Chun and Seban also determined heat transfer coefficients based upon data for heat transfer on an electrically heated tube sprayed with water on the outside. Because the heating was poor and thus the rate of evaporation low, the film thickness changed very little, and local heat transfer coefficients could be determined. In addition, the Prandtl number Pr of the liquid could be varied by a change in water temperature, between $Pr = 1.77$ and 5.7. In the turbulent region, the following equation was obtained:

$$Nu = \frac{\alpha(v_{\mathrm{L}}^2/g)^{1/3}}{\lambda_{\mathrm{L}}} = 6.616 \cdot 10^{-3}\, Re^{0.4}\, Pr^{0.65}. \tag{12.2}$$

For Reynolds numbers below the values given by Eq. (12.1), the film is laminar but wavy. As follows from investigations of Zozulja [12.2], the heat transfer on a wavy laminar film is greater than that according to the Nusselt film condensation theory. Based upon the data of Zozulja, one finds

$$\alpha = 0.876\, Re^{0.11} \alpha_0, \tag{12.3}$$

where α_0 is the local heat transfer coefficient according to the Nusselt theory, Eq. (2.11). The heat transfer coefficient α_0 can also be depicted as a function of the Reynolds number, if one enters $\delta = \lambda/\alpha_0$ for the film thickness in the earlier Eq. (2.8). One then obtains

$$Nu_0 = \frac{\alpha_0 (\nu_L^2/g)^{1/3}}{\lambda_L} = 0.693\, Re^{-1/3}. \tag{12.4}$$

Thus, Eq. (12.3), which is valid for laminar wavy falling films, is converted into

$$Nu = \frac{\alpha (\nu_L^2/g)^{1/3}}{\lambda_L} = 0.606\, Re^{-0.22}. \tag{12.5}$$

Figure 12.1 illustrates Nusselt numbers in the turbulent region according to Eq. (12.2), in the laminar wavy region according to Eq. (12.5), curve B, and according to the Nusselt theory, Eq. (12.4), curve A.

If one intersects the curves for the turbulent flow according to Eq. (12.2) with those of the laminar wavy falling film according to Eq. (12.5), then one immediately obtains the Reynolds number according to Eq. (12.1) for the transition of the laminar-wavy flow into a turbulent one. The transition is surely not as pronounced as might be expected from Fig. 12.1, but instead is gradual. The Kapitza number, which is defined as the ratio of surface force to viscous force,

$$Ka = \varrho_L \sigma^3 / g \eta_L^4,$$

is often regarded as the beginning of the formation of surface waves. According to Chun and Seban [12.1], the film becomes wavy when $Re \geqq 0.6075\, Ka^{0.09}$.

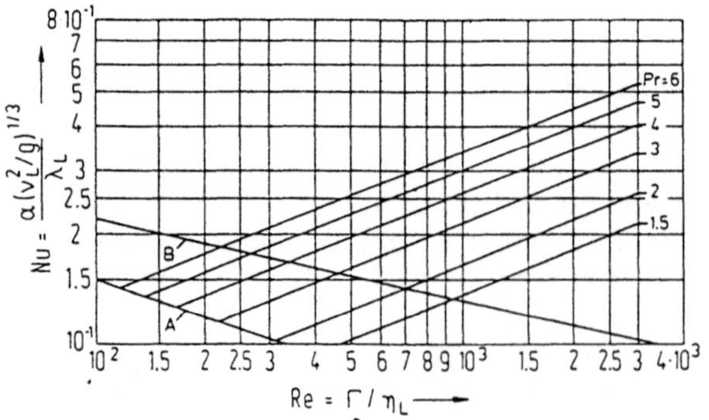

Fig. 12.1. Nusselt number for surface evaporation; to the right, ascending curves for a turbulent film according to Chun and Seban [12.1]; curve B for a laminar wavy film, according to Zozulja [12.2]; curve A, according to the Nusselt film condensation theory

Table 12.1. Kapitza and Reynolds Numbers for the Development of Surface Waves for Water

ϑ_s in °C	100	65.5	43.3	28.3
Ka	$3.185 \cdot 10^{12}$	$9.345 \cdot 10^{11}$	$2.242 \cdot 10^{11}$	$0.840 \cdot 10^{11}$
$Re = 0.6075\, Ka^{0.09}$	8.11	7.26	6.38	5.84
$Re = 0.329\, Ka^{0.125}$	14.3	12.3	10.3	9.009

This equation differs somewhat from that of Grimley [2.10], Eq. (2.34), which can also be rearranged into $Re \geqq 0.392\, Ka^{0.125}$, and which yields values of an equal order of magnitude, as is shown by Table 12.1. In view of the difficulty in determining the inception of a wavy flow, one can scarcely expect a much better correlation. A completely smooth laminar film, as required by the Nusselt film condensation theory, is not formed until below the given Reynolds numbers.

12.1.2 The Influence of Vapor Flow

As long as the vapor velocity is low, one can disregard the influence of shear stress on the liquid film. The shear stress may be disregarded as long as the Reynolds number of the vapor is smaller than 20000. Sufficiently far downstream or during evaporation in a vacuum, the quantity of vapor produced can be so large that the flow of the film is more strongly influenced by the shear stress than by gravity. For a more precise criterion for the determination of which influence is predominant, one can turn to the dimensionless characteristic quantity introduced by Wallis [4.23], Chap. 4.3,

$$j_G^* = \frac{x^* \dot{M}}{A[g d \varrho_G(\varrho_L - \varrho_G)]^{1/2}}.$$

As the flow chart of Fig. 4.12 showed, for $j_G^* \leqq 1.0$, gravity is dominant in the region where small liquid components occur in the vapor, as in the case of falling film evaporation. One can, therefore, calculate the heat transfer on the basis of the previous equations. In the case of $j_G^* \geqq 1$, however, annular flow occurs in forced convection, and one must calculate the heat transfer according to one of the basic equations for this purpose, which are yet to be discussed, for example, Eqs. (13.123) or (13.124).

12.1.3 Dryout of the Heating Surface and the Influence of Vapor Bubble Formation on Heat Transfer

In general, one attempts to avoid nucleate boiling in falling film apparatus. To be sure, the heat transfer coefficient during nucleate boiling increases with the superheating, but simultaneously liquid is dragged along by the vapor bubbles, with the result that dry spots are formed on the wall. This leads to a distinct decrease in the heat transfer, which frequently eradicates the augmentation resulting from the superheating. As yet there are no known completely valid

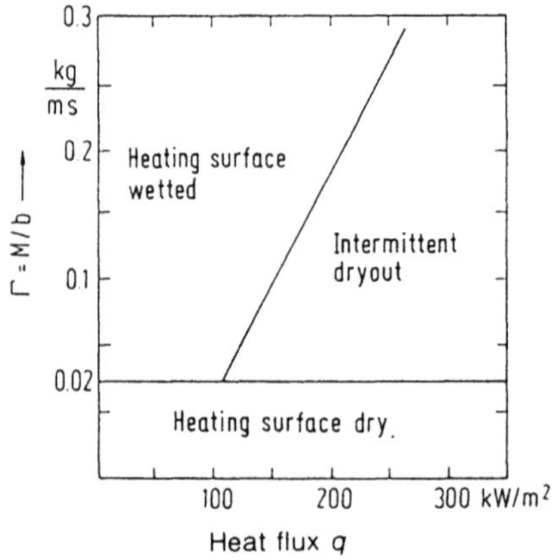

Fig. 12.2. Dryout of the heating surface, according to [12.3], during evaporation of a downward flowing water film on the outside of a vertical tube. Upward flowing vapor

equations for heat transfer during nucleate boiling in falling films. One can find equations valid for the refrigerant R-11 in a work by Struve [12.5], for alcohol-water mixtures and sugar solutions in Thoma [12.6], for water in Zaletnev et al. [12.7], and for sea water in Slesarenko [12.8].

An approximate indication when an initially saturated water film can break apart is given by Fig. 12.2, which is based upon findings of Fujita and Ueda [12.3], which have been evaluated by Palen [12.4]. Because subcooled liquids usually wet more poorly than saturated ones, one should avoid as much as possible feeding in a falling film that is too subcooled. Depending upon the size of the heat flux and the deluge flow rate $\Gamma = \dot{M}/b$ in the outlet cross section, the liquid can wet completely, or intermittent dryout can occur, or the heating surface can be completely dry. If the deluge flow rate lies below 0.01 to 0.02 kg/sm, then dryout occurs under the conditions of Fig. 12.2. In the case of saturated water, this corresponds to a Reynolds number $Re = \Gamma/\eta$ of the film between 140 and 280. According to Palen [12.4], dryout of the heating surface is avoided if

$$\frac{\Gamma\lambda}{qL} > 1.5. \tag{12.6}$$

In order to estimate whether nucleate boiling plays a role, one assumes that bubble formation prevails in the falling film and then calculates the heat transfer coefficient during nucleate boiling according to one of the known equations, for example, Eqs. (11.9) or (11.14). One then compares this heat transfer coefficient with that according to one of the previous equations for heat transfer during falling film evaporation. The predominant mechanism of heat transfer corresponds to the greater heat transfer coefficient. This is, of course, a consequence of the second law of thermodynamics, which states that of the different possible irreversible changes of condition, the one with the greater entropy production is always preferred. In our case, the mechanism with the greater heat transfer coefficient is simultaneously the one with the greater heat flux and, because the temperature differences are specified, is also the one with the greater entropy production.

One can then make approximate calculations, as though the heat transfer were determined solely by the prevailing mechanism. Actually, however, there exists a two-phase flow of liquid and vapor, and surface evaporation continues to take place. Consequently, for more precise calculations, the heat transfer coefficient must be determined by one of the equations for convective boiling yet to be treated, for example, Eqs. (13.117), (13.123), or (13.124).

13 Heat Transfer During Boiling of Pure Substances in Forced Flow

Heat transfer during boiling of a liquid in free convection is essentially determined by the difference between heating surface temperature and boiling temperature, the properties of the liquid, and the properties of the heating surface. During boiling in forced flow, the flow velocities of the vapor and the liquid phases, and the phase distribution play an additional role. As already explained in Chap. 9, the heat transfer coefficient can no longer be represented by a simple empirical correlation of the form $\alpha = cq^n$, but instead, the mass flux $\dot{m}$ and the quality x^* are additional limiting quantities, so that there are heat transfer equations of the form $\alpha = cq^n \dot{m}^s f(x^*)$. The type of such equations is essentially determined by the flow pattern, which will be discussed in this chapter.

13.1 The Different Flow Patterns

The manifold patterns of two-phase flow can be divided into definite basic types, between which, however, transitional or mixed states are possible.

13.1.1 Upward Two-Phase Flow in Vertical Tubes

The basic types of an upward two-phase flow in *vertical, unheated tubes* are shown in Fig. 13.1. In *bubble flow*, Fig. 13.1a, the gas or vapor phase is uniformly dispersed in the continuous liquid phase. Only very small bubbles are spherical; large bubbles, on the other hand, are oblate. Such a form of flow occurs, if the gas components are small. In a *plug* or *slug flow*, Fig. 13.1b, large bubbles (plugs) fill up almost the entire tube cross section. Between the plugs, liquid is interspersed with smaller bubbles (slugs).

With larger mass fluxes, the bubble structure increasingly disintegrates. There is formed a *churn flow*, Fig. 13.1c. It consists of more or less large gas or vapor fragments and has a very unstable character. This form of flow occurs especially in tubes of large diameter and at rather high pressures.

The *wispy-annular flow*, Fig. 13.1d, consists of a relatively thick liquid film on the wall, but the liquid component in the vapor or gas core of the flow is still large. The film is interspersed with small bubbles, and the liquid phase in the flow core consists of large drops, which in part have grown together to form strands of liquid. This form of flow is seen above all, when the mass fluxes are large.

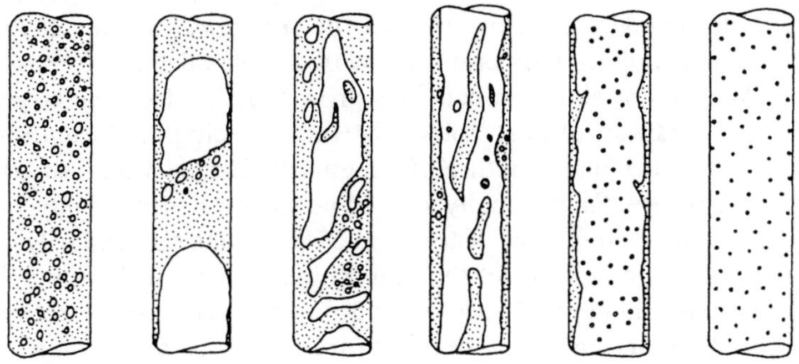

Fig. 13.1 a–f. Flow patterns in vertical, unheated tubes with upward flow. **a** bubble flow; **b** plug flow; **c** churn flow; **d** wispy-annular flow; **e** annular flow; **f** spray or drop flow

A frequently occurring form of flow is *annular flow*, Fig. 13.1e. It is characterized by the fact that the main mass of the liquid is at the wall and the gas or vapor phase, which is interspersed with drops, flows in the core of the tube with a substantially higher velocity.

As a consequence of the evaporation and especially at a high velocity of the gas or vapor, the liquid film at the wall disintegrates, and a *spray* or *drop flow*, Fig. 13.1f, is formed. This occurs particularly during evaporation under high pressures.

All these patterns of two-phase flow are observed not only in circular tubes but also in channels of a different cross section, in so far as these do not completely deviate from the circular form. They cannot, for example, have a very flat, oblate cross section.

During the evaporation of a liquid in a vertical tube, one observes the mentioned patterns of flow more or less pronounced in series, as was already shown in Fig. 9.4. The structure of such a nonadiabatic vapor-liquid flow differs normally from that of an adiabatic two-phase flow, even if the local flow parameters, such as mass flux, vapor content, etc., coincide. The causes for this are the deviation from thermodynamic equilibrium produced by the radial temperature differences, and also the deviation from hydrodynamic equilibrium. Processes which lead to a change in flow pattern, such as bubble coalescence, the dragging along of drops from the liquid film by fast flowing vapor, drop disintegration, and the like, need time. Therefore, the higher the evaporation rate, the farther removed one is from hydrodynamic equilibrium. Thus, certain flow patterns can be more pronounced in heated tubes than in unheated ones, and it is possible that other patterns do not even occur.

Figure 9.4 showed an example of different patterns of flow in a *vertical heated tube*. It is here assumed that the liquid flowing upward is initially subcooled. If the wall-adjacent layer is sufficiently superheated by external heating, then vapor bubbles are formed on the wall, which detach after reaching a critical size and diffuse to the core of the flow. As long as the core is still subcooled, the bubbles condense again (range of the *subcooled boiling*). If the core of flow has reached boiling temperature, swarms of small bubbles move about in the tube. With increasing number, they grow together forming larger bubbles. After further addition of heat, there first forms a pluglike formation (*plug flow*) or a strongly turbulent mixture of irregularly formed vapor shreds of very different sizes (*churn flow*). Finally the liquid collects increasingly on the wall (*churn-wispy and annular flow*). If the core of flow consists primarily of vapor, the vapor then tears drops out of the liquid film, because it flows much faster than the liquid (entrainment). The loss of liquid and further evaporation lead finally to a complete drying out of the film (dryout). This is the range of *spray flow*. Even before all the drops have been evaporated, the vapor temperature begins to exceed the saturated state, i.e., a thermodynamic equilibrium no longer exists between drops and vapor.

When approaching the critical state of the fluid, the picture changes. With slight contents of vapor, the flow already has a foamy appearance, because primarily small bubbles are being formed. Plug flow often does not occur at all. Especially during large mass fluxes, the drops in the vapor core frequently grow together into larger, irregular liquid formations.

13.1.2 Two-Phase Flow in Horizontal Tubes

Because of gravity, the liquid in *horizontal unheated tubes* or channels flows chiefly in the lower part of the cross section and gas or vapor in the upper part. The smaller the inertia forces are in contrast to gravity, the more the flow in horizontal tubes differs from that in vertical tubes. At very small velocities, flow patterns develop in horizontal tubes that are not seen in vertical tubes. Figure 13.2 shows characteristic forms of a horizontal two-phase flow. When the vapor bubbles collect in the upper part of the tube because of buoyant forces, one speaks of *bubble flow*, Fig. 13.2a. They can grow together into plugs and convert into *plug flow*, Fig. 13.2b.

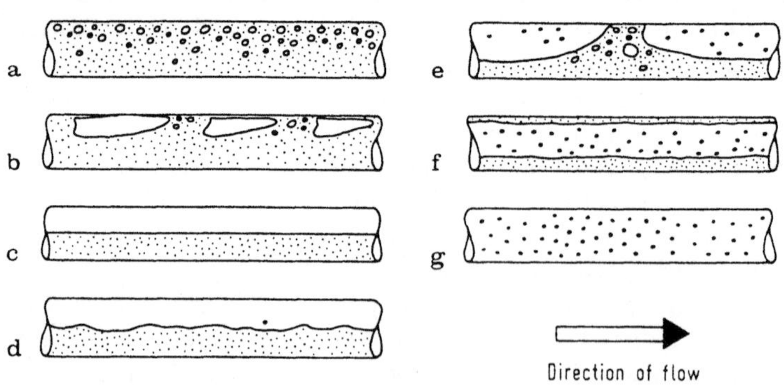

Fig. 13.2 a–g. Flow patterns in a horizontal, unheated tube. **a** bubble flow; **b** plug flow; **c** stratified flow; **d** wavy flow; **e** slug flow; **f** annular flow; **g** spray or drop flow

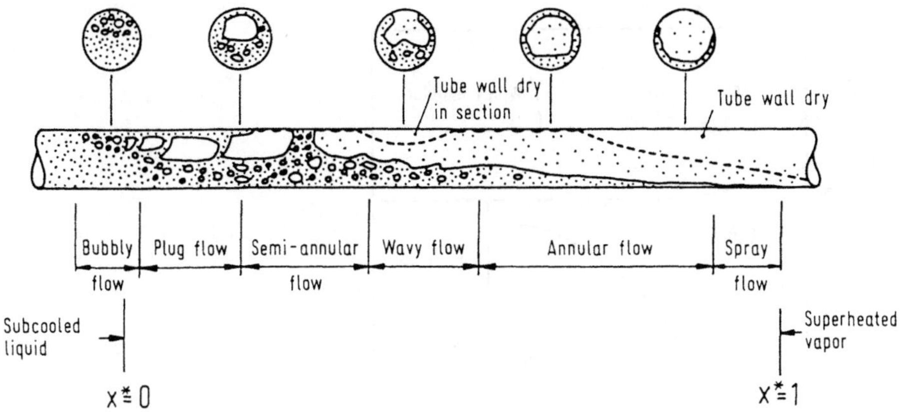

Fig. 13.3. Flow patterns in a horizontal evaporator tube, according to Collier [13.1]

At very small velocities, both phases are completely separated. This is called a *stratified flow*, Fig. 13.2c. If the gas or vapor velocity increases, then waves form on the surface of the liquid, *wavy flow*, Fig. 13.2d. The wave crests become higher and finally wet the upper canal wall in an irregular sequence, *slug flow*, Fig. 13.2e. At a yet higher velocity, *annular flow* develops, Fig. 13.2f, whose film thickness in the upper half of the tube is usually smaller than in the lower range of the tube. With a further increase in the vapor velocity, the flow converts into a *spray flow*, Fig. 13.2g. As one recognizes, the upper tube wall in the range of slug flow is alternatingly dry and wetted. In the range of annular flow, the upper canal wall dries out downstream more and more.

In *horizontal heated tubes*, the patterns of flow just discussed occur one after the other. Just as in vertical evaporator tubes, because of the radial temperature profile, no thermal equilibrium occurs. Figure 13.3 shows the different patterns of flow in horizontal evaporator tubes, assuming that the liquid enters the tube at a sufficiently small velocity below 1 m/s. The higher the mass flow rates are, the more symmetrical the phase distribution becomes and approaches that in the vertical evaporator tube.

13.2 Flow Pattern Maps

Because the different flow patterns are determined by forces between the phases, above all by the forces of inertia and gravity, it is reasonable to use diagrams, so-called flow pattern maps, to depict the boundaries between the various flow patterns in their dependence on these forces. Such a flow pattern map was first set up by Baker [13.2]. One, therefore, speaks of *Baker Maps*. These serve only for rough orientation, because the decisive forces in the different flow ranges are not known with sufficient accuracy. One must figure with rather large uncertainties, especially in two-phase flows with heat input, because most flow pattern maps were developed for adiabatic flows. Moreover, the transitions

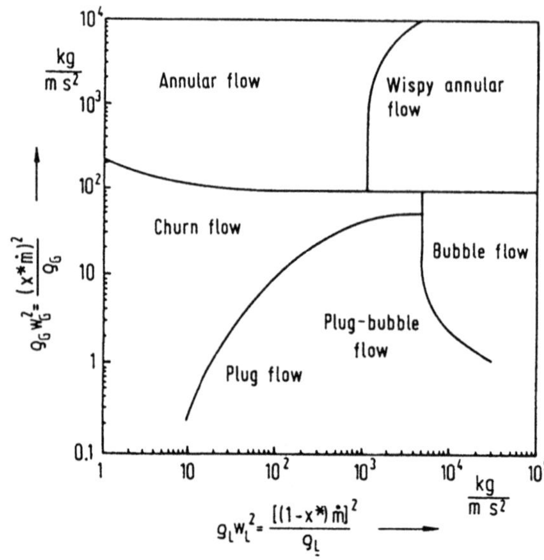

Fig. 13.4. Flow pattern diagram for vertical two-phase flows, according to Hewitt and Roberts [13.3], $1\,\mathrm{kg/ms^2} = 1\,\mathrm{Pa}$

among the flow patterns are imprecise; strict boundaries with abrupt transitions do not exist.

Figure 13.4 shows a flow pattern map, which was developed by Hewitt and Roberts [13.3] by observing the flows of air-water mixtures at atmospheric pressure and of water–water vapor mixtures at higher pressures in vertical tubes of 10 to 30 mm diameter. The momentum fluxes $\varrho_G w_G^2$ of the vapor and $\varrho_L w_L^2$ of the liquid are plotted on the ordinates. These have the dimensions of a pressure and are formed with the gas and liquid velocities, which would occur, if each of the two phases filled up the entire cross section. This yields

$$w_G = x^*\dot{m}/\varrho_G \quad \text{and} \quad w_L = (1 - x^*)\dot{m}/\varrho_L.$$

For two-phase flows in horizontal and inclined tubes, Taitel and Dukler [13.4] used models to develop a flow map, which is shown in Fig. 13.5. For the boundary between stratified annular or intermittent flow and stratified wavy flow, curve 1, the characteristic quantity

$$F_1 = \left(\frac{\varrho_G}{\varrho_L - \varrho_G}\right)^{1/2} \frac{w_G}{(dg\cos\gamma)^{1/2}} \tag{13.1}$$

is decisive, where w_G is the apparent velocity of the vapor, d the tube inner diameter, and γ the angle of inclination toward the horizontal. For the boundaries between annular-dispersed liquid and dispersed bubble or intermittent flow and between stratified wavy and stratified smooth flow, one must form the characteristic quantity

$$F_2 = \frac{\varrho_G w_G^2 w_L}{(\varrho_L - \varrho_G) g \nu_L \cos\gamma}, \tag{13.2}$$

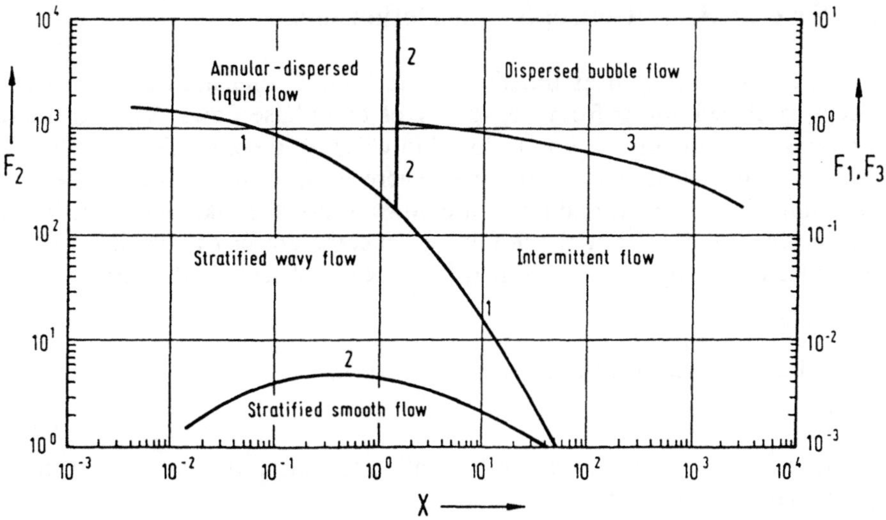

Fig. 13.5. Flow patterns in horizontal and inclined tubes, according to Taitel and Dukler [13.4]

whereas the boundary between dispersed bubble and intermittent flow, curve 3, is determined by

$$F_3 = \left[\frac{(\mathrm{d}p/\mathrm{d}z)_\mathrm{L}}{(\varrho_\mathrm{L} - \varrho_\mathrm{G}) g \cos\gamma}\right]^{1/2}, \tag{13.3}$$

where $(\mathrm{d}p/\mathrm{d}z)_\mathrm{L}$ is the friction pressure loss of the liquid phase, assuming that this alone is flowing in the tube. The Martinelli parameter X is plotted on the abscissa. It is

$$X^2 = \left(\frac{\mathrm{d}p}{\mathrm{d}z}\right)_\mathrm{L} \Big/ \left(\frac{\mathrm{d}p}{\mathrm{d}z}\right)_\mathrm{G} \tag{13.4}$$

with the friction pressure drop $(\mathrm{d}p/\mathrm{d}z)_\mathrm{L}$ of the liquid, if this alone were flowing through the tube, and with the friction pressure drop $(\mathrm{d}p/\mathrm{d}z)_\mathrm{G}$, if the gas alone were flowing through the tube. The former is

$$\left(\frac{\mathrm{d}p}{\mathrm{d}z}\right)_\mathrm{L} = -\zeta_\mathrm{L}\frac{1}{d}\frac{\varrho_\mathrm{L} w_\mathrm{L}^2}{2} \tag{13.5}$$

with the friction factor

$$\zeta_\mathrm{L} = c/Re^n \quad \text{with} \quad Re = \frac{\dot{m}(1-x^*)d}{\eta_\mathrm{L}}. \tag{13.6}$$

A corresponding equation is valid for the friction pressure drop of the gas.

13.3 Several Basic Concepts and Definitions

For the description of heat transfer during boiling, one needs several basic concepts and definitions from the theory of two-phase flows. Some of these definitions have already been introduced in Chap. 4 in connection with forced convection condensation. For our purpose here, we show in Fig. 13.6 the section of a channel in which gas and liquid are flowing. For the sake of simplicity, we show an annular flow, but the following concepts are just as valid for other flow patterns. At any cross section A, the gaseous phase fills the portion

$$\varepsilon = A_G/A, \tag{13.7}$$

and the liquid phase correspondingly the portion

$$1 - \varepsilon = A_L/A \tag{13.8}$$

because $A_G + A_L = A$. In a sufficiently small tube section Δz, these proportions do not change. Therefore,

$$\varepsilon = A_G \Delta z / A \Delta z.$$

The amount of vapor by volume in the considered tube section is thus

$$\varepsilon = V_G/V \tag{13.9}$$

and the amount of liquid by volume

$$1 - \varepsilon = V_L/V. \tag{13.10}$$

The quantity ε is called the *volumetric vapor content* or *void fraction*. To be distinguished from this is the *flowing volumetric quality*

$$\varepsilon^* = \dot{V}_G/\dot{V}, \tag{13.11}$$

which relates the volume flow of the vapor to the entire volume flow of vapor and liquid. A further concept is the *flowing mass quality* or, simply, *quality*, which represents the ratio of the mass flow rate $\dot{M}_G$ of the vapor to the entire

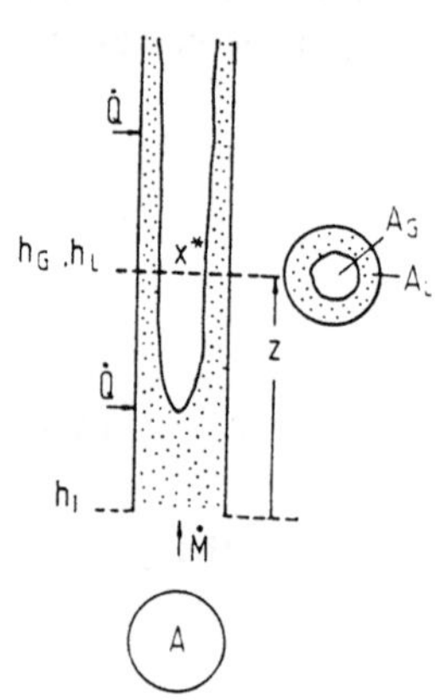

Fig. 13.6. Two-phase flow in a heated channel

mass flow rate $\dot{M} = \dot{M}_G + \dot{M}_L$

$$x^* = \dot{M}_G/\dot{M}. \tag{13.12}$$

It is

$$1 - x^* = \dot{M}_L/\dot{M}. \tag{13.13}$$

The quality is different from the static quality of the steam tables, which, as one knows, is defined by $x = M_G/M$. With the mass flow rates $\dot{M}_G$ of the gas and $\dot{M}_L$ of the liquid, one obtains the velocity of both phases in a random cross section

$$w_G = \frac{\dot{M}_G}{\varrho_G A_G} = \frac{x^* \dot{M}}{\varrho_G \varepsilon A} = \frac{x^* \dot{m}}{\varrho_G \varepsilon}, \tag{13.14}$$

$$w_L = \frac{\dot{M}_L}{\varrho_L A_L} = \frac{(1 - x^*)\dot{M}}{\varrho_L (1 - \varepsilon) A} = \frac{(1 - x^*)\dot{m}}{\varrho_L (1 - \varepsilon)}. \tag{13.15}$$

The ratio of the two velocities is called *slip* or *slip factor*. This yields

$$s = \frac{w_G}{w_L} = \frac{x^*}{1 - x^*} \frac{1 - \varepsilon}{\varepsilon} \frac{\varrho_L}{\varrho_G}. \tag{13.16}$$

Also the quality x^* and the flowing volumetric quality ε^* are connected with one another because of

$$x^* = \frac{\dot{M}_G}{\dot{M}_G + \dot{M}_L} = \frac{\dot{V}_G \varrho_G}{\dot{V}_G \varrho_G + \dot{V}_L \varrho_L} = \frac{\varepsilon^* \dot{V} \varrho_G}{\varepsilon^* \dot{V} \varrho_G + (1 - \varepsilon^*) \dot{V} \varrho_L}$$

or

$$x^* = \frac{\varepsilon^*}{\varepsilon^* + (1 - \varepsilon^*) \varrho_L / \varrho_G}. \tag{13.17}$$

The flowing volumetric quality ε^* and the void fraction ε coincide, if the slip factor becomes unity; thus $w_G = w_L$ holds, and

$$\varepsilon^* = \frac{\dot{V}_G}{\dot{V}} = \frac{w_G A_G}{w_G A_G + w_L A_L} = \frac{A_G}{A} = \varepsilon.$$

The quality x^* is usually known or is easily determined.

In unheated channels, the mass flow rates of the individual phases are generally given. In a heated channel, the quality is determined from the energy balance, Fig. 13.6. We assume here that the liquid enters subcooled into the evaporator tube. Let its specific enthalpy in the entrance cross section be h_1. It is heated by the heat transfer rate Φ to the saturation temperature and is then evaporated. If the kinetic and potential energies are disregarded, the energy balance between the entrance cross section and any other cross section at the position z gives

$$\dot{M} h_1 + \Phi = \dot{M}[x^* h_G + (1 - x^*) h_L].$$

If it is further assumed that vapor and liquid in the cross section at the position z are in thermodynamic equilibrium, then their specific enthalpies in the saturated state are to be taken at the pressure $p(z)$. In the cross section under consideration, we have $h_G = h''$ and $h_L = h'$ and $h'' - h' = \Delta h_v$ is the enthalpy of evaporation at pressure p in the cross section. Because of the assumption of thermodynamic equilibrium, we now designate the quality by x^*_{th} and call it the thermodynamic quality. This yields

$$\dot{M}h_1 + \Phi = \dot{M}(x^*_{th}\Delta h_v + h')$$

and

$$x^*_{th} = \frac{1}{\Delta h_v}\left(\frac{\Phi}{\dot{M}} + h_1 - h'\right). \tag{13.18}$$

As assumed, this equation is valid only if vapor and liquid in the cross section have the same temperature. It fails in the vicinity of the entrance cross section where the quality x^* is still very small and also if large vapor contents are present. In the vicinity of the entrance cross section, vapor bubbles can already form on the hot wall, while the core of the flow is still subcooled. Vapor and liquid thus have different temperatures. The quality is thus positive, whereas one obtains a negative value for the thermodynamic quality according to Eq. (13.18), because the liquid is still subcooled and, therefore, $\Phi + \dot{M}h_1 < \dot{M}h'$. With high qualities, one has a spray flow. Heat is transferred mainly to the vapor, which becomes superheated, although in the core of the flow liquid drops are still swept along and evaporate only gradually. The actual quality is thus still smaller than unity, although the thermodynamic quality already reaches unity according to Eq. (13.18).

In the range of intermediate qualities, where neither subcooled boiling nor spray flow is present, Eq. (13.18) nevertheless produces exact values. Figure 13.7 shows the course of the actual quality x^*_{real} over the thermodynamic quality x^*_{th}, based upon Eq. (13.18). Because no systematic investigations of the deviation from the thermodynamic quality x^*_{th} are known at present, one cannot reliably

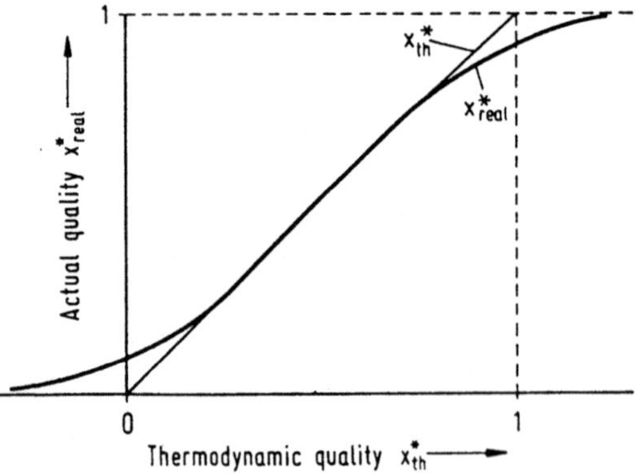

Fig. 13.7. Qualitative behavior of the actual quality x^*_{real} as a function of the thermodynamic quality x^*_{th}

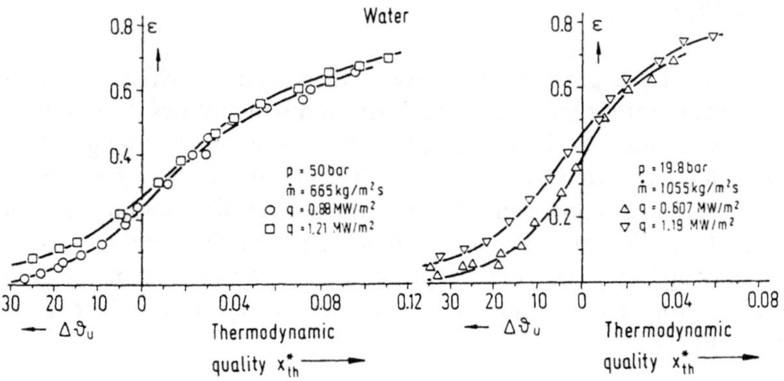

Fig. 13.8. Void fraction during subcooled boiling in two-phase flow of water, according to Rouhani [13.5]

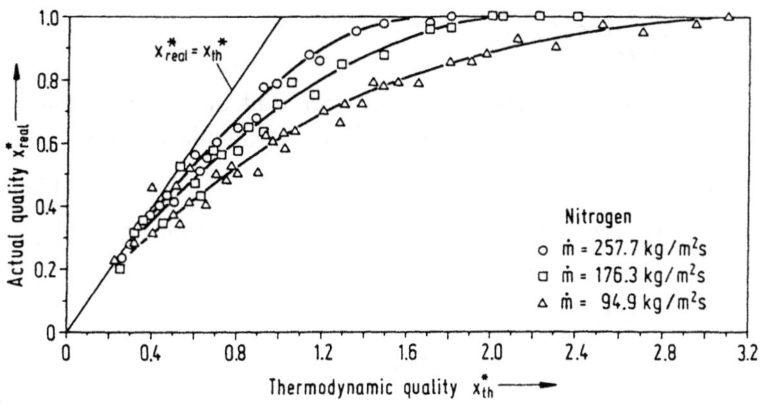

Fig. 13.9. Actual quality in a spray flow of nitrogen, according to Plummer [13.6]

state the actual quality at the limits of $x^* \to 0$ and $x^* \to 1$. The deviations from the thermodynamic quality can be substantial. As an example Fig. 13.8 shows the void fraction during subcooled boiling of water according to data of Rouhani [13.5]. At the point at which one calculates a thermodynamic quality $x_{\text{th}}^* = 0$, according to Eq. (13.18), i.e., where the liquid on the average is just reaching its saturated state, the void fraction already amounts to 20 to 50%.

It increases with decreasing pressure, and it increases with the mass flux. Also in the case of *spray flow*, i.e., with high quality, the deviation from the thermodynamic quality is substantial, because the superheated vapor still sweeps drops of liquid along. According to Fig. 13.9, which reproduces the data of Plummer [13.6] for boiling nitrogen, the last liquid drops do not disappear until the thermodynamic quality, calculated according to Eq. (13.18), lies between 1.4 and 2.6, depending upon the mass flux.

13.4 Pressure Drop

For the calculation of the pressure drop, we shall consider a channel that is inclined to the horizontal by the angle γ and through which flows a two-phase fluid, Fig. 13.10. As plotted in the figure, an element of volume of length dz is acted upon by the forces exerted by pressure, by the frictional force at the channel wall, and by gravity. We assume a steady, one-dimensional flow and a constant channel cross section. Pressure, density, and momentum flux across a cross section depend only upon the path of flow z. The balance of momentum indicates that the sum of the forces of pressure, friction, and gravity equals the change of the flow momentum:

$$\left[p - \left(p + \frac{dp}{dz}dz\right)\right]A - \tau_w U dz - \varrho g A \sin\gamma\, dz = \frac{d}{dz}(\dot{M}w)dz$$

or

$$-\frac{dp}{dz} = \tau_w \frac{U}{A} + \frac{1}{A}\frac{d}{dz}(\dot{M}w) + \rho g \sin\gamma, \tag{13.19}$$

where τ_w signifies the shear stress on the channel wall and U the channel circumference. The density of the two-phase mixture is calculated from

$$\varrho = \varepsilon\rho_G + (1-\varepsilon)\rho_L. \tag{13.20}$$

The flow momentum is composed of that of the gas and that of the liquid according to

$$\dot{M}w = \dot{M}_G w_G + \dot{M}_L w_L. \tag{13.21}$$

Together with Eqs. (13.14) and (13.15), one can also write for this

$$\dot{M}w = \frac{\dot{M}^2}{A}\left[\frac{x^{*2}}{\varepsilon\rho_G} + \frac{(1-x^*)^2}{(1-\varepsilon)\rho_L}\right]. \tag{13.22}$$

The insertion of Eqs. (13.20) and (13.22) into Eq. (13.19) yields

$$-\frac{dp}{dz} = \tau_w \frac{U}{A} + \dot{m}^2 \frac{d}{dz}\left[\frac{x^{*2}}{\varepsilon\rho_G} + \frac{(1-x^*)^2}{(1-\varepsilon)\rho_L}\right] + [\varepsilon\rho_G + (1-\varepsilon)\rho_L]g \sin\gamma. \tag{13.23}$$

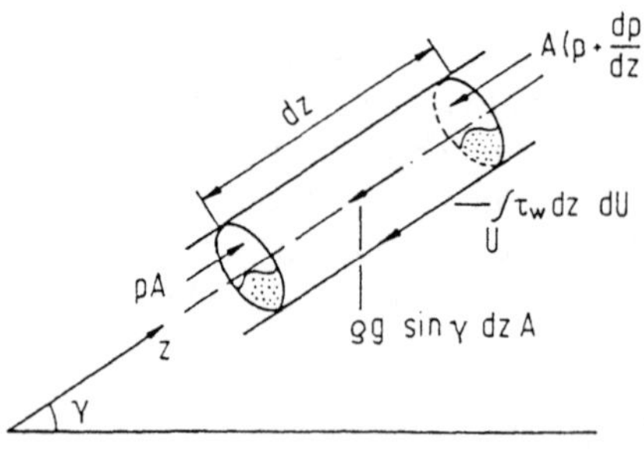

Fig. 13.10. Balance of forces on a two-phase fluid element

The total pressure drop is composed of three components: the frictional pressure drop

$$-\left(\frac{dp}{dz}\right)_{\mathrm{f}} = \tau_{\mathrm{w}} \frac{U}{A}, \tag{13.24}$$

the acceleration pressure drop

$$-\left(\frac{dp}{dz}\right)_{\mathrm{a}} = \dot{m}^2 \frac{d}{dz}\left[\frac{x^{*2}}{\varepsilon\rho_{\mathrm{G}}} + \frac{(1-x^*)^2}{(1-\varepsilon)\rho_{\mathrm{L}}}\right] \tag{13.25}$$

and the drop in pressure based upon gravity, which is also called the geodetic pressure drop,

$$-\left(\frac{dp}{dz}\right)_{\mathrm{g}} = [\varepsilon\rho_{\mathrm{G}} + (1-\varepsilon)\rho_{\mathrm{L}}]g \sin\gamma. \tag{13.26}$$

In channels with return bends and changes in cross sections, there occur additional pressure drops which we will not treat here. The reader should refer to the partinent literature [13.7].

The frictional pressure drop results from the shear stress between the flowing fluid and the channel wall. The pressure drop due to acceleration results from the change in momentum of the two phases, because in vapor-liquid flows there occurs an evaporation resulting from flashing due to loss of pressure as well as, in heated channels, an evaporation because of the addition of heat. This causes a change of mass and velocity in each phase and, consequently, also of the momentum flux of both phases. The geodetic pressure drop results from the gravitational force exerted by the fluid. It disappears in horizontal flows. The geodetic pressure drop and the acceleration pressure drop are often negligible in comparison to the frictional pressure drop. However, in heated channels with large heat and mass fluxes, the acceleration pressure drop can assume substantial values. In adiabatic two-phase flows, the acceleration pressure drop is only the result of evaporation (flashing) or, as the case may be, of the expansion of the vapor or gas phase. A simple rule for adiabatic two-phase flows in refrigeration plants states, for example, that the acceleration pressure drop does not play a role if $\Delta p_{\mathrm{f}}/p_{\mathrm{s}} < 0.2$. If this is not the case, then one determines the acceleration pressure drop according to Eq. (13.25) over the channel length between inlet cross section 1 and outlet cross section 2:

$$(p_1 - p_2)_{\mathrm{a}} = \dot{m}^2\left[\frac{x_2^{*2}}{\varepsilon_2\rho_{\mathrm{G}_2}} - \frac{x_1^{*2}}{\varepsilon_1\rho_{\mathrm{G}_1}} + \frac{(1-x_2^*)^2}{(1-\varepsilon_2)\rho_{\mathrm{L}_2}} - \frac{(1-x_1^*)^2}{(1-\varepsilon_1)\rho_{\mathrm{L}_1}}\right]. \tag{13.27}$$

Precise calculations are possible here only if a reliable equation is available for the void fraction.

If the slip factor $s = 1$ can be assumed (homogeneous flow), then according to Eq. (13.16), the void fraction becomes

$$\varepsilon = 1 \Big/ \left[\frac{\rho_{\mathrm{G}}}{\rho_{\mathrm{L}}} \frac{1-x^*}{x^*} + 1\right]. \tag{13.28}$$

In the case of complete evaporation with a change of the quality x^* between $x_1^* = 0$ and $x_2^* = 1$, $\varepsilon_1 = 0$ and $\varepsilon_2 = 1$, the acceleration pressure drop becomes

$$(p_1 - p_2)_a = \dot{m}^2 \left[\frac{1}{\rho_{G_2}} - \frac{1}{\rho_{L_1}} \right].$$

13.4.1 Frictional Pressure Drop

The frictional pressure drop generally constitutes the largest part of the total pressure drop. Yet one must rely extensively upon empirical methods for its calculation. It involves not only the transfer of momentum between fluid and wall, but also the transfer of momentum between the individual phases. The two processes cannot be measured separately and can only be estimated for simple flows. Thus, there exist only imprecise notions about the influence of the transfer of momentum between the two phases.

As a function of the quality, the frictional pressure drop does not show a linear dependence between the limiting values $x^* = 0$ and $x^* = 1$, at constant mass flux, but rather increases with increasing value x^* to a maximum between $x^* = 0.7$ and $x^* = 0.9$ and then decreases to the pressure drop of the pure vapor flow, Fig. 13.11. The maximum is all the more strongly pronounced, the greater the density difference between the phases.

Procedures for the determination of the frictional pressure drop usually start with simple models. For the most part, homogeneous flow (homogeneous distribution of the phases, $s \to 1$) or heterogeneous flow (heterogeneous distribution of the phases, $s > 1$) is assumed.

Less common are methods that are based upon definite flow patterns and are, therefore, applicable only for specific flow patterns.

For the calculation of the frictional pressure drop, it is advantageous to define several parameters that are suitable for the description of the two-phase frictional pressure drop as well as of the quality. The frictional pressure drop is often reduced to the pressure drop of the single-phase flow by using the following definitions developed by Lockhart and Martinelli [13.8]:

$$\left(\frac{dp}{dz}\right)_f = \phi_L^2 \left(\frac{dp}{dz}\right)_L \tag{13.29}$$

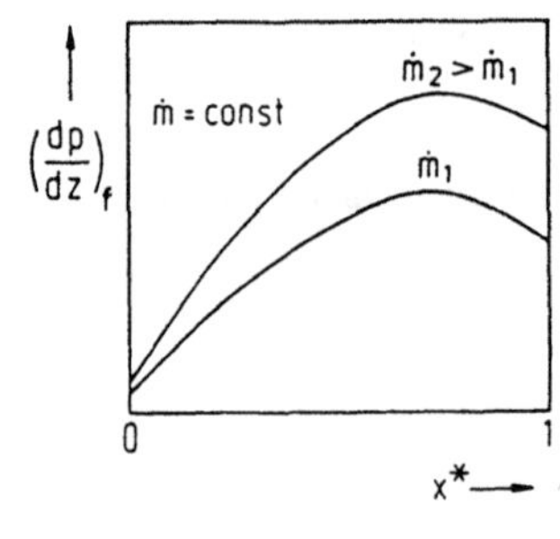

Fig. 13.11. Curve of the frictional pressure drop

or

$$\left(\frac{dp}{dz}\right)_f = \phi_G^2 \left(\frac{dp}{dz}\right)_G. \tag{13.30}$$

Here $(dp/dz)_L$ is the frictional pressure drop of the liquid and $(dp/dz)_G$ that of the vapor, assuming that each of the two phases flows alone through the channel. The factors ϕ_L^2 and ϕ_G^2 are defined by the equations above. If these factors are known, then only the pressure drops of the individual phases need be determined in order to calculate the frictional pressure drop of the two-phase flow.

The pressure drops of the two phases are determined according to known methods. Thus, for the frictional pressure drop of a liquid flow there holds

$$\left(\frac{dp}{dz}\right)_L = -\zeta_L \frac{1}{d}\frac{\rho_L w_L^2}{2} = -\zeta_L \frac{1}{d}\frac{\dot{m}_L^2}{2\rho_L}. \tag{13.31}$$

with

$$\dot{m}_L = \dot{m}(1 - x^*),$$

where d is the diameter of the flow channel and ζ_L the friction factor, which is dependent upon the Reynolds number. According to Blasius, for example, there holds

$$\zeta_L = C_1/Re_L^n, \tag{13.32}$$

where C_1 and n are functions of the state of flow and of the tube roughness. The Reynolds number is to be calculated with the assumption that the liquid is flowing alone in the tube

$$Re_L = \frac{w_L \rho_L d}{\eta_L} = \frac{\dot{m}_L d}{\eta_L} = \frac{\dot{m}(1 - x^*)d}{\eta_L}. \tag{13.33}$$

Another possibility for establishing the pressure drop of two-phase flows consists of first determining the frictional pressure drop of the gas and the liquid assuming that the entire mass of the fluid is flowing as liquid or as gas through the channel. In this case, the factors ϕ_{LO} and ϕ_{GO} are defined by

$$\left(\frac{dp}{dz}\right)_f = \phi_{LO}^2 \left(\frac{dp}{dz}\right)_{LO} \tag{13.34}$$

and

$$\left(\frac{dp}{dz}\right)_f = \phi_{GO}^2 \left(\frac{dp}{dz}\right)_{GO}. \tag{13.35}$$

This now gives

$$\left(\frac{dp}{dz}\right)_{LO} = -\zeta_{LO} \frac{1}{d}\frac{\dot{m}^2}{2\rho_L} \tag{13.36}$$

wherein the friction factor ζ_{LO} is to be determined with the Reynolds number

$$Re_{LO} = \frac{\dot{m}d}{\eta_L}. \tag{13.37}$$

13.4.2 The Homogeneous Model

In flows with a high concentration of small bubbles, $x^* \to 0$, in spray or drop flows, $x^* \to 1$, or in flows with slight differences in density, such as occur when approaching the critical state, the frictional pressure drop is described to a good approximation by the homogeneous model. In these cases, less accurate results are obtained with the heterogeneous model.

Calculations according to the homogeneous model are made just as in the case of single-phase flows, using however, suitably defined mean property values. For the frictional pressure drop of a homogeneous two-phase flow, there holds

$$\left(\frac{dp}{dz}\right)_f = -\zeta \frac{1}{d}\frac{\rho w^2}{2} = -\zeta \frac{1}{d}\frac{\dot{m}^2}{2\rho}. \tag{13.38}$$

Assuming equal velocity of liquid and vapor, one obtains from Eq. (13.16) the void fraction as

$$\varepsilon = 1 \bigg/ \left(1 + \frac{\rho_G}{\rho_L}\frac{1-x^*}{x^*}\right) \tag{13.39}$$

and, thus, the density of the homogeneous flow is

$$\rho = \varepsilon\rho_G + (1-\varepsilon)\rho_L = 1 \bigg/ \left(\frac{x^*}{\rho_G} + \frac{1-x^*}{\rho_L}\right). \tag{13.40}$$

The friction factor is to be calculated as in the case of the single-phase flow, for example, according to Blasius

$$\zeta = c_1/Re^n. \tag{13.41}$$

The dynamic viscosity of the homogeneous two-phase flow is now to be entered into the Reynolds number $Re = \dot{m}d/\eta$. In the literature [13.9–13.11], usually empirical assumptions have been suggested which contain as limiting cases $\eta(x^* = 0) = \eta_L$ and $\eta(x^* = 1) = \eta_G$.

According to McAdams et al. [13.9], there holds

$$\frac{1}{\eta} = \frac{x^*}{\eta_G} + \frac{1-x^*}{\eta_L}, \tag{13.42}$$

according to Cicchitti et al. [13.10]

$$\eta = x^*\eta_G + (1-x^*)\eta_L \tag{13.43}$$

and according to Dukler et al. [13.11]

$$\eta = \rho\left[x^* \frac{\eta_G}{\rho_G} + (1 - x^*)\frac{\eta_L}{\rho_L}\right]. \tag{13.44}$$

With Eq. (13.40) one obtains from Eq. (13.38) the frictional pressure drop of the homogeneous two-phase flow as

$$\left(\frac{dp}{dz}\right)_f = -\zeta \frac{1}{d}\frac{\dot{m}^2}{2}\left(\frac{x^*}{\rho_G} + \frac{1 - x^*}{\rho_L}\right). \tag{13.45}$$

On the other hand, assuming that only liquid is flowing in the tube, the pressure drop is

$$\left(\frac{dp}{dz}\right)_L = -\zeta_L \frac{1}{d}\frac{\dot{m}^2(1 - x^*)^2}{2\rho_L}. \tag{13.46}$$

According to the Blasius relation, the friction factor ζ_L of the liquid flow is

$$\zeta_L = c/Re_L^n \quad \text{and} \quad Re_L = \dot{m}(1 - x^*)d/\eta_L. \tag{13.47}$$

With Eqs. (13.46), (13.38), and (13.29), one finds the factor ϕ_L^2 for the homogeneous two-phase flow as

$$\phi_L^2 = \frac{\zeta}{\zeta_L}\left[1 + x^*\left(\frac{\rho_L}{\rho_G} - 1\right)\right]\cdot(1 + x^*)^{-2}. \tag{13.48}$$

Here one can eliminate the friction factor by using the Blasius relation, Eqs. (13.41) and (13.47), and then obtains

$$\phi_L^2 = (1 + x^*)^{n-2}\left[1 + x^*\left(\frac{\rho_L}{\rho_G} - 1\right)\right]\Big/\left[1 + x^*\left(\frac{\eta_L}{\eta_G} - 1\right)\right]^n, \tag{13.49}$$

where n is the Blasius exponent, Eq. (13.41), or as the case may be, Eq. (13.47).

13.4.3 The Heterogeneous or Separated Model

In contrast to the homogeneous model, in the heterogeneous model the two phases are said to flow separately from one another and to have different velocities, so that a slip exists between the phases.

13.4.3.1 The Procedure of Lockhart and Martinelli

An especially simple and frequently used procedure was developed by Lockhart and Martinelli [13.8]. It is based upon data for water-air and oil-air mixtures in horizontal tubes at low pressures. The calculation procedure has, however, proven valid for vertical upward two-phase flows of single component and multi-component mixtures. The basic idea of the Lockhart–Martinelli procedure is that, as discussed before, the frictional pressure drop of a two-phase flow can be determined by using a correction factor from the frictional pressure drop of the single phases. From this, the two-phase multipliers ϕ_L^2 and ϕ_G^2 were defined according to Eqs. (13.29) and (13.30).

One puts

$$\frac{\phi_G^2}{\phi_L^2} = X^2 = \frac{(dp/dz)_L}{(dp/dz)_G} \tag{13.50}$$

with $(dp/dz)_L$ according to Eq. (13.46) and

$$\left(\frac{dp}{dz}\right)_G = -\zeta_G \frac{1}{d} \frac{\dot{m}^2 x^{*2}}{2\rho_G}. \tag{13.51}$$

From Eqs. (13.46) and (13.51), follows

$$X^2 = \frac{(dp/dz)_L}{(dp/dz)_G} = \frac{\zeta_L}{\zeta_G}\left(\frac{1-x^*}{x^*}\right)^2 \frac{\rho_G}{\rho_L}. \tag{13.52}$$

The quantity X defined by Eq. (13.50) is usually characterized as the *Lockhart–Martinelli parameter.*

It assumes different values, depending upon whether the type of flow in the two phases is laminar or turbulent. The following combinations, which shall be characterized by indices with respect to the value X, are possible:

Flow State of the Phases	Indices of X
gas laminar, liquid laminar	ll
gas laminar, liquid turbulent	lt
gas turbulent, liquid laminar	tl
gas turbulent, liquid turbulent	tt.

If one assumes the validity of the Blasius relation for the friction factor

$$\zeta_L = c/Re_L^n = c\Big/\left[\frac{\dot{m}(1-x^*)d}{\eta_L}\right]^n \tag{13.53}$$

and

$$\zeta_G = c/Re_G^n = c\Big/\left(\frac{\dot{m}x^* d}{\eta_G}\right)^n \tag{13.54}$$

there follows according to Eq. (13.52)

$$X_{tt}^2 = \left(\frac{1-x^*}{x^*}\right)^{2-n}\left(\frac{\eta_L}{\eta_G}\right)^n \frac{\rho_G}{\rho_L}. \tag{13.55}$$

For turbulent flow in technically smooth tubes, the exponent n lies between 0.2 and 0.25. With $n = 0.2$, this becomes

$$X_{tt} = \left(\frac{1-x^*}{x^*}\right)^{0.9}\left(\frac{\eta_L}{\eta_G}\right)^{0.1}\left(\frac{\rho_G}{\rho_L}\right)^{0.5} \tag{13.56}$$

Lockhart and Martinelli then assumed that each of the two factors Φ_G and Φ_L can be depicted as a function of the parameter X. In Fig. 13.12 are given the values Φ_L and Φ_G derived by Lockhart and Martinelli for the different types of flow as functions of the parameter X.

The curves are reproduced well by the following equations:

$$\Phi_L^2 = 1 + \frac{C}{X} + \frac{1}{X^2}, \tag{13.57}$$

and

$$\Phi_G^2 = 1 + CX + X^2. \tag{13.58}$$

The values in the following table are to be used for the constant C.

Constant C in the Eqs. (13.57) and (13.58):

Gas/Vapor	Liquid	Index	C
laminar	laminar	ll	5
laminar	turbulent	lt	10
turbulent	laminar	tl	12
turbulent	turbulent	tt	20

Laminar flow is assumed if $Re < 1000$, and turbulent flow occurs if $Re > 2000$. No certain prediction can be made for the transition region $1000 < Re < 2000$. To be safe, one should assume the presence of turbulent flow.

The procedure of Lockhart and Martinelli is especially simple and easy to carry out. To be sure, it is surpassed in accuracy in certain parameter regions by other procedures, but independent of application, it renders satisfactory results in pressure drop within a range of uncertainty of approximately $\pm 50\%$. Greater deviations are to be expected for tube diameters of $d > 0.1$ m. Also to be considered is the fact that the two-phase multipliers Φ_L and Φ_G were determined from measurements at low pressures. For high pressure steam-water flows, the method of Martinelli and Nelson [13.12], which is to be treated in the following chapter, is preferable. However, in the case of large mass fluxes $\dot{m} > 1500$ kg/m^{2}s, one must expect rather large errors, when using this procedure. The method of Baroczy [13.13], which is also to be treated later and which is valid for substances other than steam-water flows, has proven here to be more precise.

13.4.3.2 The Procedure of Martinelli and Nelson

Because the Lockhart–Martinelli multipliers were determined chiefly from experiments at atmospheric pressure, Martinelli and Nelson [13.12] extended the region of validity of the procedure up to the critical pressure by using data for steam-water flow in evaporator tubes. They assumed turbulent flow for both phases and then determined an equation between Φ_L and X_{tt} close to the critical pressure, where the differences between the viscosities and densities of both phases were very slight. In addition, they adopted the value $\Phi_L(X_{tt})$ for atmos-

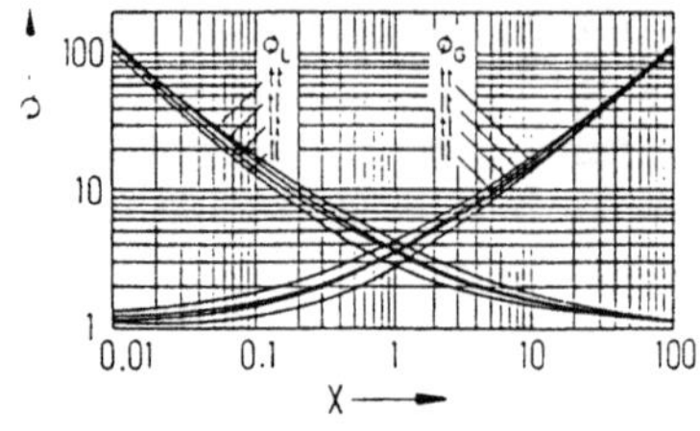

Fig. 13.12. Derivation of the frictional pressure drop according to Lockhart and Martinelli [13.8]

ϕ_L, ϕ_G: Two-phase multiplier according to Eqs. (13.29) and (13.30)

X: Lockhart–Martinelli parameter according to Eq. (13.46)

phases were very slight. In addition, they adopted the value $\Phi_L(X_{tt})$ for atmospheric pressure from Lockhart and Martinelli according to Fig. 13.12. From test data, they then determined the pattern of the function $\Phi_L(X_{tt})$ for different pressures between both limiting values. Finally, the group of functions obtained in this manner was converted with the pressure p as a parameter, and the already defined parameter Φ^2_{LO} was established:

$$\left(\frac{dp}{dz}\right)_f = \Phi^2_{LO}\left(\frac{dp}{dz}\right)_{LO}, \tag{13.59}$$

where $(dp/dz)_{LO}$ is the pressure drop of the entire mass flow rate, which is calculated with the properties of the liquid

$$\left(\frac{dp}{dz}\right)_{LO} = -\zeta_{LO}\frac{1}{d}\frac{\dot{m}^2}{2\rho}. \tag{13.60}$$

The friction factor $\zeta_{LO}(Re_{LO})$ depends upon the Reynolds number $Re_{LO} = \dot{m}d/\eta_L$.

Together with Eq. (13.46) there follows

$$\frac{(dp/dz)_L}{(dp/dz)_{LO}} = \frac{\Phi^2_L}{\Phi^2_{LO}} = \frac{\zeta_L}{\zeta_{LO}}(1-x^*)^2. \tag{13.61}$$

Assuming that the Blasius friction factor is valid,

$$\zeta_L = c \bigg/ \left(\frac{\dot{m}d(1-x^*)}{\eta_L}\right)^n$$

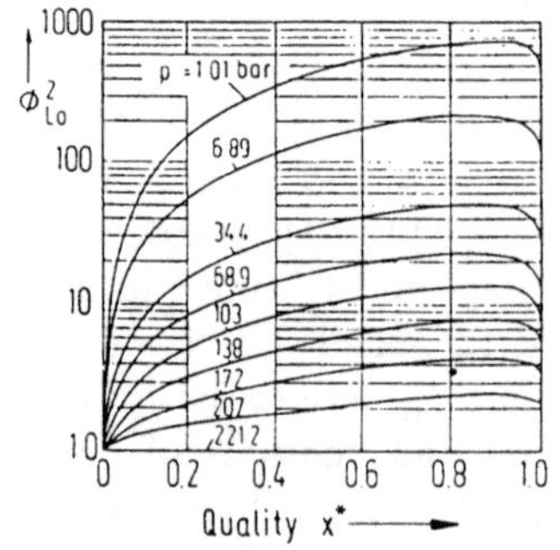

Fig. 13.13. Derivation of the frictional pressure drop in steam-water flows according to Martinelli and Nelson [13.12]

ϕ_{LO}: Two-phase multiplier according to Eq. (13.34)

and

$$\zeta_{LO} = c \bigg/ \left(\frac{\dot{m}d}{\eta_L}\right)^n$$

there results from Eq. (13.61) the relation between Φ_L and Φ_{LO}

$$\Phi_L^2 = \Phi_{LO}^2 (1 - x^*)^{2-n}. \tag{13.62}$$

The findings of Martinelli and Nelson are depicted in Fig. 13.13. A comparison with test data shows that the procedure becomes unreliable only at very high mass fluxes $\dot{m} > 1500\,\mathrm{kg/m^2\,s}$.

13.4.3.3 The Procedure of Baroczy and Baroczy–Chisholm

In the procedure of Baroczy [13.13], the mass flux and the type of fluid as well as the pressure are allowed for in the parameters. As an independent variable, the following parameters are introduced:

$$\frac{\Phi_{GO}^2}{\Phi_{LO}^2} = \frac{1}{\Gamma_0^2} = \frac{\rho_G}{\rho_L}\left(\frac{\eta_L}{\eta_G}\right)^{0.2}. \tag{13.63}$$

The quantities Φ_{GO} and Φ_{LO} are defined here by Eqs. (13.34) and (13.35). The index of the viscosity ratio in Eq. (13.63) follows from the assumption that in the Blasius equation for n the value 0.2 can be entered for turbulent flow in smooth tubes.

Also, according to Baroczy, the frictional pressure drop is to be determined using the two-phase multiplier Φ_{LO}^2, which is depicted in Fig. 13.14 as a function of $1/\Gamma_0^2$ with the quality x^* as parameter. If only liquid $x^* = 0$ is flowing, then according to Eq. (13.34), $\Phi_{LO}^2 = 1$, whereas during pure vapor flow, $x^* = 1$, $\Phi_{GO}^2 = 1$ and $\Phi_{LO}^2 = \Gamma_0^2$.

Figure 13.14 is valid only for a mass flux of $\dot{m} = 1356\,\mathrm{kg/m^2s}$. The conversion to other mass fluxes is made possible by a diagram, Fig. 13.15, which was also

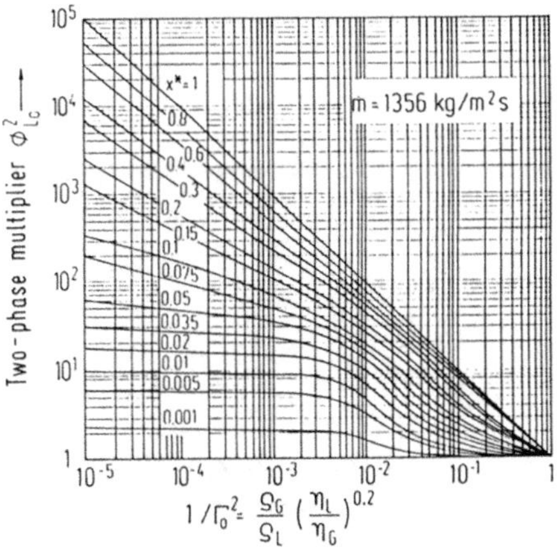

Fig. 13.14. Derivation of the frictional pressure drop according to Baroczy [13.13]

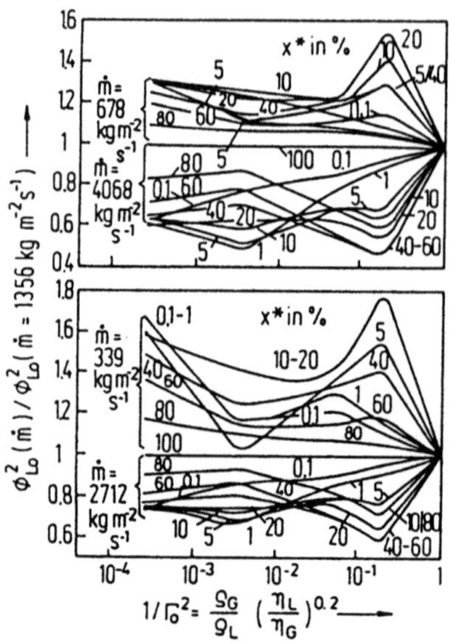

Fig. 13.15. Correction factors according to Baroczy [13.13]

developed by Baroczy, and from which correction factors for other values of $\dot{m}$ as a function of x^* can be taken. Although these curves were calculated empirically and run unsystematically, most test results are reproduced more accurately than by the previously treated procedures.

Because it is difficult to determine by interpolation the correction factor for other values of the mass flux, Chisholm [13.14] reproduced the values of Baroczy by an empirical equation

$$\Phi_{LO}^2 = 1 + (\Gamma_0^2 - 1)[Bx^{*0.9}(1 - x^*)^{0.9} + x^{*1.8}] \tag{13.64}$$

with Γ_0 according to Eq. (13.63). The parameter B is dependent upon the combination of property values Γ_0 and upon the mass flux, in accordance with the following table.

Table 13.1. Values for the Parameter B in Eq. (13.64)

Γ_0	$\dot{m}$ in kg/m²s	B
$\leqq 9.5$	< 500	4.8
	500 to 1900	$2400/\dot{m}$
	$\geqq 1900$	$55/\dot{m}^{0.5}$
9.5 to 28	$\leqq 600$	$520/(\Gamma_0 \dot{m}^{0.3})$
	> 600	$21/\Gamma_0$
$\geqq 28$	—	$15000/(\Gamma_0^2 \dot{m}^{0.5})$

In addition, Chisholm slightly modified the values of Baroczy, because the comparison with test data in certain regions according to Baroczy yielded somewhat too small frictional pressure drops.

13.4.3.4 The Procedure of Theissing

The goal of the procedure of Theissing [13.15] was not only to calculate the pressure drop of two-phase flows in straight channels but also to calculate the drop in return bends, valves, packed columns, heat exchangers, etc. and to develop a procedure that is valid for immiscible and non-Newtonian liquids. The procedure is like that of Lockhart and Martinelli. It contains as limiting cases the pressure drop of the single-phase flow and fulfills in addition the requirement that the value of the single-phase flow be yielded in the case of identical property values of the phases and for vanishing interfacial tension, at the critical point, for example. The increase of the pressure drop by momentum exchange between the phases is allowed for by an interaction parameter φ.

An approximation equation developed on the basis of these assumptions contained additionally three empirical constants, which were so developed that the pressure drop of a gas-liquid flow in the tube agreed with that according to Lockhart and Martinelli.

As a result, Theissing obtained the following empirical equation for the frictional pressure drop:

$$\Delta p_{\mathrm{f}} = \left[\Delta p_{\mathrm{LO}}^{1/n\varphi} \left(\frac{\dot{M}_{\mathrm{L}}}{\dot{M}} \right)^{1/\varphi} + \Delta p_{\mathrm{GO}}^{1/n\varphi} \left(\frac{\dot{M}_{\mathrm{G}}}{\dot{M}} \right)^{1/\varphi} \right]^{n\varphi}. \tag{13.65}$$

Δp_{LO} and Δp_{GO} signify the pressure drops of liquid and gas flows, respectively, at the total mass flow rate $\dot{M} = \dot{M}_{\mathrm{L}} + \dot{M}_{\mathrm{G}}$.

The exponents φ and n are to be evaluated as follows: There is

$$\varphi = 3 - 2\left(\frac{2\sqrt{\rho_{\mathrm{L}}/\rho_{\mathrm{G}}}}{1 + \rho_{\mathrm{L}}/\rho_{\mathrm{G}}} \right)^{0.7/n} \tag{13.66}$$

and

$$n = \frac{n_1 + n_2(\Delta p_{\mathrm{G}}/\Delta p_{\mathrm{L}})^{0.1}}{1 + (\Delta p_{\mathrm{G}}/\Delta p_{\mathrm{L}})^{0.1}} \tag{13.67}$$

with

$$n_1 = \frac{\ln(\Delta p_{\mathrm{L}}/\Delta p_{\mathrm{LO}})}{\ln \dot{M}_{\mathrm{L}}/\dot{M}}, \quad n_2 = \frac{\ln \Delta p_{\mathrm{G}}/\Delta p_{\mathrm{GO}}}{\ln \dot{M}_{\mathrm{G}}/\dot{M}}. \tag{13.68}$$

Δp_{L} and Δp_{G} are the pressure drops of liquid and gas flows, respectively, at the mass flow rates $\dot{M}_{\mathrm{L}}$ and $\dot{M}_{\mathrm{G}}$.

A review using 176 test data for gas-liquid flows in a rough tube yielded a mean quadratic deviation in the pressure drop of 23%. This procedure thus yields values at least equal in accuracy in the tested region as does that of Baroczy and Chisholm. It is, however, more valid in general. The pressure drop in a gas-liquid flow in packed columns is also satisfactorily reproduced. A comparison with test data for three-phase flows and non-Newtonian liquids is still unavailable.

13.4.3.5 The Procedure of Chawla

Whereas the previously treated procedures of the Lockhart-Martinelli type are not based upon assumptions concerning the form of flow, Chawla [13.16] assumes an annular flow and determines the two-phase frictional pressure drop based upon the exchange of momentum between the two phases and between the fluid and the tube wall. Applying the Blasius friction factor, the frictional pressure drop of the gaseous core flow is

$$\left(\frac{dp}{dz}\right)_G = -\zeta_G \frac{\rho_G w_G^2}{2d_G} = -\frac{0.3164}{(\rho_G w_G d_G/\eta_G)^{1/4}} \frac{\rho_G^2 w_G^2}{2d_G \rho_G}. \tag{13.69}$$

Here, an equivalent tube diameter according to

$$A_G = d_G^2 \frac{\pi}{4}$$

is ascribed to the gas cross section A_G. With $\varepsilon = A_G/A = d_G^2/d^2$, Eq. (13.69) converts into

$$\left(\frac{dp}{dz}\right)_G = -\frac{0.3164\varepsilon^{1/8}}{(\rho_G w_G \varepsilon d/\eta_G)^{1/4}} \frac{\rho_G^2 w_G^2}{2\varepsilon^{1/2} d \rho_G}.$$

Here is, according to Eq. (13.14)

$$\rho_G w_G \varepsilon = x^* \dot{m},$$

and thus

$$\left(\frac{dp}{dz}\right)_G = -\frac{0.3164}{(\dot{m}d/\eta_G)^{1/4}} \frac{\dot{m}^2 x^{*7/8}}{2d\rho_G} \frac{1}{\varepsilon^{19/8}}. \tag{13.70}$$

The volumetric content ε of the vapor is obtained from Eq. (13.16)

$$\frac{1}{\varepsilon} = 1 + \frac{1-x^*}{x^*} s \frac{\rho_G}{\rho_L}.$$

Thus, Eq. (13.70) converts into

$$\left(\frac{dp}{dz}\right)_G = -\frac{0.3164}{(\dot{m}d/\eta_G)^{1/4}} \frac{\dot{m}^2 x^{*7/8}}{2d\rho_G} \left(1 + \frac{1-x^*}{x^*} s \frac{\rho_G}{\rho_L}\right)^{19/8}. \tag{13.71}$$

Chawla used Eq. (13.71) established for the frictional pressure drop of the gas core flow for the calculation of the total pressure drop. The slip factor is thus no longer a true velocity ratio but rather a quantity allowing for the exchange of momentum between the two phases, the influence of the different forms of flow of the two phases, and the influence of the roughness of the heating surface. In order to be able to encompass these influences and to determine the frictional pressure drop $(dp/dz)_f$, the slip factor s is replaced by a two-phase flow parameter ε_C

$$\left(\frac{dp}{dz}\right)_f = -\frac{0.3164}{(\dot{m}d/\eta_G)^{1/4}}\frac{\dot{m}^2 x^{*7/8}}{2d\rho_G}\left(1+\frac{1-x^*}{x^*}\varepsilon_C\frac{\rho_G}{\rho_L}\right)^{19/8}. \tag{13.72}$$

Using test data for technically rough tubes with relative roughnesses

$$10^{-6} < k/d < 10^{-3}$$

there resulted

$$\frac{1}{\varepsilon_C^3} = \frac{1}{\varepsilon_{C_1}^3} + \frac{1}{\varepsilon_{C_2}^3} \tag{13.73}$$

with

$$\log \varepsilon_{C_1} = 0.960 + \log B$$

and

$$\log \varepsilon_{C_2} = (0.168 - 0.055 \log(k/d)) \log B - 0.67$$

with

$$B = \frac{1-x^*}{x^*}(Re_L Fr_L)^{-1/6}\left(\frac{\rho_L}{\rho_G}\right)^{-0.9}\left(\frac{\eta_L}{\eta_G}\right)^{-0.5},$$

where Re_L is the Reynolds number and Fr_L is the Froude number of the liquid phase:

$$Re_L = \frac{\dot{m}(1-x^*)d}{\eta_L}, \quad Fr_L = \frac{\dot{m}^2(1-x^*)^2}{\rho_L^2 g d}.$$

For the range of validity of his method of calculation, Chawla gives the following values:

$$0.006\,m \leqq d \leqq 0.154\,m;\ 5.9\cdot 10^{-6} < k/d < 6.8\cdot 10^{-2};\ 10^2 \leqq Re_L < 3\cdot 10^5;$$
$$10^{-5} \leqq Fr_L \leqq 10^2;\ 30 \leqq \rho_L/\rho_G \leqq 850;\ 40 \leqq \eta_L/\eta_G \leqq 7000;\ 10^{-4} \leqq x^* \leqq 0.98.$$

For the VDI Wärmeatlas [13.17], Chawla has recently worked out a modification of this computation procedure, which is valid not only for annular flows but also for dispersed flows.

13.4.3.6 The Equation of Lombardi and Pedrocchi

Based upon an investigation of the determining values of influence for the frictional pressure drop in vertical two-phase flows, Lombardi and Pedrocchi [13.18] found a very simple empirical correlation for practical use:

$$\left(\frac{dp}{dz}\right)_f = -k\dot{m}^n\sigma^{0.4}/(d^{1.2}\rho^{0.866}) \tag{13.74}$$

where ϱ is the density of the homogeneous two-phase mixture according to

Eq. (13.40). In Eq. (13.74), σ signifies the surface tension and d the diameter of the tube.

The constant k and the exponent n assume for flow in tubes the values $k = 0.83$ and $n = 1.4$. The values $k = 0.213$ and $n = 1.6$ are valid for longitudinal flow in tube bundles as well as for annular channels. These values are to be entered in SI units ($\dot{m}$ in kg/m^{2}s, ρ in kg/m^3, σ in N/m, d in m; one obtains $(dp/dz)_f$ in N/m^2). Equation (13.74) is based upon test data in the following regions:

$$500\,\mathrm{kg/m^2s} \leqq \dot{m} \leqq 5000\,\mathrm{kg/m^2s};\, 0.005\,\mathrm{m} \leqq d \leqq 0.025\,\mathrm{m};$$

$$20 \cdot 10^{-3}\,\mathrm{N/m} \leqq \sigma \leqq 80 \cdot 10^{-3}\,\mathrm{N/m};$$

$$0.01 < x^* < 0.98 \text{ and } 15 < \rho_L/\rho_G < 100.$$

The equation reproduces approximately 87% of the test data used for the determination of the constants and exponents with a relative error of $\pm 18\%$.

13.4.3.7 Summarizing Remarks on the Calculation of the Frictional Pressure Drop

In addition to those discussed, there are numerous further procedures for the calculation of the frictional pressure drop. By using banks of test data, one has tried, therefore, to compare the accuracy of reproduction of the individual procedures. With the aid of such statistical investigations one can determine which procedures yield the smallest error of deviation; one should check from case to case, however, whether a procedure, which reproduces the entirety of the data rather poorly, is not actually more accurate for the special case in hand. Therefore, in the treatment of the procedures discussed in detail, reference is always made to their specific range of applicability.

As comparative studies [13.19 to 13.21] have shown, all of the known data collectively are best reproduced on the average by the procedures of Baroczy, Baroczy and Chisholm, and Lombardi and Pedrocchi; in every case the mean quadratic deviation in pressure drop is 30 to 35%. One must also bear in mind that comparative data frequently deviate from one another by more than $\pm 20\%$. This is due primarily to the fact that the frictional pressure drop is most often calculated by subtracting the acceleration pressure drop and the geodetic pressure drop from the total pressure drop. However, the acceleration and geodetic pressure drops can already be compromised by significant errors, because the void fraction is not known precisely.

A considerable improvement in the reproduction of test data by equations with a large range of validity is, therefore, scarcely to be expected. Only in narrowly defined parameter regions can more precise results be achieved by special flow models.

13.5 The Different Regions of Heat Transfer

We shall consider a vertical evaporator tube, which is uniformly heated over the entire length and to which subcooled liquid is supplied from below. Let the heat flux q be low and the tube long enough that the supplied liquid can be completely evaporated. Figure 13.16 once again shows the different flow patterns. In addition, the various regions of heat transfer and the liquid and wall temperatures are plotted.

As long as the wall temperature remains below the temperature necessary for the formation of vapor bubbles, heat is transferred by single-phase, forced flow. If the wall is sufficiently superheated, vapor bubbles can form there, although the liquid core is still subcooled. One speaks of the region of subcooled boiling. The wall temperature is nearly constant in this region and is a few degrees Kelvin above the saturation temperature. By definition, the transition to nucleate boiling is put at that point, at which the liquid on the average reaches the saturation temperature and the thermodynamic quality becomes $x^* = 0$. Actually at this point, as indicated in Fig. 13.16, because of the radial temperature profile, the liquid in the core is still subcooled, while simultaneously vapor is already present, so that the mean enthalpy is equal to that of the saturated liquid. As already explained in section 13.3, saturation temperature in the core does not occur until further downstream from the point $x^* = 0$.

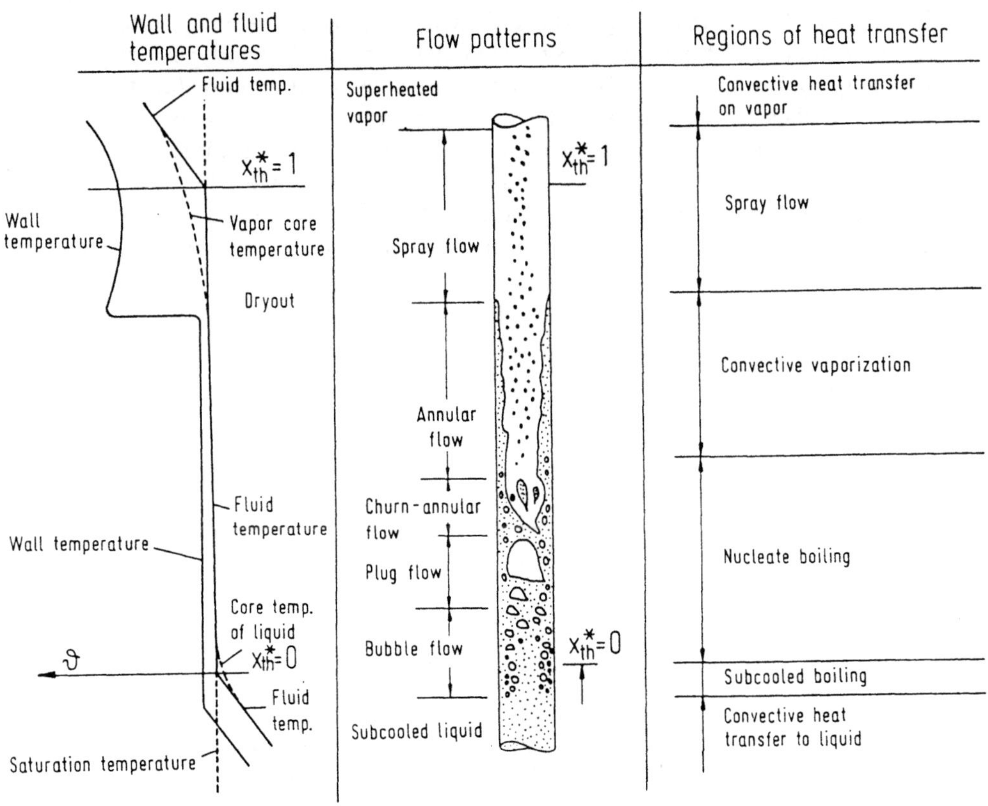

Fig. 13.16. Wall and fluid temperatures, flow patterns, and the accompanying heat transfer ranges in a vertical, heated tube

In the region of nucleate boiling, the heat transfer is determined chiefly by the formation of vapor bubbles and only to a lesser degree by convection. This region encompasses the actual bubble, plug, churn-annular, and parts of annular flow. The quality of the flow increases constantly downstream and at sufficiently high quality converts into an annular flow with a liquid film at the wall and vapor in the core, which entrains drops of liquid. It is characteristic that vapor bubbles are formed in the entire region of the nucleate boiling at the wall. In the region of annular flow, however, the liquid film becomes so thin and its thermal resistance so small, that the liquid near the wall is no longer sufficiently superheated and the bubble formation at the wall is suppressed to a great extent. The heat is conducted mainly by the liquid film, which evaporates at its surface. Heat is transferred by *convective vaporization.*

As soon as the liquid film at the wall is completely evaporated, the temperature of a wall heated with a constant heat flux increases. The transition is called dryout. One comes into the region of spray flow, which is followed after the complete evaporation of all the liquid droplets by a region, in which heat is transferred by convection to the vapor.

The heat transfer regions can be of very different extent, depending upon the size of the heat and mass fluxes, as is seen in Fig. 13.17. There, on the right, the different heat transfer regions are drawn in on a vertical evaporator tube. The dependence of the heat transfer coefficient on the thermodynamic quality is shown by the curve q_1. In the region of subcooled boiling (a), the heat transfer coefficient increases with the quality, because a part of the formed vapor bubbles in the core condense again and the liquid is thereby set into oscillatory motion, which leads to an increase in the heat transfer with the quality. The coefficient remains constant once established boiling is reached. In region (b), the liquid reaches on the average saturation temperature. Bubbles formed at

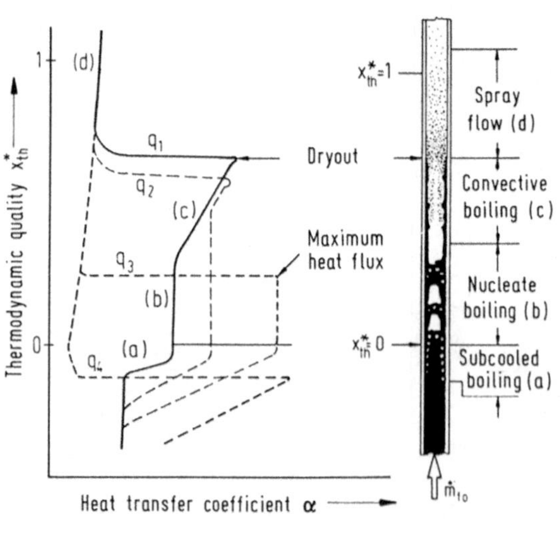

Fig. 13.17. Trends of the heat transfer coefficient in a vertical evaporator tube, $\dot{m} = \text{const}$, $q = \text{const}$ with $q_1 < q_2 < q_3 < q_4$

the wall no longer condense, and the heat transfer coefficient is essentially independent of the vapor content. In the region of convective boiling (c) scarcely any more bubbles are formed at the wall; heat is conducted by the liquid film, and this evaporates increasingly downstream. The film thickness decreases, and, thus, the heat transfer coefficient increases with the quality. Shortly before the heating surface dries out, the heat transfer coefficient reaches its greatest value and then drops off quickly, because the wall temperature rises after the dryout. In the adjoining region of spray flow (d), the vapor is laden with liquid drops, which increasingly evaporate and increase the flow velocity. The heat transfer coefficient, therefore, increases in the region of spray flow with the quality, and, after complete evaporation, converts into the almost constant coefficient of the convective heat transfer to pure vapor.

The curve of q_2 shows the influence of a larger heat flux. The subcooled boiling begins earlier, and the heat transfer coefficient in the region of nucleate boiling is larger. It remains unchanged in the region of convective vaporization, and the heating surface dries out at a lower quality. A further increase of the heat flux, curve q_3, also results in the subcooled boiling setting in earlier and the heat transfer in the region of nucleate boiling being larger. The bubble density adjacent to the wall can then become so great that the heat transfer sinks drastically before reaching the region of convective vaporization. As in the case of boiling in free convection, the supplied heat flux at a definite point of the tube coincides with the peak heat flux, which in this case is dependent upon the vapor content. Subsequently, the heat transfer coefficient decreases, and the temperature rises rapidly. Upon further increase of the heat flux, curve q_4, the maximum heat flux can occur even in the region of subcooled boiling. It is thus seen that low heat transfer coefficients occur after the surpassing of the peak heat flux at either low quality or high quality, i.e., in the region of film boiling and in the region of spray flow.

The previous considerations are valid when assuming a constant mass flux. In the region of subcooled boiling and of nucleate boiling, the heat transfer coefficient is only weakly dependent upon the mass flux. In the region of convective boiling, the heat transfer coefficient increases with the mass flux, Fig. 9.5. The dryout already sets in when the quality is rather small. In the following sections, the different heat transfer regions will be treated individually.

13.6 Subcooled Boiling

If a liquid with less than the boiling temperature enters a heated channel, Fig. 13.18, the heat is initially transferred by forced convection. The wall temperature and the mean liquid temperature increase. The transferred heat flow rate is calculated for the flow in a tube from

$$q d\pi z = \dot{M} c_{pL}[\vartheta_L(z) - \vartheta_1] = \alpha d\pi z[\vartheta_w - \vartheta_L(z)], \tag{13.75}$$

if $\vartheta_L(z)$ is the mean liquid temperature at a position z of the tube. This yields the wall temperature

$$\vartheta_w(z) = \vartheta_L(z) + \frac{q}{\alpha} = \vartheta_1 + q\left(\frac{d\pi z}{\dot{M}c_{pL}} + \frac{1}{\alpha}\right). \tag{13.76}$$

In the single-phase region of flow the wall temperature increases linearly with the tube length, if the heat flux is constant. "Subcooled boiling" sets in, if the wall in comparison to the saturation temperature is sufficiently superheated, so that vapor bubbles form. The wall temperature may fall if only very small cavities are available for nucleation; see Fig. 13.18. Immediately before the inception of subcooled boiling, let the wall be superheated by $\Delta\vartheta = \vartheta_w - \vartheta_s$ in comparison to the saturation temperature, Fig. 13.18. The length z_1 of the tube section in which there is only single-phase flow is given by

$$\vartheta_w = \vartheta_s + \Delta\vartheta = \vartheta_1 + q\left(\frac{d\pi z_1}{\dot{M}c_{pL}} + \frac{1}{\alpha}\right) \tag{13.77}$$

with the initial subcooling $\Delta\vartheta_u = \vartheta_s - \vartheta_1$ at

$$z_1 = \left(\frac{\Delta\vartheta_u + \Delta\vartheta}{q} - \frac{1}{\alpha}\right)\frac{\dot{M}c_{pL}}{d\pi}. \tag{13.78}$$

The position z_2 at which the mean liquid temperature is equal to the saturation temperature $\vartheta_L(z_2) = \vartheta_s$ results from Eq. (13.75) at

$$.z_2 = \frac{\Delta\vartheta_u \dot{M}c_{pL}}{d\pi q}. \tag{13.79}$$

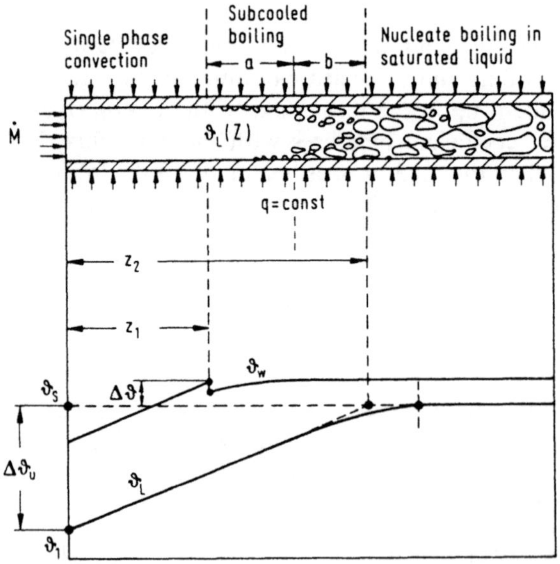

Fig. 13.18. Wall and mean liquid temperature during subcooled boiling. Range a: partial subcooled boiling, heat transfer determined by convection; range b: completely developed subcooled boiling, heat transfer determined by bubble formation

There is subcooled boiling along the section

$$z_2 - z_1 = \left(\frac{1}{\alpha} - \frac{\Delta\vartheta}{q}\right)\frac{\dot{M}c_{pL}}{d\pi}. \tag{13.80}$$

This does not occur, if the mean liquid temperature in the region of the single-phase convection is already equal to the saturation temperature. This gives $q = \alpha(\vartheta_w - \vartheta_s)$ or, according to Eq. (13.80), $z_2 - z_1 = 0$.

In the region of subcooled boiling, the heat transfer is significantly better than in the region of single-phase convection. The vapor bubbles formed at the wall transport latent energy into the colder core flow, in which they condense. In addition, the convection in the liquid is strengthened by the rapid growth and condensation of bubbles. As long as there are few bubbles present, a substantial part of the heat is transferred by forced convection to the liquid between the bubbles. One has a region partially with nucleate boiling and partially with convective heat transfer. Further downstream, the bubble density becomes greater and the heat transfer is determined to a great extent by the formation of vapor bubbles, as is shown in Fig. 13.18. The vapor bubbles glide along the wall and condense in the core of the flow until saturation temperature is reached in the core. This region is then called region of completely developed subcooled boiling.

13.6.1 Inception of Subcooled Boiling

For vapor bubbles to be able to form at the wall, the wall temperature must be greater than the boiling temperature. According to Eq. (13.76), there thus holds

$$\vartheta_w = \vartheta_1 + q\left(\frac{d\pi z}{\dot{M}c_{pL}} + \frac{1}{\alpha}\right) > \vartheta_s \tag{13.81}$$

or, if one introduces the subcooling $\vartheta_s - \vartheta_1 = \Delta\vartheta_u$ in the inlet cross section,

$$\vartheta_w - \vartheta_1 = q\left(\frac{d\pi z}{\dot{M}c_{pL}} + \frac{1}{\alpha}\right) > \vartheta_s - \vartheta_1 = \Delta\vartheta_u. \tag{13.82}$$

From this equation results that position z, at which the wall temperature begins to exceed the saturation temperature. The more the liquid is subcooled, the further downstream this position lies. It shifts at the specified subcooling downstream, if the supplied heat flux is small and the mass flow rate is large.

If the heat flux q is plotted versus the subcooling, Fig. 13.19, then one obtains a straight line with the slope

$$\frac{q}{\Delta\vartheta_u} = \left(\frac{d\pi z}{\dot{M}c_{pL}} + \frac{1}{\alpha}\right)^{-1}. \tag{13.83}$$

On it lie all the points at which the wall temperature is equal to the saturation temperature. Below the straight line is the region, in which the added heat flux is not sufficient or the mass flow rate is too large in order to cause the wall

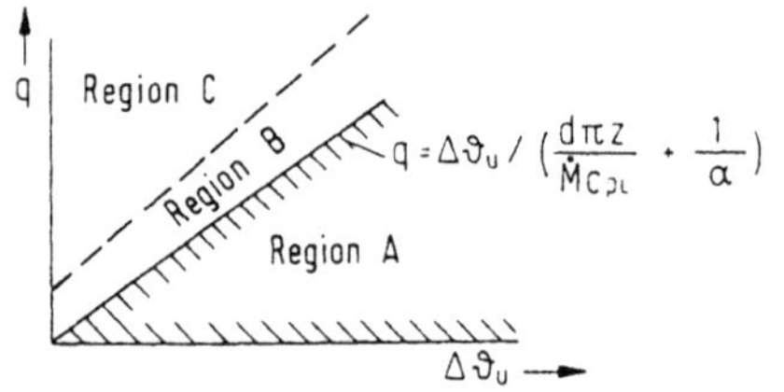

Fig. 13.19. Bubble formation as a function of the subcooling $\Delta\vartheta_u$ in the inlet cross section. Region A: no bubble formation; region B: partial subcooled boiling; region C: completely developed subcooled boiling

temperature to exceed the boiling temperature, region A in Fig. 13.19. Above the straight line, the added heat fluxes are so large and the mass flow rates so small, that boiling can set in. To be sure, a slight exceeding of the boiling temperature by the wall temperature does not lead to an intensive, i.e., fully developed nucleate boiling. Vapor bubbles form only at single nucleation sites of the wall, and the heat transfer is determined for the most part by single-phase convection in the liquid, region B in Fig. 13.19. Not until sufficiently large heat fluxes, region C, does the wall temperature lie clearly above the boiling temperature. Enough nucleation sites are activated, and the heat transfer is determined mainly by the nucleate boiling.

According to Fig. 13.18, boiling first sets in when the wall is superheated by a temperature $\Delta\vartheta$ in comparison to the boiling temperature. In the following, this excess temperature is to be determined; with it the wall temperature, and, because of Eq. (13.78), also the position z_1 at the inception of boiling are fixed. We shall consider a single vapor bubble, Fig. 13.20, which has formed at the heated wall and extends into the temperature boundary layer. Let the bubble still be so small that one may approximate the temperature of the liquid in the vicinity of the bubble by a straight line

$$\vartheta_L(z) = \vartheta_w - \frac{qz}{\lambda_L}. \tag{13.84}$$

In order that the bubble tip not condenses, the liquid there must be superheated, as the considerations concerning the origin and the growth of vapor bubbles showed, Sect. 11.1, because a curved surface is in equilibrium only with a superheated liquid. According to Hsu's work for pool boiling [13.22], it is assumed that the superheating of the liquid at the bubble tip must be at least as large as the superheating of a bubble which is in equilibrium with a liquid of constant temperature. We have already calculated this in Chap. 10. It is

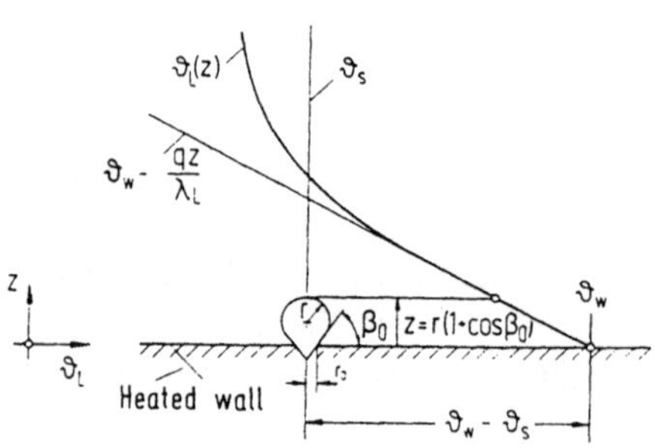

Fig. 13.20. Temperatures in a subcooled liquid

derived from Eq. (10.12) for a spherically shaped bubble of the radius r from

$$r = \frac{2\sigma T_s}{\varrho'' \Delta h_v \Delta\vartheta}(1 + \Delta\vartheta\omega^*),$$

in which the factor

$$\omega^* = \frac{\Delta\varrho}{\varrho'\sigma}\frac{d}{d\vartheta}\left(\frac{\varrho'\sigma}{\Delta\varrho}\right) = \frac{d}{d\vartheta}\ln\left(\frac{\varrho'\sigma}{\Delta\varrho}\right)$$

(SI unit 1/K) is to be formed at the saturation temperature T_s of the flat phase interface associated with the pressure p_0. The superheat $\Delta\vartheta$ is the difference between the liquid temperature ϑ_L and the saturation temperature ϑ_s with a flat phase interface. The superheat is given by

$$\Delta\vartheta = \frac{2\sigma T_s}{\varrho'' \Delta h_v r}\frac{1}{1 - \dfrac{2\sigma T_s}{\varrho'' \Delta h_v r}\omega^*},$$

or we have

$$\vartheta(r) = \vartheta_s + \frac{2\sigma T_s}{\varrho'' \Delta h_v r}\frac{1}{1 - \dfrac{2\sigma T_s}{\varrho'' \Delta h_v r}\omega^*}. \tag{13.85}$$

At the bubble tip of a spherical bubble of the radius r sitting on the wall in accordance with Fig. 13.20, according to Hsu, the temperature of the liquid $\vartheta_L(z)$ according to Eq. (13.84) must be at least equal to the temperature $\vartheta(r)$ according to Eq. (13.85), which is necessary for the existence of a bubble. Here we assume that a spherical bubble of the radius r grows according to Fig. 13.20 out of a depression of mouth radius r_0. The wetting angle of the sitting bubble shall be β_0. From the wall, the bubble tip is at a distance of

$$z = (1 + \cos\beta_0)r = c_1 r \tag{13.86}$$

and it is

$$r_0 = \sin\beta_0 \cdot r = c_2 r. \tag{13.87}$$

If the bubble assumes a hemisphere shape we have $\beta_0 = \pi/2$, $c_1 = 1$ and $c_2 = 1$.

In Fig. 13.21 are depicted the temperatures $\vartheta_L(z)$ according to Eq. (13.84) and the temperatures $\vartheta(r)$ in conformity with Eq. (13.85). With greater heat flux $q_2 > q_1$ and at a maintained wall temperature ϑ_w, the difference between the wall and liquid temperatures at a definite point increases. For a heat flux q_1, according to Fig. 13.21 the liquid temperature $\vartheta_L(z)$ is larger than the necessary minimum temperature $\vartheta(r)$ only between points A and B. A bubble of the radius $r > r_A$ would extend with its tip into a liquid that is too cold and could not continue to grow. A smaller bubble of the radius $r < r_B$ needs for further growth a greater superheating than is present in the liquid and would therefore condense. Thus only bubbles of a radius of

$$r_B \leqq r \leqq r_A$$

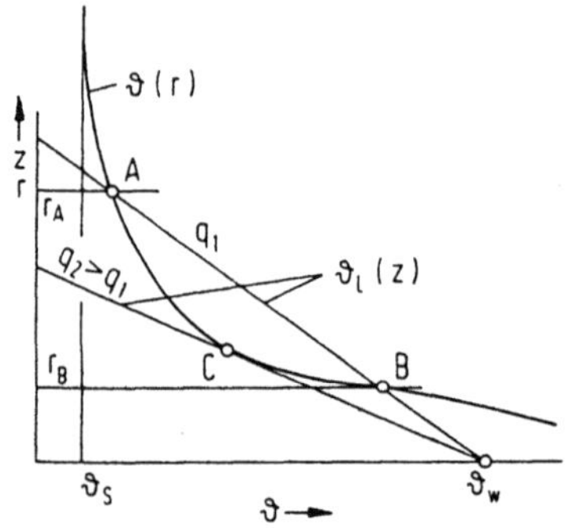

Fig. 13.21. Conditions for the growth of bubbles in a subcooled liquid

are viable. With increasing heat flux, the interval r_A, r_B of the viable bubbles becomes ever smaller until at a definite heat flux, in Fig. 13.21 the heat flux q_2, the points A and B coincide in point C, because the straight line $\vartheta_L(z)$ just touches the curve $\vartheta(r)$. Then, only a bubble of a quite definite *critical radius* r_c is still viable. Smaller bubbles condense again, because they need a higher superheating than is present in the liquid. Larger bubbles cannot continue to grow, because the liquid is too cold at the bubble tip. If the heat flux were to be increased beyond q_2, then for a given wall temperature the straight line $\vartheta_L(z)$ would lie below the curve $\vartheta(r)$. The liquid would be superheated in a thin wall-adjacent layer, in which no bubbles could be formed, because the smallest viable bubble is larger than the superheated liquid layer.

At the point of contact C, let the bubble radius be r_c. We stipulate $z = c_1 r_c$, for the thickness of the superheated liquid layer in conformity with Eq. (13.86), see Fig. 13.20. There then holds in the point of contact $z = c_1 r_c$

$$\vartheta_L = \vartheta \quad \text{and} \quad \frac{d\vartheta_L}{dz} = \frac{d\vartheta}{dr}. \tag{13.88}$$

We calculate the critical bubble radius by first equating the derivatives. This then yields

$$\frac{q}{\lambda_L} = \frac{1}{a_0\left(\frac{r_c}{a_0} - \omega^*\right)^2} \tag{13.89}$$

with the abbreviation

$$a_0 = \frac{2\sigma T_s}{\varrho'' \Delta h_v} \tag{13.89a}$$

and ω^*, from which one obtains the critical radius

$$r_c = a_0\left[\omega^* + \left(\frac{\lambda_L}{q a_0}\right)^{1/2}\right]. \tag{13.90}$$

There are fundamentally two solutions for r_c, because one would have to write $\pm(\lambda_L/qa_0)^{1/2}$ in Eq. (13.90). Because ω^* at ambient temperature is of the order of magnitude of 10^{-2} to $10^{-3}\,K^{-1}$ and $(\lambda_L/qa_0)^{1/2}$ is of the order of magnitude of 10^1 to $10^2\,K^{-1}$, a minus sign before the term $(\lambda_L/qa_0)^{1/2}$ would lead to negative bubble radii, which is physically impossible. The expression $(\lambda_L/qa_0)^{1/2}$ must, therefore, be placed in Eq. (13.90) with a positive sign. Written out, the equation for the critical bubble radius reads

$$r_c = \frac{2\sigma T_s}{\varrho'' \Delta h_v}\left[\frac{d}{d\vartheta}\ln\frac{\varrho'\sigma}{\Delta\varrho} + \left(\frac{\lambda_L}{qa_0}\right)^{1/2}\right]. \tag{13.91}$$

Because of Eq. (13.87), the critical size of pore, which is minimally necessary for bubble formation, is found to be

$$(r_0)_c = \sin\beta_0 r_c = c_2 r_c. \tag{13.92}$$

The expression (13.91) deviates from the value for the critical bubble radius found by Davis and Anderson [13.23], because these authors assumed in their derivations that the vapor pressure curve for the curved phase interface of the vapor bubble would run parallel to the vapor pressure curve of the flat phase interface, which strictly speaking is not accurate, as has been shown by the observations in Sect. 10.1; one should also remember in this regard Fig. 10.3. Equation (13.91) can be further simplified, if one considers that even at high pressures and large heat fluxes there is

$$\left(\frac{\lambda_L}{qa_0}\right)^{1/2} \gg \frac{d}{d\vartheta}\ln\left(\frac{\varrho'\sigma}{\Delta\varrho}\right).$$

For example, in water of 198.33 bar ($\vartheta_s = 365\,°C$) at a heat flux of $10^6\,W/m^2$ there is

$$\left(\frac{\lambda_L}{qa_0}\right)^{1/2} = 5.33\,K^{-1} \quad \text{and} \quad \frac{d}{d\vartheta}\ln\left(\frac{\varrho'\sigma}{\Delta\varrho}\right) = -0.12\,K^{-1}.$$

Thus, there holds to a good approximation

$$r_c = \left(\frac{2\sigma T_s \lambda_L}{\varrho'' \Delta h_v q}\right)^{1/2} = \left(\frac{a_0 \lambda_L}{q}\right)^{1/2}, \tag{13.93}$$

which agrees with the approximate equation found by Davis and Anderson [13.23].

By equating the temperatures at the point $z = c_1 r_c$ and $r = r_c$ one obtains from Eqs. (13.84) and (13.85)

$$\vartheta_w - \frac{qc_1 r_c}{\lambda_L} = \vartheta_s + \frac{a_0}{r_c}\frac{1}{1 - \frac{a_0}{r_c}\omega^*},$$

with a_0 according to Eq. (13.89a). From this follows

$$(\vartheta_w - \vartheta_s)_c = c_1 \frac{q r_c}{\lambda_L} + \frac{a_0}{r_c} \frac{1}{1 - \frac{a_0}{r_c}\omega^*}. \tag{13.94}$$

With the aid of Eq. (13.90) one determines from this the superheating of the wall necessary for bubble formation

$$(\vartheta_w - \vartheta_s)_c = c_1 \frac{q a_0}{\lambda_L}\omega^* + (c_1 + 1)\left(\frac{q a_0}{\lambda_L}\right)^{1/2}. \tag{13.95}$$

Because in this the first term on the right side can be disregarded in contrast to the second one, except for conditions close to the critical point, one determines the superheating for the inception of boiling, valid to a good approximation, to be

$$(\vartheta_w - \vartheta_s)_c = (c_1 + 1)\left(\frac{q a_0}{\lambda_L}\right)^{1/2}. \tag{13.96}$$

According to Eqs. (13.92) and (13.93), the critical size of pore minimally necessary for the formation of bubbles equals

$$(r_0)_c = c_2 r_c = c_2 \left(\frac{a_0 \lambda_L}{q}\right)^{1/2}.$$

The best agreement with data were with the analytical predictions of Bergles and Rohsenow [13.24], who approximated the barely viable bubbles situated at the cavity mouths by hemispheres. Then the wetting angle is $\beta_0 = \pi/2$, $c_1 = 1 + \cos\beta_0 = 1$ and $c_2 = \sin\beta_0 = 1$. The necessary minimum superheating for bubble growth in subcooled liquids is determined to be

$$(\vartheta_w - \vartheta_s)_c = 2\left(\frac{2 q \sigma T_s}{\varrho'' \Delta h_v \lambda_L}\right)^{1/2} \tag{13.97}$$

and the mouth radius of the smallest pores necessary for bubble formation to be

$$(r_0)_c = \left(\frac{2\sigma T_s \lambda_L}{\varrho'' \Delta h_v q}\right)^{1/2}. \tag{13.98}$$

Instead of Eq. (13.97), for water Bergles and Rohsenow derived a dimensional equation that accounts for the variable slope of the saturation line

$$(\vartheta_w - \vartheta_s)_c = \frac{5}{9}\left[\left(\frac{q}{1120}\right)^{0.463} p^{-0.535}\right]^{p^{0.0234}} \tag{13.97a}$$

which is valid in the pressure region

$$1.03\,\text{bar} \leqq p \leqq 138\,\text{bar}$$

with q in W/m^2, p in bar, and $(\vartheta_w - \vartheta_s)_c$ in K. The values calculated from this agree quite well with those from Eq. (13.97). Instead of the minimal superheating of the wall $(\vartheta_w - \vartheta_s)_c$, one can also state the highest admissible subcooling $(\vartheta_s - \vartheta_L(z))_c = \Delta\vartheta_u(z)$ of the liquid. For boiling to set in, this must not be surpassed. There must be

$$\vartheta_w - \vartheta_s \geqq (\vartheta_w - \vartheta_s)_c$$

with $(\vartheta_w - \vartheta_s)_c$ from Eq. (13.97). Valid also, therefore, is

$$\vartheta_w - \vartheta_s - (\vartheta_L(z) - \vartheta_s) \geqq (\vartheta_w - \vartheta_s)_c - (\vartheta_L(z) - \vartheta_s)$$

or

$$\vartheta_w - \vartheta_L(z) \geqq (\vartheta_w - \vartheta_s)_c + \Delta\vartheta_u .$$

With

$$q = \alpha(\vartheta_w - \vartheta_L(z))$$

the admissible subcooling follows at

$$\Delta\vartheta_u(z) \leqq \frac{q}{\alpha} - (\vartheta_w - \vartheta_s)_c . \tag{13.99}$$

According to the above considerations, two conditions must be fulfilled in order to have bubble formation in a subcooled liquid. The superheating of the wall must exceed certain minimal values in accordance with Eq. (13.97) and the heating surface must be covered with pores of a minimal size, Eq. (13.98). For water at 10 bar, $\vartheta_s = 179.88\,°\text{C}$, one obtains for a heat flux of $2\,\text{MW/m}^2$ the necessary minimal superheating at which boiling sets in from Eq. (13.97) $(\vartheta_w - \vartheta_s)_c = 6.63\,\text{K}$ (according to Bergles and Rohsenow, Eq. (13.97a), there results $(\vartheta_w - \vartheta_s)_c = 5.88\,\text{K}$) and one finds the smallest pore radius necessary for bubble formation from Eq. (13.98) as $(r_0)_c = 1.11\,\mu\text{m}$. If a sufficient number of pores of this size are present, then the formation of bubbles sets in at the stated superheating. If the heating surface is very smooth, or if the larger cavities have been flooded by a highly wetting liquid, the wall must be more strongly superheated in order for bubbles to form.

If the mouth radius of the largest pores present in sufficient number is known, the required superheating can be calculated. By multiplication of Eqs. (13.97) and (13.98) there results

$$(\vartheta_w - \vartheta_s)_c = \frac{1}{(r_0)_c} \frac{4\sigma T_s}{\varrho'' \Delta h_v} . \tag{13.100}$$

It should be stressed that the above equations lead only to rough approximations, because they were derived using very far-reaching assumptions. It was assumed, for example, that the superheating around a spherical bubble in the interior of a liquid is to coincide with that of the liquid of a bubble growing on the wall. It was further assumed concerning this bubble that it should be a hemisphere. As Bankoff [13.25] showed, liquids which wet especially well penetrate into the normally required pores with mouth radii $(r_0)_c$ between 1 and 5 μm, with the result that these pores cannot contribute to nucleation. The first vapor bubbles do not form until the wall reaches a sufficiently high superheat, and this can lie far above the calculated values. The vapor bubbles can stimulate bubble formation further downstream, so that there lower superheats are sufficient. As of now, the processes associated with this have by no means been adequately researched.

13.6.2 Partial Subcooled Boiling

After the beginning of the subcooled boiling at a point z_1, which can be calculated approximately from Eq. (13.78) with the aid of $\Delta\vartheta = \vartheta_w - \vartheta_s$ according to Eq. (13.97), only a few bubbles are formed at first, so that a substantial part of the heat flux is still transferred convectively to the liquid between the bubbles. One designates this process as partial subcooled boiling. With increasing wall temperature, the bubble density increases and the component of the convectively transferred heat flux decreases. With further increase of the bubble density, the component of the convectively transferred heat flux becomes negligible. One finds oneself in the region of the fully developed subcooled boiling, in which the flow velocity and the subcooling play almost no role, in contrast to partial subcooled boiling.

These phenomena can be made clear, if, following a suggestion of Collier [13.1], one considers the heat flux given off by the wall at a definite point z

$$q = \alpha(\vartheta_w - \vartheta_L(z)). \tag{13.101}$$

Assuming the mean liquid temperature $\vartheta_L(z)$ as given, we shall investigate the heat flux dependence on the wall temperature, Fig. 13.22. In the region of forced convection without bubble formation, the heat transfer coefficient α is practically constant. The heat flux increases linearly with the wall temperature. After the boiling temperature has been sufficiently exceeded $\vartheta_w > \vartheta_s$, the transferred heat flux increases more than proportionally with the wall temperature. In Fig. 13.22, the curve for nucleate boiling is simultaneously drawn in with a hatched line, in

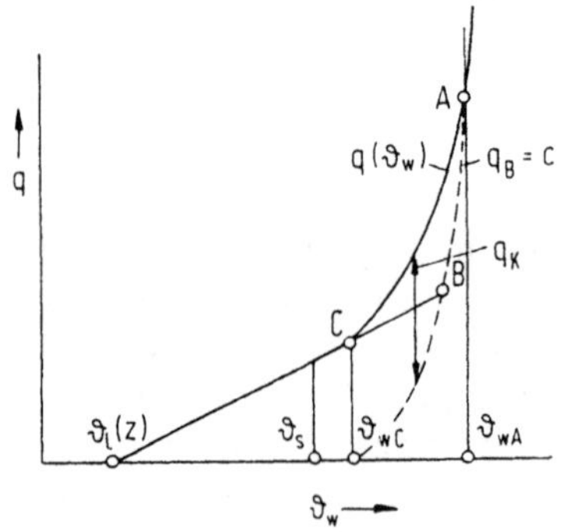

Fig. 13.22. Heat flux at a particular axial location z as a function of the wall temperature during subcooled boiling

keeping with $q = q_B = c(\vartheta_w - \vartheta_s)^n$. At a sufficiently high wall temperature, the two curves approach one another because the component of the heat flux transferred by convection in $q(\vartheta_w)$ becomes negligible. As shown, the difference between the curves represents the convective component of the heat flux q_K. By partial subcooled boiling is meant the region of the heat flux below the point of contact A in Fig. 13.22, where the component q_K is finite, whereas above the point of contact one speaks of fully developed subcooled boiling, because the component q_K is negligibly small. The convective component thus has the behavior depicted in Fig. 13.23. In the region of single-phase flow, the heat flux transferred by convection increases with the wall temperature. In the region of subcooled boiling, it decreases again, because with increasing wall temperature, the heating surface is more densely covered with bubbles. Therefore, less heat can be transferred by convection. Procedures for the calculation of the heat transfer subdivide, according to Fig. 13.22, the heat flux into a component resulting from bubble formation q_B and a convective component q_K

$$q = q_B + q_K. \tag{13.102}$$

According to the *suggestion of Bowring* [13.26], the individual heat fluxes in this equation are calculated in the following manner. As long as the wall temperature is smaller than the necessary minimum temperature according to Eq. (13.97), or the subcooling is greater than the highest permissible one according to the right side of Eq. (13.99), then the equations of the single-phase heat transfer are valid, because no viable bubbles are yet present. If the wall temperature is sufficiently great or the subcooling sufficiently small, then the convective component can be computed from

$$q_K = \alpha(\vartheta_s - \vartheta_L(z)) = \dot{m} c_{pL}(\vartheta_L(z) - \vartheta_1) \tag{13.103}$$

where ϑ_1 is the inlet temperature and α the heat transfer coefficient during single-phase convection. Because this heat transfer coefficient is practically temperature-independent, the suggestion of Bowring results in assuming the convective component q_K as constant for a prescribed value $\vartheta_L(z)$, as shown by the hatched line in Fig. 13.23. Above the temperature ϑ_{wA}, the component q_K disappears. Then the equations for fully developed subcooled boiling, which are yet to be explained, apply. The component q_B is computed from one of the

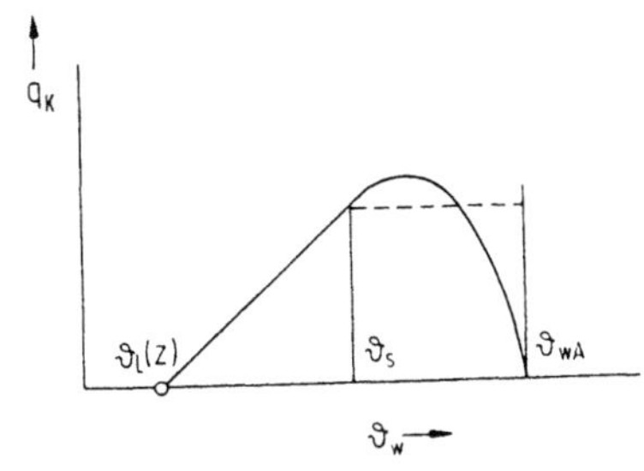

Fig. 13.23. Convective portion of the heat flux at a definite point z in dependence upon the wall temperature

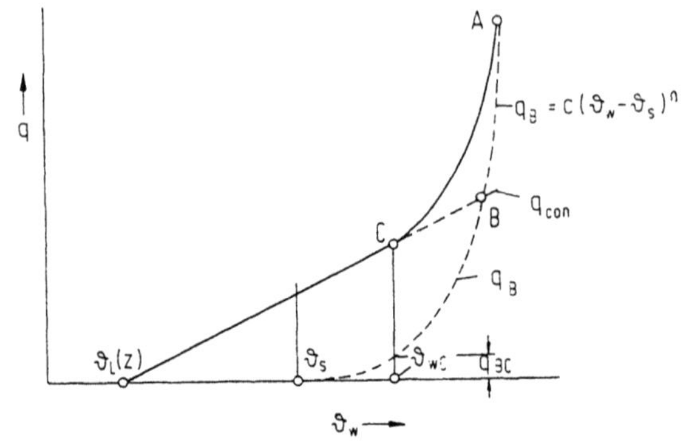

Fig. 13.24. Heat flux during subcooled boiling; procedure of Bergles and Rohsenow [13.28]

equations for the heat transfer during nucleate boiling in free convection, Sect. 11.4.

After checking data, Engelberg-Forster and Grief [13.27] suggested the simple assumption for the calculation of the heat flux in the point A of Fig. 13.22

$$q_A = 1.4 q_B \tag{13.104}$$

where the heat flux q_B is obtained according to Fig. 13.22 in the point of intersection B of the straight line for the single-phase convective heat transfer

$$q = \alpha(\vartheta_w - \vartheta_L(z))$$

and of the curve for $q_B = c(\vartheta_w - \vartheta_s)^n$. If both curves are extended so that they intersect, one obtains the values q_B, ϑ_{wB} and then from Eq. (13.104) also the temperature ϑ_{wA}.

According to the more accurate *method of Bergles and Rohsenow* [13.28], the heat flux is evaluated in the region of partial subcooled boiling

$$q = (q_{con}^2 + q_b^2)^{1/2}. \tag{13.105}$$

The convective component q_{con} and the nucleate boiling component q_b are evaluated in a manner different from that suggested by Bowring. As shown in Fig. 13.24, the convective component is that during single-phase convective heat transfer

$$q_{con} = \alpha(\vartheta_w - \vartheta_L(z))$$

with the heat transfer coefficient α for single-phase forced convection. As shown in Fig. 13.24, the component q_b is equal to the heat flux q_B for fully developed subcooled boiling minus the component q_{BC} of the heat flux at inception of nucleate boiling

$$q_b = q_B - q_{BC}. \tag{13.106}$$

Until the inception of nucleate boiling in point C, this component is equal to zero. With the aid of Eq. (13.106), one rearranges the assumption of Bergles and Rohsenow, Eq. (13.105), into

$$q = q_{\text{con}} \left[1 + \left(\frac{q_{\text{BC}}}{q_{\text{con}}} \right)^2 \left(1 - \frac{q_{\text{B}}}{q_{\text{BC}}} \right)^2 \right]^{1/2} . \tag{13.107}$$

13.6.3 Fully Developed Subcooled Boiling

After the wall temperature has exceeded the value ϑ_{wA}, Fig. 13.22, one finds oneself in the region of fully developed subcooled boiling. The heat transfer is determined solely by the nucleate boiling. The flow velocity and the subcooling have a negligibly small influence on the wall temperature, which is dependent mainly upon the heat flux and the pressure. The heating surface has less influence than during boiling in free convection, because the transferred heat fluxes and also the wall temperatures are larger, with the result that smaller nucleation sites become activitated than during pool boiling. In the absence of other information, one can use the equations for the heat transfer during boiling in free convection, Sect. 11.4, for the calculation of the heat transfer during fully developed subcooled boiling in forced flow.

The numerous data on heat transfer during subcooled boiling of water have been described by Jens and Lottes [13.29] and later by Thom et al. [13.30] by means of empirical correlations. Thom published the equation

$$\Delta\vartheta = \vartheta_{\text{w}} - \vartheta_{\text{s}} = 22.65 q^{0.5} \text{e}^{-p/87} \tag{13.108}$$

in which the heat flux q is to be entered in MW/m^2, the pressure p in bar and $\Delta\vartheta$ is obtained in K. The equation is based upon data on vertical, electrically heated steel or nickel tubes with inner diameters of from 3.63 to 5.74 mm; the pressures were between 7 and 172 bar, the water temperatures between 115 and 340 °C, the mass fluxes between 11 and $1.05 \cdot 10^4\ \text{kg/m}^2\text{s}$ and the heat fluxes reached values up to $12.5\ \text{MW/m}^2$. During the experiments the water flowed upward.

13.7 Nucleate Boiling in a Saturated Liquid

By definition, nucleate boiling in a saturated liquid, the so-called *saturated boiling*, begins when the quality, assuming thermodynamic equilibrium, Eq. (13.18), becomes $x_{\text{th}}^* = 0$. The mean liquid temperature is then equal to the saturation temperature. Actually, however, at the inception of saturated boiling, the liquid core is still subcooled and the wall-adjacent liquid layer is superheated. Although bubbles have already formed in the region of subcooled boiling, the bubble formation can be depressed again, even if the mean temperature approaches the saturation temperature. That is the case, when as a result of better heat transfer the heat removed by two-phase flow is equal to that supplied by the wall. Thus saturation boiling sets in only if the heat supplied by the wall is greater than the heat which the two-phase flow can remove without additional formation of bubbles. For this a sufficiently large heat flux is necessary. According to the explanations of the previous section, the wall temperature must exceed certain minimal values. Because of Eq. (13.97) there must be

$$\vartheta_w - \vartheta_s \geqq 2\left(\frac{2q\sigma T_s}{\varrho''\Delta h_v \lambda_L}\right)^{1/2}.$$

If one enters the heat transfer coefficient

$$\alpha_{2Ph} = q/(\vartheta_w - \vartheta_s)$$

of the two-phase flow, the supplied heat flux must be

$$q \geqq \frac{8\sigma T_s}{\varrho''\Delta h_v \lambda_L}\alpha_{2Ph}^2. \tag{13.109}$$

Here α_{2Ph} is the heat transfer coefficient of a two-phase flow without vapor formation. As already explained, this heat transfer coefficient increases with the mass flux and also with the quality. At the prescribed heat flux the temperature difference $\vartheta_w - \vartheta_s$ then decreases. In case it becomes too small, no vapor bubbles can form at the prescribed heat flux. The heat flux supplied at the wall can be transferred to the two-phase flow without formation of vapor bubbles. The region of subcooled boiling converts immediately into that of convective boiling, without passing through the region of saturated nucleate boiling. In order to determine whether the imbalance, Eq. (13.109), is fulfilled and whether saturated boiling occurs, it is sufficient to calculate approximately the heat transfer coefficient α_{2Ph}. This can be accomplished, for example, by means of the equation of Dengler and Addams [13.31]

$$\frac{\alpha_{2Ph}}{\alpha_L} = 3.5\left(\frac{1}{X_{tt}}\right)^{0.5}, \tag{13.110}$$

in which α_L is the single-phase heat transfer coefficient, assuming the total flow to be liquid. The quantity X_{tt} is the Lockhart–Martinelli parameter used earlier

$$X_{tt} = \left(\frac{\varrho_G}{\varrho_L}\right)^{0.5}\left(\frac{\eta_L}{\eta_G}\right)^{0.1}\left(\frac{1-x^*}{x^*}\right)^{0.9}.$$

The preceding considerations presuppose that the cavity distribution is sufficient to start bubble formation. In case the mouth radii of the pores are smaller than is necessary on the basis of Eq. (13.98), it is expedient to estimate first the necessary minimum superheating $(\vartheta_w - \vartheta_s)_c$ according to Eq. (13.100) based upon the size of pores present and compare it with the superheat $\vartheta_w - \vartheta_s = q/\alpha_{2Ph}$, which would set in in the two-phase flow without bubble formation. If it is smaller than the necessary minimum superheat, bubble formation is no longer possible. Conversely, if a greater superheating than is required for bubble formation were to set in, a two-phase flow without bubble formation could not exist.

Adjoining the region of subcooled boiling is the region of saturated boiling. For the calculation of the heat transfer, one assumes as in the case of boiling in a subcooled liquid, that the heat transfer is determined to a great extent by the nucleate boiling. In addition, the convection has an influence, even if it is slight, on the heat transfer. Models and empirical equations start with boiling in free convective flow and allow for the influence of forced flow by means of an additional term. They are based on the idea that bubble growth is only slightly influenced by the flow as long as bubbles are smaller than the superheated boundary layer. The heat transfer coefficient is composed, therefore, of two components

$$\alpha_{2Ph} = \alpha_B + \alpha_K, \tag{13.111}$$

where α_B is the heat transfer coefficient during boiling in free convection, Sect. 11.4, and α_K is the heat transfer coefficient during forced flow, as if only liquid were flowing through the tube. One calculates the latter from the McAdams equation for the heat transfer of a single-phase forced flow

$$Nu = \frac{\alpha_K d}{\lambda_L} = 0.023\, Re^{0.8} Pr^{0.4}$$

with the Reynolds number $Re = \dot{m}_L d/\eta_L = \dot{m}(1 - x^*)d/\eta_L$ and the Prandtl number $Pr = \nu_L/a_L$ of the liquid. In Eq. (13.111), the component α_K of the heat transfer by forced flow is usually only a small percentage of the entire heat transfer. This method of calculation does not take into consideration the influence of the quality on the heat transfer in the region of saturated boiling, even if it is only small. In order to encompass it, Chawla [13.32] has suggested on the basis of experiments with refrigerants an equation of the form

$$\frac{\alpha_{2Ph}}{\alpha_B} = 29\, Re^{-0.3} Fr^{0.2} \tag{13.112}$$

with the Reynolds number of the liquid phase

$$Re = \dot{m}(1 - x^*)d/\eta_L$$

and the Froude number

$$Fr = \dot{m}^2(1 - x^*)^2/(\varrho_L^2 g d).$$

The heat transfer coefficient α_B is once again to be determined from the equations for the heat transfer during boiling in free convective flow, Sect. 11.4. The measurements of Chawla, which led to Eq. (13.112), were carried out with horizontal tubes. As Stephan and Auracher have shown [13.33], Eq. (13.112) also reproduces measurements with vertical tubes. Recent, as yet unpublished, data of Auracher have shown that Eq. (13.112) overevaluates the influence of the mass flux $\dot{m}$ and of the quality x^* for larger values of these parameters. Equation (13.112) should be used only if

$$1 \leqq 29\, Re^{-0.3} Fr^{0.2} \leqq 1.2. \tag{13.113}$$

is valid for the right side.[1] The investigations of Auracher and also an evaluation

[1] Personal communication of H. Auracher

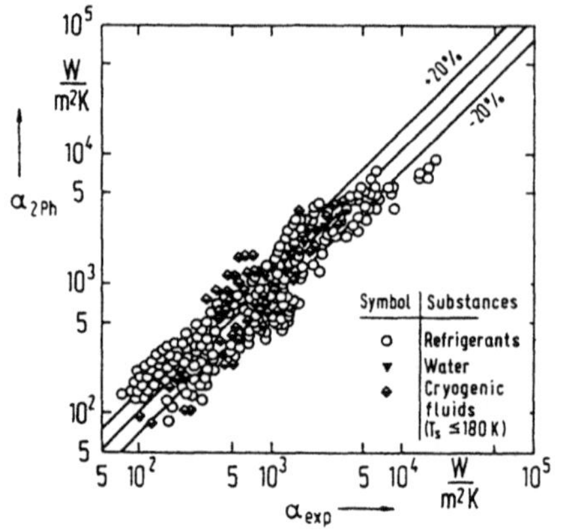

Fig. 13.25. Heat transfer coefficients α_{2Ph} calculated according to Eq. (13.112) as compared to data α_{exp}

by Steiner [13.34] of the previously known data indicate that the exponent of the heat flux in the region of saturated boiling is more strongly dependent upon pressure than during nucleate boiling and especially at higher pressures is smaller than the corresponding exponent during nucleate boiling. The spread of the test data is, however, still so considerable, that there is substantial uncertainty in these equations. It is to be recommended, therefore, that until more precise results are available, the heat transfer for vertical tubes be calculated according to Eq. (13.112) with allowance for the region of validity. An approximate impression of the size of the error to be expected for different substance groups is given by Fig. 13.25 taken from a work of Stephan and Auracher [13.33].

If vapor and liquid flow downward in a vertical tube, the relative velocity between vapor and liquid is diminished as a result of buoyancy. This leads to a reduction of the heat transfer coefficient. According to measurements of Pujol [13.35] for the refrigerant R-113, the heat transfer coefficient α_{down} of the downward flow is related to that of the upward flow α_{up} by

$$\alpha_{down} = 0.75\alpha_{up}. \tag{13.114}$$

The value of α_{up} can be calculated, for example, from Eq. (13.112).

Although no distinction was made between measurements on vertical and *horizontal tubes* when establishing Eq. (13.112), recent data of Steiner and colleagues [among others, 13.34], have confirmed the expectation that vapor and liquid are always distributed unevenly in a horizontal tube, if very high flow velocities are disregarded. One cannot calculate, therefore, the heat transfer during saturated boiling in vertical and horizontal tubes according to common equations. The upper side of the horizontal tube is more poorly wetted than is the lower side. This effect is especially noticeable, if the tube wall is thin and poorly heat-conducting, because the temperatures around the tube wall can be less well equalized than in thick-walled tubes that conduct heat well. Heat transfer coefficients are, therefore, variable in the flow direction and around the circumference. Local heat transfer measurements have not yet been made. Only

Table 13.2. Values K_1 for different refrigerants, according to Slipčević [13.36]

Refri-gerant	K_1 in $W^{0.3}m^{0.1}s^{0.1}kg^{-0.1}K^{-1}$ at a saturation temperature in °C						
	−50	−40	−30	−20	−10	0	10
R-11	–	–	–	0.082	0.085	0.088	0.091
R-12	0.105	0.112	0.118	0.123	0.129	0.134	0.138
R-13	0.139	0.152	0.170	0.193	0.218	0.247	–
R-13B1	0.113	0.119	0.127	0.135	0.145	0.156	0.171
R-21	–	0.080	0.084	0.089	0.093	0.098	0.104
R-22	0.116	0.122	0.128	0.134	0.141	0.149	0.158
R-40	–	–	–	0.161	0.164	0.167	0.171
R-113	–	–	–	–	–	0.069	0.073
R-114	–	–	–	–	–	0.095	0.097

heat transfer coefficients averaged around the circumference have been measured by several authors. Steiner [13.34] has evaluated these and his own measurements and has developed equations for the calculation of the heat transfer for horizontal tubes with a good thermal conductivity and for those with a poor thermal conductivity. Mention should be made here of the survey in the VDI-Wärmeatlas [13.34]. There one also finds data on the heat transfer during saturated boiling in inclined and helical tubes.

A simple, empirical correlation for the mean heat transfer coefficient during saturated boiling of refrigerants in tubes has been published by Slipčević [13.36]:

$$\bar{\alpha}_{2Ph} = K_1 \frac{\dot{m}^{0.1} \dot{q}^{0.7}}{(A_a/A_i)^{0.7} d^{0.5}}. \tag{13.115}$$

The property-dependent quantity K_1 is given in Table 13.2 for different refrigerants. In Eq. (13.115), the ratio A_a/A_i of the outer to the inner tube surface still appears, because the equation builds upon data on the heat transfer during evaporation on the outside of tubes and thus had to be converted to heat transfer coefficients for the inside. The inner tube diameter is indicated by d.

13.8 Convective Boiling

Adjoining the region of saturated boiling is that of convective boiling or forced convective vaporization. The heat transfer from the wall to the fluid improves there with increasing quality. The thermal resistance of the boundary layer decreases in comparison to the thermal resistance of nucleate boiling. The wall temperature decreases also, compare Fig. 13.16, so that the number of vapor bubbles formed at the wall decreases or bubbles are no longer formed at all. The heat transfer is determined predominantly, or exclusively, by evaporation at the phase interface between the liquid located at the wall and the vapor of the core flow.

13.8.1 Vertical Tubes

For the description of the heat transfer, Chen [13.37] has suggested a procedure similar to that for the calculation of the heat transfer in the region of saturated boiling. His equations are, therefore, valid for the region of saturated boiling and that of convective boiling. As in the case of saturated boiling, it is assumed that the heat transfer coefficient is composed of two components which are independent of one another, that for the bubble formation α'_B and that for the convection α'_K

$$\alpha_{2Ph} = \alpha'_B + \alpha'_K. \tag{13.116}$$

One reduces the component α'_B originating from the bubble formation to the heat transfer coefficient α_B for nucleate boiling in free convection. Because in a forced flow the temperature rise in the boundary layer is steeper than during nucleate boiling in free convection, more heat is removed from the wall by conduction and the bubble formation, in comparison with that in free convection, is partially suppressed. Chen allowed for this influence by a correction factor $S \leqq 1$ (suppression factor), with which the heat transfer coefficient α_B during nucleate boiling in free convection is to be multiplied, $\alpha'_B = S\alpha_B$.

The factor S for a vanishingly small mass flux approaches unity, then the heat transfer coefficient α'_B agrees with that during nucleate boiling, and it approaches zero for a very large mass flux, because then the heat transfer is determined exclusively by the convective component α'_K in Eq. (13.116).

The convective component α'_K of the heat transfer coefficient is related to the heat transfer for the single-phase liquid. Because of the more rapidly flowing vapor in the core of the flow and also because of the vapor bubbles still present, one must augment the heat transfer coefficient α_K of the single-phase forced liquid flow by an enhancement factor $F \geqq 1$,

$$\alpha'_K = F\alpha_K.$$

The factor F is determined mainly by the shear stress exerted by the vapor on the liquid and, as Chen has shown, can be expressed by the Martinelli parameter

$$X_{tt} = \left(\frac{\varrho_G}{\varrho_L}\right)^{0.5} \left(\frac{\eta_L}{\eta_G}\right)^{0.1} \left(\frac{1-x^*}{x^*}\right)^{0.9}.$$

Equation (13.116) now reads

$$\alpha_{2Ph} = S\alpha_B + F\alpha_K. \tag{13.117}$$

The heat transfer coefficient α_B for nucleate boiling was obtained by Chen from the pool boiling equation of Forster and Zuber [13.38]

$$\alpha_B = 0.00122 \left[\frac{\lambda_L^{0.79} c_{pL}^{0.45} \varrho_L^{0.49}}{\sigma^{0.5} \eta_L^{0.29} (\varrho_G \Delta h_v)^{0.24}}\right] \Delta\vartheta^{0.24} \Delta p^{0.75}. \tag{13.118}$$

To be placed into this equation are the property values in SI units, m, kg, N, and K, and also the pressure difference Δp in N/m^2 which belongs to the temperature difference $\Delta\vartheta$ corresponding to the vapor pressure curve. The heat transfer coefficient is obtained in W/m^2K. The equation reproduces test data for water,

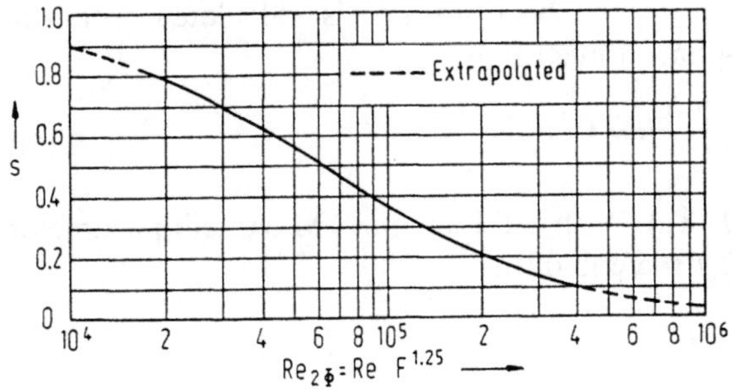

Fig. 13.26. Correction factor S in Eq. (13.117), according to Chen [13.37]

ethyl alcohol, n-pentane and benzene as far as such were known when the equation was developed in 1966. Since then, however, more precise measurements with a much larger number of substances have been carried out. One obtains, therefore, more precise results, when one determines the heat transfer coefficient α_B according to more recent equations. Jallouk [13.39], for example, was able to reproduce his measurements on the refrigerant R-114 better, when he computed the heat transfer coefficient α_B not according to the equation of Forster and Zuber, but rather according to an equation of Rohsenow and calculated with it the heat transfer coefficient α_{2Ph}. Thus it seems more expedient to calculate α_B with the aid of one of the application equations of Sect. 11.4.

The factor S in Eq. (13.117) depends upon the mass flux of the liquid and can be expressed by its Reynolds number

$$Re = \dot{m}(1 - x^*)d/\eta_L.$$

It is, however, not independent of the velocity and temperature field of the boundary layer, because the latter is decisively determined by the vapor flow and, thereby, is also dependent upon the factor F. Figure 13.26 shows the factor S determined by Chen. It is

$$S = f(ReF^{1.25}) = f(Re_{2\Phi}),$$

$$S = \frac{1}{1 + 2.53 \cdot 10^{-6} Re_{2\Phi}^{1.17}} \qquad (13.119)$$

with $Re_{2\Phi} = ReF^{1.25}$.

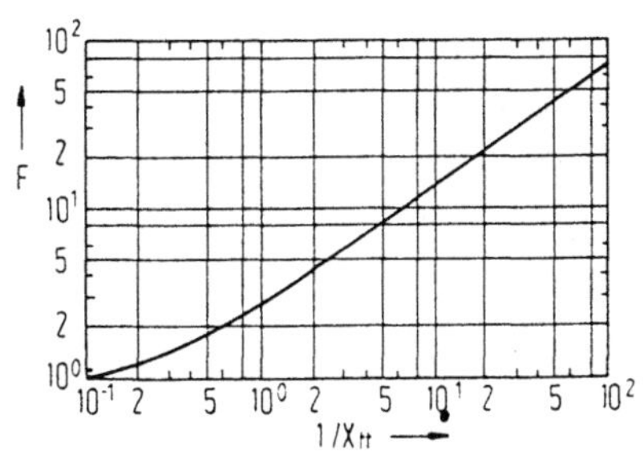

Fig. 13.27. Correction factor F in Eq. (13.117), according to Chen [13.37]

The convective contribution to the heat transfer is calculated from the McAdams equation for single-phase forced flow

$$\alpha_K = \frac{\lambda_L}{d} 0.023 \left[\frac{\dot{m}(1-x^*)d}{\eta_L} \right]^{0.8} Pr^{0.4}. \tag{13.120}$$

Figure 13.27 shows the factor $F(X_{tt})$. With the Lockhart–Martinelli parameter, this curve can be represented by the equation

$$F = 1 \quad \text{for} \quad \frac{1}{X_{tt}} \leqq 0.1 \quad \text{and}$$

$$F = 2.35 \left(\frac{1}{X_{tt}} + 0.213 \right)^{0.736} \quad \text{for} \quad \frac{1}{X_{tt}} > 0.1. \tag{13.121}$$

The correlation of Chen was tested in experiments for convective boiling in vertical canals and until now has been much used for the calculation of heat transfer coefficients. Recent investigations of Gungor and Winterton [13.40], in which the correlation was compared with around 4300 test data for different substances, showed that, depending upon the substance and the range of its condition, the heat transfer coefficients computed were too large by the factor of 1.2 to 1.9. A better agreement with the test data was obtained, when one computed the factors S and F in Eq. (13.117) according to the following correlations

$$S = 1 + 1.15 \cdot 10^{-6} F^2 Re^{1.17}, \tag{13.119a}$$

$$F = 1 + 2.4 \cdot 10^4 Bo^{1.16} + 1.37 X_{tt}^{-0.86} \tag{13.121a}$$

with $Bo = q/\dot{m}\Delta h_v$, $Re = \dot{m}(1-x^*)/\eta_L$, and the Martinelli parameter X_{tt}. The mean deviation from the test data equals $\pm 21.4\%$. The heat transfer coefficient α_B during nucleate boiling should be calculated either according to Eq. (11.9) or the very simple, but physically scarcely plausible, dimensional equation of Cooper [13.41]

$$\alpha_B = 55 q^{2/3} (p/p_{cr})^a (-\log p/p_{cr})^{-0.55} M^{-0.5}$$

with

$$a = 0.12 - 0.2 \log R_p$$

whereas the heat transfer coefficient α_K is obtained from Eq. (13.120). In the equation of Cooper, M designates the molar mass in kg/kmol, q is the heat flux in W/m^2, R_p is the depth of surface smoothness in μm and one obtains α_B in W/m^{2}K.

For *convective boiling in annuli heated from one wall* only, an equivalent diameter d_e must be introduced, which is given by

$$d_e = 4A/U_w$$

for a slot width over 4 mm and

$$d_e = 4A/U_h$$

for a clearance under 4 mm. A is the flow cross section, U_w is the wetted perimeter, and U_h the heated perimeter.

Procedure during Computation and Further Application Equations

For the calculation of heat transfer coefficients according to Eq. (13.117) of Chen, it is expedient to begin with the following steps:

—One calculates first the Lockhart–Martinelli parameter

$$X_{tt} = \left(\frac{\varrho_G}{\varrho_L}\right)^{0.5} \left(\frac{\eta_L}{\eta_G}\right)^{0.1} \left(\frac{1 - x^*}{x^*}\right)^{0.9}.$$

—One finds the heat transfer coefficient α_B during nucleate boiling for the prescribed heat flux q from the application equations given in Sect. 11.4 or, following the suggestion of Gungor and Winterton, from the equation of Cooper. The factors S and F are calculated from Eqs. (13.119a) and (13.121a).
—With these values, the heat transfer coefficient α_{2Ph} according to Eq. (13.117) is fixed and the wall temperature $\vartheta_w = \vartheta_s + q/\alpha_{2Ph}$ can be determined.

This computation procedure must be modified for *subcooled boiling*, because the driving temperature differences for the nucleate boiling and the forced convection are different. The transferred heat flux is then

$$q = \alpha_K(\vartheta_w - \vartheta_L) + S\alpha_B(\vartheta_w - \vartheta_s),$$

where ϑ_L is the mean fluid temperature. The enhancement factor F does not appear in this equation, because, on average, vapor is not produced over a cross section. The correction factor S for nucleate boiling is still to be calculated according to Eqs. (13.119a) and (13.121a). The mean deviation between test data and the heat fluxes calculated according to this procedure amounts to $\pm 25\%$.

Another group of correlations presupposes that the ratio α_{2Ph}/α_K of the heat transfer coefficients of the two-phase flow and that of the single-phase liquid flow can be represented by the Lockhart–Martinelli parameter alone or by it and the boiling number

$$Bo = q/(\dot{m}\Delta h_v). \qquad (13.122)$$

In this, the boiling number

$$Bo = \frac{q}{\Delta h_v}\frac{1}{\dot{m}} = \varrho_G w_G \frac{1}{\varrho w}$$

can be interpreted as the ratio of the mass flux of the generated vapor perpendicular to the wall to the entire mass flux parallel to the wall. Such correlations are of the form

$$\frac{\alpha_{2Ph}}{\alpha_K} = A\left(\frac{1}{X_{tt}}\right)^b \qquad (13.123)$$

or

$$\frac{\alpha_{2Ph}}{\alpha_K} = M\left[Bo \cdot 10^4 + N\left(\frac{1}{X_{tt}}\right)^n\right]^m. \tag{13.124}$$

The empirical constants A, b and M, N, n, m of this equation are compiled for different fluids in Tables 13.3 and 13.4. In equations of the type of Eq. (13.123), the heat transfer coefficient α_{2Ph} of the two-phase flow is independent of the heat flux. In equations of the type of Eq. (13.124), it is proportional to the heat flux in the region in which bubble formation is decisive, i.e., the term $Bo \cdot 10^4$ dominates

Table 13.3. Numerical values for the constants A and b in Eq. (13.123)

Fluid	A	b	Authors
Water, forced convection, upward	3.0	0.5	Dengler and Addams [13.31]
Different liquids, free convection	3.4	0.45	Guerrieri and Talty [13.42]
Water, forced convection, upward	2.5	0.75	Schrock and Grossman [13.43]
Water, forced convection, upward	2.9	0.66	Bennett et al. [13.44]
Water, forced convection, downward	2.721	0.581	Wright et al. [13.45]
n-Butanol, forced convection, downward	7.55	0.328	Somerville [13.46]
Water, forced convection, upward	2.167	0.699	Collier et al. [13.47]
R-113, forced convection, upward and downward	4.0	0.37	Pujol and Stenning [13.48]

Table 13.4. Numerical values for the constants in Eq. (13.124)

Fluid	M	N	n	m	Authors
Water, forced convection, upward	0.739	1.5	2/3	1	Schrock and Grossman [13.43]
Water, forced convection, downward	1.48	1.5	2/3	1	Sani [13.49]
Water, forced convection, downward	1.39	1.5	2/3	1	Wright et al. [13.45]
n-Butanol, forced convection, downward	2.45	1.5	2/3	1	Somerville [13.46]
R-12, R-22, forced convection, horizontal	1.91	1.5	2/3	0.6	Chaddock [13.50]
R-113, forced convection, upward	0.9	4.45	0.37	1	Pujol and Stenning [13.48]
R-113, forced convection, downward	0.53	7.55	0.37	1	Pujol and Stenning [13.48]

on the right side, if one disregards the findings of Chaddock [13.50]. In Eq. (13.124), the mass flux on the left side is contained in the heat transfer coefficient α_K and is actually $\alpha_K \sim \dot{m}^{0.8}$; in the boiling number Bo, the mass flux appears in the denominator, so that the heat transfer coefficient α_{2Ph} in the region of large values of the boiling number is scarcely dependent upon the mass flux any more. None of these findings correctly reproduces the physical facts of the case. Therefore, the stated equations should be used only for the investigated substances and for the pressure region, which lies in the mentioned works between 1 bar and roughly 5 bar, and which covered in the experiments of Schrock and Grossman [13.43] solely the pressure region from 2.9 to 34.8 bar.

An empirical equation, which reproduces the data of numerous authors of the USSR, has been formulated by Borishanskij and colleagues [13.51]. It describes measurements on water and air-water mixtures and reads

$$\frac{\alpha_{2Ph}}{\alpha^*} = \left[1 + 7 \cdot 10^{-9} \left(\frac{\alpha_B}{\alpha^*} \right)^2 \left(1 + \frac{\varrho' - \varrho''}{\varrho''} x^* \right)^{3/2} Bo^{-3/2} \right]^{1/2} \tag{13.125}$$

with

$$\alpha^* = [\alpha_K^2 + (0.7\alpha_B)^2]^{1/2}$$

and the heat transfer coefficient for nucleate boiling of water

$$\alpha_B = 0.625(p_s^{0.14} + 8.95 \cdot 10^{-14} p_s^2) q^{0.7}.$$

In these equations, the heat flux q is to be stated in W/m^2 and the saturation pressure p_s in N/m^2. One obtains the heat transfer coefficients in W/m^2K. x^* is the mean quality of the considered tube section $x^* = (x_1^* + x_2^*)/2$, if x_1^* is the quality in the inlet cross section and x_2^* that in the outlet cross section. The equation reproduces test data on water and on water-air mixtures of many authors of the USSR in the region

$$0.05 \leqq x^* \leqq 0.9$$
$$8.1 \cdot 10^4\,W/m^2 \leqq q \leqq 5.6 \cdot 10^6\,W/m^2$$
$$2\,bar \leqq p \leqq 170\,bar$$
$$5\,mm \leqq d \leqq 34\,mm$$

with a mean error of about $\pm 30\%$. Figure 13.28, which was taken from the work of Borishanskij, gives an approximate idea of the accuracy. It further shows that in the region of smaller values of the flow parameter

$$\left(\frac{\alpha_B}{\alpha^*} \right)^{4/3} \left(1 + \frac{\varrho' - \varrho''}{\varrho''} x^* \right) Bo^{-1}$$

or of larger values of the boiling number $Bo = q/(\dot{m}\Delta h_v)$, the ratio α_{2Ph}/α^* approaches unity. The heat transfer is then determined for the most part by bubble

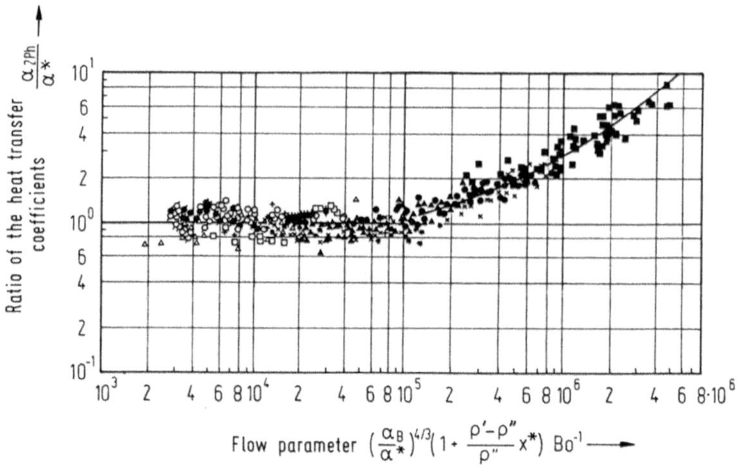

Fig. 13.28. Correlation of data, according to Borishanskij et al. [13.51]

formation. For sufficiently large values of

$$\left(\frac{\alpha_B}{\alpha^*}\right)^{4/3}\left(1+\frac{\varrho'-\varrho''}{\varrho''}x^*\right)Bo^{-1} > 5\cdot 10^{-4},$$

i.e., of large values of the flow parameter or of small values of the boiling number, the heat is removed predominantly by convection.

For practical considerations, it is difficult to decide which of the many known equations one should use. It is recommended that the heat transfer coefficient be computed according to several equations, because one thus gets an idea of the uncertainty and can adapt the heating surface accordingly. In many boilers, the overall heat transfer coefficient is not strongly influenced by the heat transfer coefficient on the boiling side. In such boilers, this makes the uncertainty in the heat transfer coefficient tolerable.

13.8.2 Horizontal Tubes

Under the influence of gravity, the liquid collects for the most part on the lower part of the tube, whereas the upper side is wetted poorly or not at all, with the result that during heating with a constant heat flux higher wall temperatures occur at the top than at the base. The temperature differences are equalized more or less well depending upon the thermal conductivity of the tube wall. As already explained in connection with saturated boiling, the heat transfer coefficient averaged over the circumference is thus dependent upon the thermal conductivity of the tube wall. As in the case of condensation in horizontal tubes, the heat transfer coefficient averaged over the circumference is composed of the heat transfer coefficient α_b of the wetted lower arc section, compare Fig. 4.13, and of the heat transfer coefficient α_{dr} of the upper poorly wetted or dry section in accordance with

$$\bar{\alpha} = \frac{\varphi}{\pi}\bar{\alpha}_b + \left(1 - \frac{\varphi}{\pi}\right)\bar{\alpha}_{dr}. \tag{13.126}$$

Unfortunately, the dependence of $\bar{\alpha}_b$, $\bar{\alpha}_{dr}$ and also the wetting angle φ on heat and mass fluxes and on the property values are not known. At present, there are not yet available sufficiently reliable and proven methods of computation for heat transfer during boiling in forced flow in horizontal tubes. It has been noted that the models applied to condensation are not applicable here. First assumptions and approximate values for the wetting angle φ are found in recent works, the findings of which Steiner has summarized [13.34].

On the basis of an evaluation of experiments with water at pressures of 1 bar to 165 bar, with R-11, R-12, R-22, and R-113, and with cyclohexane, Shah [13.52, 13.53] has developed a simple equation for practical use, which also served as the basis for the equation for heat transfer during condensation in horizontal tubes, Eq. (4.50). In it the McAdams equation for the heat transfer of a single-phase turbulent tube flow is expanded by a function of the quality, of the density ratio ϱ_L/ϱ_G of liquid and gas, and of the Froude number

$$Fr = \dot{m}^2/(\varrho_L^2 g d).$$

The heat transfer coefficient averaged over the circumference is expressed by

$$\frac{\alpha_{2Ph}}{\alpha_K} = 3.9 Fr^{0.24}\left(\frac{x^*}{1 - x^*}\right)^{0.64}\left(\frac{\varrho_L}{\varrho_G}\right)^{0.4} \tag{13.127}$$

with the heat transfer coefficient α_K of the single-phase liquid flow according to the McAdams equation

$$\alpha_K = \frac{\lambda_L}{d} 0.023\left(\frac{\dot{m}(1 - x^*)d}{\eta_L}\right)^{0.8} Pr^{0.4}.$$

The equation is valid, assuming that a stratified flow has formed in the horizontal tube. According to the data of Shah, that is the case if the Froude number is

$$Fr = \dot{m}^2/(\varrho_L^2 g d) < 0.04. \tag{13.128}$$

For Froude numbers above 0.04, the tube is completely wetted, with the result that an annular flow forms, which can be computed according to the equations for vertical tubes which have already been discussed. Shah recommends computing the heat transfer coefficient for the annular flow in the horizontal tube also from Eq. (13.127) and replacing there the factor

$3.9 Fr^{0.24}$ by the value 1.8

so that Eq. (13.127) for $Fr \geqq 0.04$ converts into

$$\frac{\alpha_{2Ph}}{\alpha_K} = 1.8\left(\frac{x^*}{1 - x^*}\right)^{0.64}\left(\frac{\varrho_L}{\varrho_G}\right)^{0.4} \tag{13.129}$$

The evaluation of Eq. (13.127) was based upon measurements in the region of $3 \cdot 10^{-4} \leqq Fr \leqq 0.04$. In addition, Shah recommends Eq. (13.127) or, as the case may be, Eq. (13.129) only if the boiling number exceeds certain minimum values $Bo \geqq 10^{-4}$. Otherwise, the test data will be too inaccurately reproduced by the empirical equation.

For the calculation of heat transfer coefficients during *convective boiling in horizontal tubes*, using Eq. (13.128), one must first check whether stratified flow is present. If this is the case, one can compute according to Eq. (13.127) of Shah; otherwise it is recommended that one calculates according to the equations for heat transfer in vertical tubes, for example, according to Eq. (13.117) of Chen.

If one puts the heat transfer coefficient α_K of the single-phase liquid flow into Eq. (13.127), one finds the following dependency upon the quality

$$\alpha_{2Ph} = K(1 - x^*)^{0.16} x^{*0.64} \tag{13.130}$$

with

$$K = \frac{\lambda_L}{d} 0.023 \left(\frac{\dot{m} d}{\eta_L}\right)^{0.8} Pr^{0.4} 3.9 Fr^{0.24} \left(\frac{\varrho_L}{\varrho_G}\right)^{0.4}, \tag{13.131}$$

which, in keeping with the empirical nature of this equation, no longer accurately reproduces the limiting cases $x^* = 0$ and $x^* = 1$. By differentiation of Eq. (13.130) with respect to the quality, one finds a maximum of the circumference-averaged heat transfer coefficient at the point $x^* = 0.8$, in agreement with data of Chawla [13.32]. It is easy to visualize that such a maximum occurs, because, with increasing quality, the heat transfer coefficient of the two-phase flow at first increases due to the increasing vapor shear stress. At sufficiently high quality, total dryout occurs and the low thermal conductivity of the vapor again leads to a decrease of the heat transfer coefficient. Figure 13.29 shows, for example, the pattern of the heat transfer coefficient α_{2Ph} as a function of the quality, as measured by Chawla.

The mean heat transfer coefficient over a longitudinal section $\Delta L = L_2 - L_1$ is frequently of interest for practical computations. One obtains it from Eq. (13.130) by integration

$$\alpha_{2Ph} = \frac{K}{\Delta L} \int_{z=L_1}^{L_2} (1 - x^*)^{0.16} x^{*0.64} \mathrm{d}z.$$

Assuming that one can regard the quality in the considered section ΔL as a linear function of the length

$$x^* = \frac{x_2^* - x_1^*}{\Delta L}(z - L_1) + x_1^*$$

then, after series expansion of the expression $(1 - x^*)^{0.16} = 1 - 0.16 x^*$, while disregarding the terms of higher order in x^*, the mean heat transfer coefficient

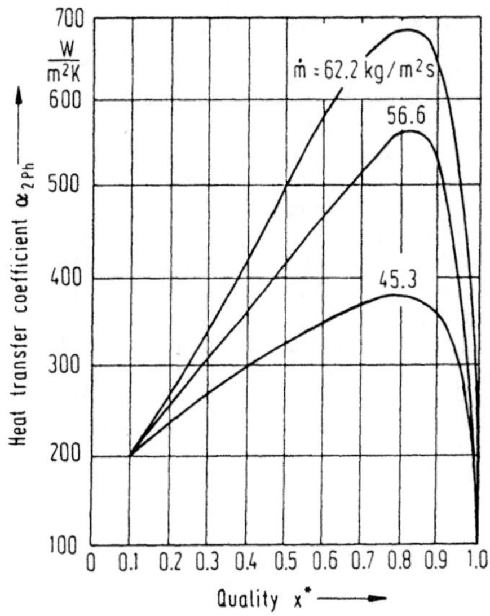

Fig. 13.29. Heat transfer coefficient α_{2Ph} as a function of quality, according to Chawla [13.32]. Horizontal tube, heated with $q = 1740\,\mathrm{W/m^2}$, R-11 at an evaporation temperature of 10 °C

is found to be

$$\bar{\alpha}_{2Ph} = \frac{K}{x_2^* - x_1^*}\left(\frac{x^{*1.64}}{1.64} - 0.16\frac{x^{*2.64}}{2.64}\right)_{x_1^*}^{x_2^*}$$

with K according to Eq. (13.131). Then,

$$\frac{\bar{\alpha}_{2Ph}}{\alpha_K(x^* = 0)} = 3.9 Fr^{0.24}\left(\frac{\varrho_L}{\varrho_G}\right)^{0.4}\frac{1}{x_2^* - x_1^*}(0.61x^{*1.64} - 0.06x^{*2.64})_{x_1^*}^{x_2^*} \tag{13.132}$$

with $\alpha_K(x^* = 0)$ from the equation of McAdams.

According to the investigations of Gungor and Winterton [13.40], the presently known data of heat transfer during convective boiling in horizontal tubes can be reproduced very well, if one slightly modifies the procedure of Chen for vertical tubes, which was described in the previous section. One merely has to multiply the correction factor S of Eq. (13.119a) by

$$S_1 = Fr^{1/2}$$

and the enhancement factor F of Eq. (13.121a) by

$$F_1 = Fr^{0.1 - 2Fr}$$

if the Froude number $Fr = \dot{m}^2/(\varrho_L^2 g d)$ is smaller than 0.05, i.e., if gravity causes an unequal distribution of the liquid film over the tube circumference.

Slipčević [13.36] has published another practical empirical equation for the calculation of the mean heat transfer coefficient for convective boiling of

Table 13.5. Values K_2 for different refrigerants, according to Slipčević [13.36]

Refrigerants	K_2 in $\mathrm{Ws^{1.4}\,m^{1.3}\,kg^{-1.4}\,K^{-1}}$ at saturation temperature in °C						
	−50	−40	−30	−20	−10	0	10
R-11	–	–	–	0.835	0.640	0.490	0.380
R-12	0.525	0.399	0.310	0.256	0.194	0.156	0.126
R-13	0.113	0.090	0.073	0.062	0.053	0.045	–
R-13B1	0.148	0.115	0.092	0.077	0.065	0.055	0.046
R-21	–	1.748	1.290	0.949	0.705	0.557	0.440
R-22	0.635	0.470	0.351	0.272	0.215	0.169	0.138
R-40	–	–	–	1.167	0.925	0.710	0.540
R-113	–	–	–	–	–	0.451	0.296
R-114	–	–	–	–	–	0.208	0.163

refrigerants. It has the form

$$\bar{\alpha}_{2\mathrm{Ph}} = K_2 \dot{m}^{1.4}/d^{0.5} \tag{13.133}$$

and is valid for complete evaporation of the refrigerant at the outlet cross section. The value K_2 is property-dependent and is given in Table 13.5 for a series of refrigerants.

Together with Eq. (13.115) for saturated boiling of refrigerants, one can also calculate the mean heat transfer coefficient for complete evaporation of refrigerants. The equation that yields the greater heat transfer coefficient is always to be used. As is recognized from a comparison of Eqs. (13.115) and (13.133), saturated boiling is present if there is

$$\dot{m} < \left(\frac{K_1}{K_2}\right)^{0.769} q^{0.538}.$$

Otherwise one must use Eq. (13.133) for convective boiling.

13.9 Critical Boiling Conditions

As already presented in connection with boiling in free convection or pool boiling, the heat transfer coefficient decreases sharply after surpassing a peak heat flux. The liquid at the wall is displaced by vapor. If, for instance, heat is supplied to the heated surface by electrical or nuclear heating, or by radiation, then the wall temperature increases suddenly after reaching the peak heat flux. If, on the other hand, as in a heat exchanger or condenser, the wall temperature is imposed by another fluid, the reduction in the heat transfer coefficient leads to a drastic drop in the heat flux. This phenomenon is summarized by the concept of *critical boiling conditions*. By this is generally meant the reduction of the heat transfer coefficient after surpassing a critical heat flux. In the case of boiling under forced convection, this heat flux is generally not identical with the peak heat flux during pool boiling and can be caused by

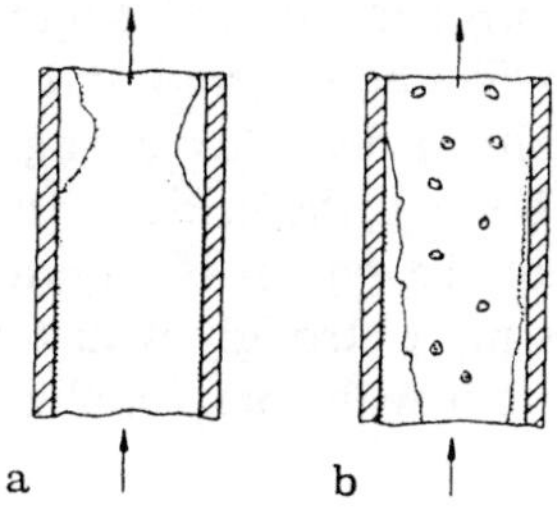

Fig. 13.30 a, b. Types of boiling crisis. **a** film boiling; **b** dryout of heating surface

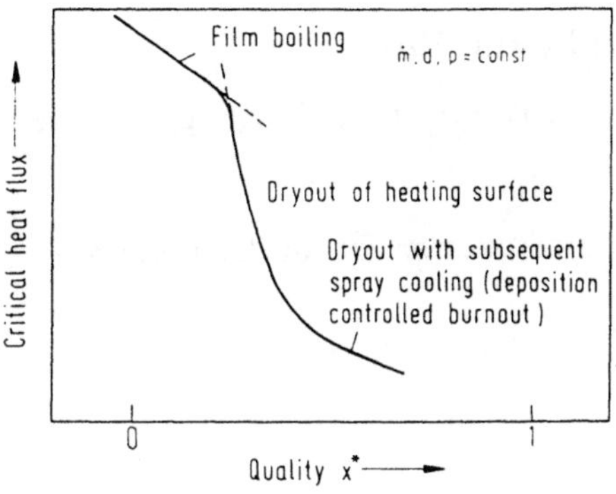

Fig. 13.31. Critical heat flux as a function of the quality

different mechanisms, as will be explained in the following. We, therefore, speak generally of critical heat flux in contrast to peak heat flux during nucleate boiling in free convection.

The phenomena, which can be observed in free convection boiling, also occur in a forced flow, but are more complicated there. Generally speaking, two different types of boiling crisis can be distinguished:

a) *Film boiling* occurs when there is a small amount of vapor by volume. The liquid forms a continuous phase and after the critical heat flux is reached, a vapor film is formed at the heated wall which separates the liquid from the wall. The high thermal resistance of the vapor film leads to a drop in the heat flux, in case the wall temperature is imposed, or to a rise in the wall temperature, if the heat flux is applied. The smaller the amount of vapor by volume, the greater the critical heat flux.
b) With a large amount of vapor by volume, annular flow occurs. At the wall there is mainly liquid, and in the core the vapor forms the continuous phase. After the critical heat flux is reached, the liquid film at the wall is dispersed, and the wall is covered by vapor. One speaks of *dryout of the heated surface*.

If the vapor content is sufficiently large, the heated surface already dries out during very small heat fluxes. Downstream, small drops of liquid from the core of the flow can reach the heated surface, which evaporate only partially because of the small heat flux. This type of dryout with adjoining spray cooling of the wall by drops of liquid is often called "deposition controlled dryout". Figure 13.30 depicts the two types of the boiling crisis, namely film boiling and dryout of the heated surface. Figure 13.31 shows how the critical heat flux changes qualitatively with the quality. In the region of film boiling, the critical heat flux decreases very close to linear with the quality. In the region of dryout, many data suggest that the critical heat flux decreases more strongly with the quality, and only at large qualities in the region of dryout with spray cooling, does it decrease more weakly with the quality.

13.9.1 Limiting Values for the Critical Heat Flux

A lower limiting value for the critical heat flux is found from the energy balance

$$q d\pi z = \dot{M} c_{pL}(\vartheta_L(z) - \vartheta_1) = \alpha d\pi z(\vartheta_w - \vartheta_L(z)),$$

where ϑ_1 is the temperature in the inlet cross section. From this follows

$$\vartheta_w = \vartheta_L(z) + \frac{q}{\alpha} = \vartheta_1 + \frac{q d\pi z}{\dot{M} c_{pL}} + \frac{q}{\alpha}$$

or

$$q = \frac{\vartheta_w - \vartheta_1}{\dfrac{d\pi z}{\dot{M} c_{pL}} + \dfrac{1}{\alpha}}.$$

Because, after reaching the critical heat flux, the wall temperature clearly lies above the saturation temperature, one obtains as the lower limiting value for the critical heat flux

$$q_{cr} > \frac{\vartheta_s - \vartheta_1}{\dfrac{d\pi z}{\dot{M} c_{pL}} + \dfrac{1}{\alpha}} = \frac{\Delta\vartheta_u}{\dfrac{d\pi z}{\dot{M} c_{pL}} + \dfrac{1}{\alpha}}, \tag{13.134}$$

where $\Delta\vartheta_u$ is the subcooling of the liquid in the inlet cross section.

From this, for example, one finds that for water with an initial subcooling of $\Delta\vartheta_u = 50\,\text{K}$, a mass flux of $\dot{m} = 1000\,\text{kg/m}^2\text{s}$ and a specific heat of $c_{pL} = 4.186\,\text{kJ/kgK}$, which boils in a 1 m long tube of $d = 25\,\text{mm}$ inner diameter, assuming a heat transfer coefficient of $\alpha = 10\,000\,\text{W/m}^2\text{K}$, the critical heat flux must be

$$q_{cr} > 3.62 \cdot 10^5\,\text{W/m}^2.$$

If, on the other hand, the critical heat flux is achieved, not all of the liquid is evaporated yet. The upper limit is thus the critical heat flux that is necessary in order to evaporate all the liquid. This is found again from an energy balance

$$q d\pi z = \dot{M}(h'' - h_1) = \dot{M}(\Delta h_v + h' - h_1).$$

If one substitutes

$$\dot{M}(h' - h_1) = \dot{M}\bar{c}_{pL}(\vartheta_s - \vartheta_1) = \dot{M}\bar{c}_{pL}\Delta\vartheta_u,$$

where $\bar{c}_{pL}$ is the mean specific heat between the temperature ϑ_1 in the inlet cross section and the boiling temperature ϑ_s, one finds because of $q_{cr} < q$

$$q_{cr} < \frac{\dot{M}\Delta h_v}{d\pi z}\left(1 + \frac{\bar{c}_{pL}\Delta\vartheta_u}{\Delta h_v}\right). \tag{13.135}$$

With the data of the above example and an evaporation enthalpy of 2100 kJ/kg, one calculates

$$q_{cr} < 1.4 \cdot 10^7 \text{ W/m}^2.$$

so that, for the example, the critical heat flux must lie between $3.62 \cdot 10^5$ and $1.4 \cdot 10^7$ W/m². In view of the large number of empirical methods for the calculation of the critical heat flux and of the many limiting values that it depends upon, for the sake of accurate computation, it is always recommended that one estimates the interval in which the critical heat flux must lie.

13.9.2 Experimental Results for the Critical Heat Flux

With the development of boiling water reactors in the last twenty-five years, the critical heat flux has been much investigated. Most experiments have been conducted with water and refrigerants in vertical, electrically heated tubes. Nevertheless, there are very few models for determining the dependence of the critical heat flux on the major physical parameters. This results from the fact that the processes are very complicated, that experiments cover only a limited parameter region, and that the measurements are so difficult that there are often substantial errors in measurement.

We will report in the following several important dependencies that can be recognized on the basis of the experiments up to now. As the experiments have demonstrated, with a vertical flow in a uniformly electrically heated tube, the critical heat flux is dependent upon the mass flux $\dot{m}$, the heated length z, the local quality $x^*(z)$, the tube diameter d and the pressure p. The property values are dependent on pressure and temperature, and, if there is phase equilibrium in a cross section, they are established by the specification of pressure or temperature. Then, there holds for a particular fluid

$$q_{cr} = q_{cr}(\dot{m}, x^*, z, d, p). \tag{13.136}$$

The quality, Eq. (13.18),

$$x^* = \frac{1}{\Delta h_v}\left(\frac{\Phi}{\dot{M}} + h_1 - h'\right)$$

can be written as

$$x^* = \frac{1}{\Delta h_v}\left(\frac{q4z}{\dot{m}d} - \bar{c}_{pL}\Delta\vartheta_u\right). \tag{13.137}$$

because of

$$h_1 - h' = \bar{c}_{pL}(\vartheta_1 - \vartheta_s) = -\bar{c}_{pL}\Delta\vartheta_u$$

and

$$\Phi/\dot{M} = qd\pi z/(\dot{m}d^2\pi/4).$$

An equation equivalent to Eq. (13.136) thus reads

$$q_{cr} = q_{cr}(\dot{m}, \Delta\vartheta_u, z, d, p), \tag{13.138}$$

where $\Delta\vartheta_u$ is approximately equal to the subcooling in the inlet cross section. Finally, it is

$$x^* = \frac{h(z) - h_1}{\Delta h_v}$$

and, consequently, Eq. (13.136) can also be represented by

$$q_{cr} = q_{cr}(\dot{m}, h, z, d, p). \tag{13.139}$$

13.9.2.1 The Dependence of the Critical Heat Flux on the Mass Flux and Subcooling

We will first consider the dependence of the critical heat flux on the mass flux $\dot{m}$ and the initial subcooling $\Delta\vartheta_u$ at a definite point z of a tube with the diameter d at the assumed pressure p of the fluid. As Fig. 13.32 shows, the critical heat

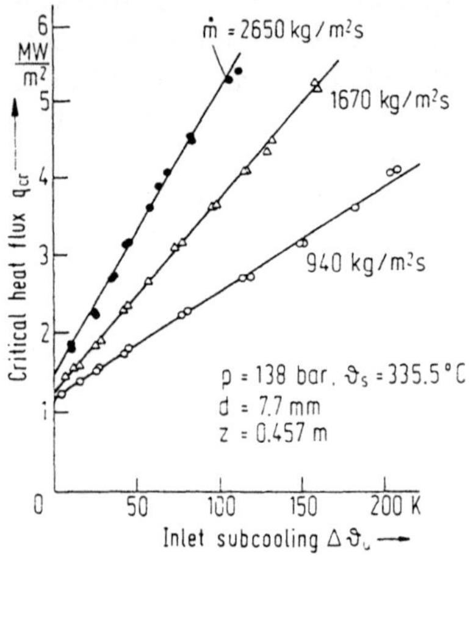

Fig. 13.32. Influence of subcooling on the critical heat flux of water, according to [13.54]

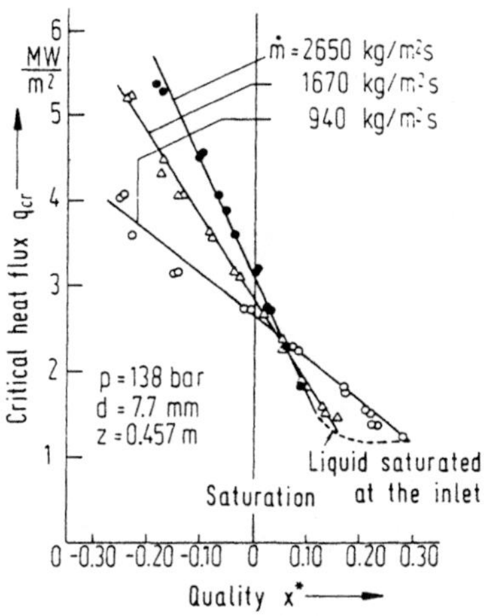

Fig. 13.33. Influence of quality on the critical heat flux of water, according to [13.49]

flux increases almost linearly with the inlet subcooling and it increases for a specified inlet subcooling with the mass flux.

The dependence of the critical heat flux upon the mass flux and the quality

$$q_{cr} = q_{cr}(\dot{m}, x^*)$$

at fixed values of z, d, p is more complicated. One finds this dependence, if one calculates the quality x^* according to Eq. (13.137) for the values in Fig. 13.32. As a result, one obtains Fig. 13.33. As this shows, in the range of subcooled boiling, $x^* < 0$, and at the specified quality, the critical heat flux increases with the mass flux. This corresponds to the results of Fig. 13.32. In the region of saturated boiling, $x^* > 0$, the curves $\dot{m} = \text{const}$ intersect, with the result that at sufficiently large quality, despite an increasing mass flux, the critical heat flux decreases again. This phenomenon has been confirmed by the data of Silvestri [13.55], but as of yet has not been satisfactorily explained.

13.9.2.2 The Dependence of the Critical Heat Flux on the Subcooling and Heated Length

We assume the mass flux $\dot{m}$, the tube diameter d and the pressure p of the fluid as fixed. According to data of Lee and Obertelli [13.56, 13.57], Fig. 13.34, the critical heat flux decreases with increasing quality. The dependence on the tube length is, however, slight. At the specified values of $\dot{m}, p, d$, the critical heat flux is determined mainly by the quality, so that Eq. (13.136), which is valid for a definite substance, can be simplified to

$$q_{cr} = q_{cr}(\dot{m}, x^*, d, p). \tag{13.140}$$

As the figure shows, given other parameters fixed, a definite heat flux is associated with a definite quality. The boiling length Δz at the critical heat flux conditions,

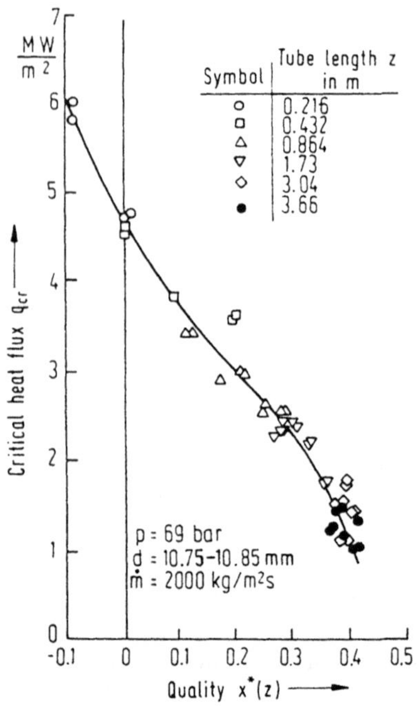

Fig. 13.34. Influence of tube length and of outlet quality on the critical heat flux, according to [13.56, 13.57]

calculated from the point $x^* = 0$ at which the mean liquid temperature is equal to the saturation temperature, can be determined from an energy balance:

$$q_{cr} d\pi \Delta z = \dot{m} \frac{d^2 \pi}{4} x^* \Delta h_v,$$

from which follows

$$\Delta z = \frac{\dot{m} d}{4 q_{cr}} x^* \Delta h_v. \tag{13.141}$$

We designate Δz as the length of boiling. Now the critical heat flux for a definite fluid can also be written, according to Eq. (13.140), in the form

$$q_{cr} = q_{cr}(\dot{m}, \Delta z, d, p). \tag{13.142}$$

It also follows from Eq. (13.140) that the quality upon reaching the critical heat flux is given by

$$x^*_{cr} = x^*_{cr}(\dot{m}, q_{cr}, d, p) \tag{13.143}$$

or because of Eq. (13.137) by

$$x^*_{cr} = x^*_{cr}(\dot{m}, \Delta z, d, p). \tag{13.144}$$

From this results the fact that 'the liquid portion that can be evaporated at the critical heat flux depends upon the boiling length, Fig. 13.35.

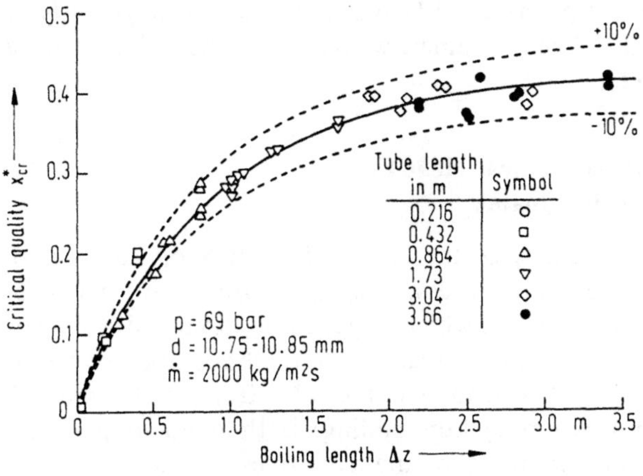

Fig. 13.35. Critical quality of water as a function of the boiling length, according to [13.56, 13.57]

13.9.2.3 The Dependence of the Critical Heat Flux on the Subcooling and the Tube Diameter

As Fig. 13.36 shows, at constant inlet subcooling, the critical heat flux increases with the tube diameter, if the other parameters of mass flux $\dot{m}$ and length z are kept constant. In the region of small tube diameters $d \leqq 12.8$ mm, under the conditions of Fig. 13.36, the critical heat flux rises proportionally with the inlet

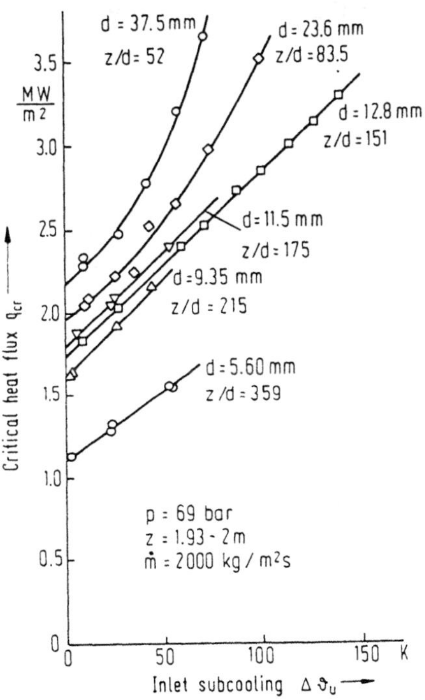

Fig. 13.36 Critical heat flux of water as a function of the inlet subcooling and the tube diameter, according to [13.56, 13.58]

subcooling; for larger tube diameters $d > 12.8$ mm the rise is stronger. Also at the specified quality and other fixed parameters, the critical heat flux increases with increasing tube diameter.

13.9.2.4 The Dependence of the Critical Heat Flux on the Pressure and the Quality

Figure 13.37 shows the dependence of the critical heat flux of water in the region of film boiling upon the relative pressure p/p_{cr} according to Eq. (13.150) of Doroshchuk and colleagues [13.59], which is yet to be discussed. It is valid for film boiling. Film boiling no longer occurs for qualities above $x^* = 0.5$. The heating surface dries out. The dotted curve $x^* = 0.5$, which would be calculated from Eq. (13.150) assuming film boiling, is then no longer valid. Instead, the critical heat flux adopts the smaller values of the solid curve for $x^* = 0.5$. These are calculated according to Eq. (13.151), which is yet to be discussed. For vanishing quality $x^* = 0$ the critical heat flux decreases sharply above the relative pressure $p/p_{cr} = 0.136$ ($p = 30$ bar). One has found by measurements that at pressures less than about $p/p_{cr} = 0.136$, the critical heat flux passes through a maximum and then decreases with increasing pressure. In several cases [13.60, 13.61], one has even observed a second maximum at pressures between 100 and 200 bar, which is the more pronounced, if the subcooling and the mass flux are raised and the ratio of the tube length to the diameter is lowered. If the entrance temperature is kept constant, curve $\vartheta_1 = 174\,°C$ in Fig. 13.37, then the subcooling in the inlet cross section increases

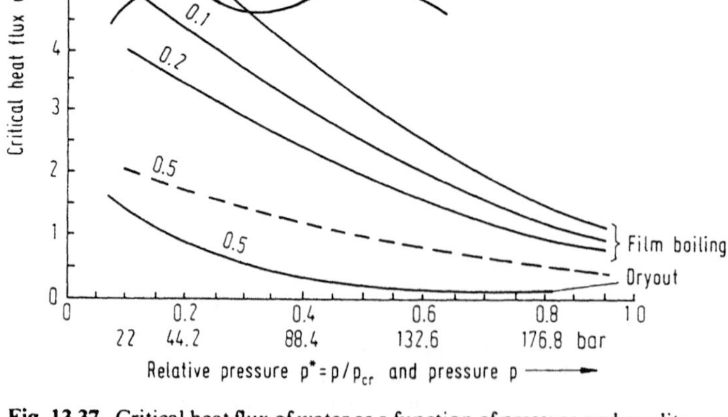

Fig. 13.37. Critical heat flux of water as a function of pressure and quality, according to Eqs. (13.150) and (13.151). $\vartheta_1 = 174\,°C$, according to Macbeth [13.60]

with the pressure of the system. The second maximum at higher pressures is amplified. As a whole, however, the critical heat flux along the curve $\vartheta_1 = 174\,°\mathrm{C}$ is more weakly dependent upon the pressure than at constant inlet subcooling, as is shown by the other curves in Fig. 13.37.

13.9.3 Application Formulas for the Critical Heat Flux

13.9.3.1 Vertical Tubes

Because of the great practical significance of the critical heat flux, especially for the design of boiling water reactors, numerous empirical correlations have also been developed on the basis of many experiments. Many of them are valid only over narrow ranges of parameters. Only a few equations are based upon models dealing with the type of the critical condition, i.e., based upon film boiling or the dryout of the heated surface. Most equations are of a purely empirical nature and do not differentiate between the types of critical heat flux. To these equations belongs the frequently used equation of Thomson and Macbeth [13.61], which is valid over a wide region of parameters. It was later simplified for practical purposes by Bowring [13.62]. As already presented, the critical heat flux corresponding to Eq. (13.140)

$$q_{\mathrm{cr}} = q_{\mathrm{cr}}(\dot{m}, x^*, d, p)$$

is dependent upon the local quality $x^*(z)$. A linear dependence upon the local quality $x^*(z)$ is assumed, which is not particularly accurate for high qualities, as is shown, for example, in Fig. 13.34. Therefore, let there be

$$q_{\mathrm{cr}} = A' - B'x^*, \tag{13.145}$$

where for a definite substance the values A', B' are still dependent upon the mass flux, the tube diameter, and the pressure: $A'(\dot{m}, d, p)$ and $B'(\dot{m}, d, p)$.

With the aid of Eq. (13.137) one can also write for this

$$q_{\mathrm{cr}} = A' - \frac{B'}{\Delta h_{\mathrm{v}}}\left(\frac{q_{\mathrm{cr}} 4z}{\dot{m}d} - \bar{c}_{\mathrm{pL}}\Delta\vartheta_{\mathrm{u}}\right).$$

Solving explicitly for the critical heat flux, one obtains

$$q_{\mathrm{cr}} = \frac{\dfrac{A'}{B'}\dfrac{\Delta h_{\mathrm{v}}\dot{m}d}{4} + \dfrac{\dot{m}d}{4}\bar{c}_{\mathrm{pL}}\Delta\vartheta_{\mathrm{u}}}{\dfrac{\Delta h_{\mathrm{v}}\dot{m}d}{4B'} + z}$$

or

$$q_{\mathrm{cr}} = \frac{A\dfrac{\Delta h_{\mathrm{v}}\dot{m}d}{4} + \dfrac{\dot{m}d}{4}\bar{c}_{\mathrm{pL}}\Delta\vartheta_{\mathrm{u}}}{B\dfrac{\Delta h_{\mathrm{v}}\dot{m}d}{4} + z}. \tag{13.146}$$

Here the values A and B are functions of mass flux, pressure, and diameter, which can be written for water, according to the equations of Bowring [13.62], in the following form:

$$A = 2.317 \frac{F_1(p)}{1.0 + 0.0143 F_2(p) d^{1/2} \dot{m}}, \tag{13.147}$$

$$B = 0.308 \frac{F_3(p)/\Delta h_v}{1.0 + 0.347 F_4(p)(\dot{m}/1356)^n}. \tag{13.148}$$

The tube diameter d is to be entered in m, the entire mass flux $\dot{m}$ in kg/m²s, the evaporation enthalpy Δh_v in J/kg, the mean specific heat $\bar{c}_{pL}$ of the liquid between entrance temperature ϑ_1 and saturation temperature ϑ_s in J/kgK, and the critical heat flux q_{cr} is obtained in W/m². The exponent n is pressure dependent and is given by

$$n = 2.0 - 0.00725 p \tag{13.149}$$

with the system pressure p in bar. The pressure functions F_1 to F_4 are given in Table 13.6. The empirical correlation is based upon measurements with water in the following ranges:

$$2\,\text{bar} \leqq p \leqq 190\,\text{bar}$$

$$0.002\,\text{m} \leqq d \leqq 0.045\,\text{m}$$

$$0.15\,\text{m} \leqq z \leqq 3.7\,\text{m}$$

$$136\,\text{kg/m}^2\text{s} \leqq \dot{m} \leqq 18600\,\text{kg/m}^2\text{s}.$$

The root mean square error is given at 7%, and 95% of the test data lie in a confidence interval of ± 14%.

By a comparison of roughly 3000 series of experiments, including those from the USSR, Drescher and Köhler [13.63] have shown that in the region of *film boiling in water*, an equation developed by Doroshchuk et al. [13.59], yields the most precise values. It reads

$$q_{cr} = (10.3 - 17.5 p^* + 8 p^{*2}) \left(\frac{8 \cdot 10^{-3}}{d}\right)^{1/2} \left(\frac{\dot{m}}{1000}\right)^{0.68 p^* - 1.2 x^* - 0.3} e^{-1.5 x^*}. \tag{13.150}$$

In it, the tube inner diameter d is to be entered in m and the mass flux $\dot{m}$ in kg/m²s; $p^* = p/p_{cr}$ is the pressure relative to the critical pressure. One obtains q_{cr} in MW/m². The equation reproduces test data in the following regions:

$$29\,\text{bar} \leqq p \leqq 196\,\text{bar corresponding to } 0.13 \leqq p^* \leqq 0.89$$

$$500\,\text{kg/m}^2\text{s} \leqq \dot{m} \leqq 5000\,\text{kg/m}^2\text{s}$$

$$0\,\text{K} \leqq \Delta\vartheta_u \leqq 75\,\text{K}$$

$$4 \cdot 10^{-3}\,\text{m} \leqq d \leqq 25 \cdot 10^{-3}\,\text{m}.$$

Table 13.6. Pressure functions in Eqs. (13.147) and (13.148) for calculation of the critical heat flux of water, according to Bowring [13.62]

p bar	F_1	F_2	F_3	F_4
1	0.478	1.782	0.400	0.0004
5	0.478	1.019	0.400	0.0053
10	0.478	0.662	0.400	0.0166
15	0.478	0.514	0.400	0.0324
20	0.478	0.441	0.400	0.0521
25	0.480	0.403	0.401	0.0753
30	0.488	0.390	0.405	0.1029
35	0.519	0.406	0.422	0.1380
40	0.590	0.462	0.462	0.1885
45	0.707	0.564	0.538	0.2663
50	0.848	0.698	0.647	0.3812
60	1.043	0.934	0.890	0.7084
68.9	1.000	1.000	1.000	1.000
70	0.984	0.995	1.003	1.030
80	0.853	0.948	1.033	1.322
90	0.743	0.903	1.060	1.647
100	0.651	0.859	1.085	2.005
110	0.572	0.816	1.108	2.396
120	0.504	0.775	1.129	2.819
130	0.446	0.736	1.149	3.274
140	0.395	0.698	1.168	3.760
150	0.350	0.662	1.186	4.277
160	0.311	0.628	1.203	4.825
170	0.277	0.595	1.219	5.404
180	0.247	0.564	1.234	6.013
190	0.220	0.534	1.249	6.651
200	0.197	0.506	1.263	7.320

The standard deviation between the calculated and the measured heat flux is given at 16%. By iteration, one can easily determine from Eq. (13.150) the critical quality at the prescribed critical heat flux.

Data for the critical heat flux of water in the region of the dryout of the heating surface could be reproduced best according to the investigation of Drescher and Köhler by an equation of Kon'kov [13.64]. It reads

$$q_{cr} = c_1 x^{*-0.8} \dot{m}^{-2.664} d^{-0.56} e^{c_2 p}. \tag{13.151}$$

In it, the tube diameter d is to be entered in mm, the mass flux $\dot{m}$ in kg/m^2s, and the pressure p in bar. One obtains the critical heat flux in MW/m^2. The values c_1 and c_2 are dependent upon the pressure region and are given in Table 13.7. The equation is valid in the region

$$200\,kg/m^2s \leqq \dot{m} \leqq 5000\,kg/m^2s,$$

$$4\,mm \leqq d \leqq 32\,mm.$$

At the prescribed critical heat flux q_{cr}, the critical quality can be calculated easily from Eq. (13.151).

Table 13.7. Constants c_1 and c_2 in Eq. (13.151) for calculation of the critical heat flux

Pressure range		
4.9 bar $\leqq p <$ 29.4 bar	$1.8447 \cdot 10^5$	0.1372
29.4 bar $\leqq p <$ 98 bar	$2.0048 \cdot 10^7$	−0.0204
98 bar $\leqq p <$ 196 bar	$1.1853 \cdot 10^9$	−0.0636

As Fig. 13.31 showed, the critical heat flux decreases in the region of the dryout of the heating surface more strongly with the quality than in the region of film boiling. If one imagines the two curves for film boiling and for dryout of the heating surface in Fig. 13.31 as extended beyond their point of intersection (hatched line), then it is recognized that the smaller value is the critical heat flux. For practical calculations, one, therefore, determines the critical heat flux according to Eqs. (13.150) and (13.151). The smaller value is decisive. With it one can simultaneously decide which type of boiling crisis is occurring.

Both equations presuppose steady flow just as Eq. (13.146) does. If the mass flux, for example, is variable as the result of pressure fluctuations in the system, then the boiling crisis can already begin with much smaller mass fluxes and qualities than with steady values.

For low pressures $p < 5$ bar and small mass fluxes $\dot{m} \leqq 300\ \mathrm{kg/m^2s}$, Alad'yev et al. [13.65] have published a simple correlation

$$q_{cr} = k\left(\frac{\dot{m}}{L/d}\right)^{0.8} (1 - 2x_E^*) \qquad (13.152)$$

with $k = 460$ for water according to Thomson and Macbeth [13.61]. It is valid for the ratio of tube length L to inner diameter d

$$10 \leqq L/d \leqq 100$$

and for subcooled boiling with qualities (thermodynamic qualities) in the inlet cross section

$$-0.2 \leqq x_E^* < 0.$$

The mass flux $\dot{m}$ is to be entered in $\mathrm{kg/m^2s}$. One obtains the critical heat flux in $\mathrm{kW/m^2}$.

13.9.3.2 Horizontal and Inclined Tubes

In horizontal or inclined tubes, the influence of gravity is negligible in comparison to the inertia forces only at extremely high mass fluxes over 4000 to 6000 $\mathrm{kg/m^2s}$, so that the inclination of the tube has no influence on the boiling crisis. With smaller mass fluxes, the boiling crisis begins at the upper side of the tube, whereas the tube below can still be wetted by liquid.

In order to estimate the ratio of inertia to gravity forces, let us consider a gas volume V_G moving in a liquid in a horizontal or inclined tube. Let the gas velocity be w_G, and let the gas occupy a cross section A_G. The inertia force on the gas volume is then

$$F_T = \dot{M}_G w_G = \dot{m}_G A_G w_G = \dot{m}_G^2 A_G / \varrho_G.$$

Simultaneously, buoyant forces affect the gas volume

$$F_A = V_G(\varrho_L - \varrho_G)g.$$

The ratio of the two forces is

$$\frac{F_T}{F_A} = \frac{\dot{m}_G^2/\varrho_G}{(\varrho_L - \varrho_G)g} \frac{A_G}{V_G}.$$

We now assume that the gas substantially fills the tube cross section. Then there is

$$\frac{A_G}{V_G} = \frac{d\pi L}{\frac{d^2\pi}{4} L} = \frac{4}{d} \quad \text{and thus}$$

$$\frac{F_T}{F_A} = \frac{\dot{m}_G^2/\varrho_G}{(\varrho_L - \varrho_G)g} \frac{4}{d}.$$

Instead of which one can also consider the quantity

$$\left(\frac{F_T}{F_A}\right)^{1/2} = \frac{\dot{m}_G/\varrho_G^{1/2}}{[(\varrho_L - \varrho_G)gd]^{1/2}} \cdot 2.$$

We assume now that gas and liquid are in phase equilibrium, and may, therefore, set $\varrho_G = \varrho''$, $\varrho_L = \varrho'$. In addition, upon inception of the boiling crisis, $\dot{m}_G = x_{cr}^* \dot{m}$. The thus obtained Froude number

$$Fr = x_{cr}^* \dot{m} \left[\frac{1}{\varrho''(\varrho' - \varrho'')gd}\right]^{1/2} \tag{13.153}$$

represents the ratio of inertia to buoyant forces in a two-phase flow.

For $Fr > 7$, the forces of inertia are so great in comparison to the forces of gravity that no noticeable stratification occurs [13.66]. The slope of the tube then has no influence on the location of the boiling crisis, and the equations for the vertical tubes can be used. With smaller Froude numbers, especially from $Fr \leqq 3$, the stratification is very pronounced, and the influence of the slope of the tube on the boiling crisis must be considered. In horizontal tubes, the upper side dries out even at small qualities, whereas the tube below remains wetted until complete evaporization. Because of the heat flow from the unwetted upper side to the lower side, the temperature rise at the inception of the boiling crisis is smaller in a horizontal tube than in a vertical tube. Figure 13.38 shows the pattern of wall and mean fluid temperature in a horizontal tube. The wall temperature passes through two relative maxima. Already in the region of the subcooled boiling $x^* < 0$, so much vapor collects on the upper side of the tube,

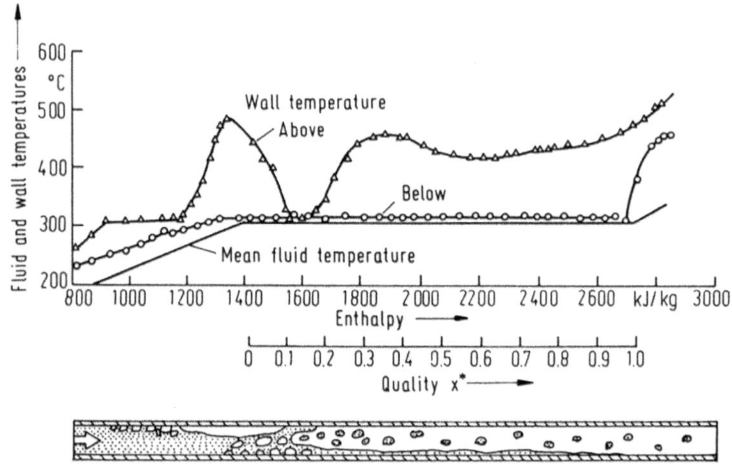

Fig. 13.38. Wall and mean fluid temperatures in a horizontal, uniformly heated tube, according to [13.72]. $p = 100$ bar, $\dot{m} = 500\,\text{kg/m}^2\text{s}$, $q = 300\,\text{kW/m}^2$, $d = 24.3$ mm, fluid: water

that the tube dries out on top and the wall temperature rises. Further downstream, with increasing quality, the fluid velocity also increases. The flow changes from stratified to slug flow or even to annular flow. The upper side is again wetted, so that the wall temperature falls, until at a sufficiently high quality the upper side once again dries out, and the wall temperature climbs again. It falls off somewhat in the adjoining region of spray flow, and finally rises downstream to the degree that the liquid disappears.

Until now, measurements for the critical heat flux in horizontal or inclined tubes [13.67] have been limited mostly to a narrow region of parameters, so that at present there are no correlations for them as there are for vertical tubes. The same is true for curved or helical tubes. Mention should be made of the articles in the VDI-Wärmeatlas [13.68].

13.9.3.3 Nonuniform Heating Around the Tube Circumference

If evaporator tubes are heated on one side, as for example, the tubes of a boiler heated by radiation, then the heat flux is not uniformly distributed around the circumference. The maximum of the heat flux is then clearly larger than the mean heat flux computed from the supplied heat flow rate and the surface area, Fig. 13.39. Usually, one can approximate the heat flux on the inside of the tube to a good approximation by a cosine profile

$$\frac{q}{\bar{q}} = \frac{1 - q_{\max}/\bar{q}}{\dfrac{2}{\pi} - 1}(\cos\varphi - 1) + \frac{q_{\max}}{\bar{q}}, \qquad (13.154)$$

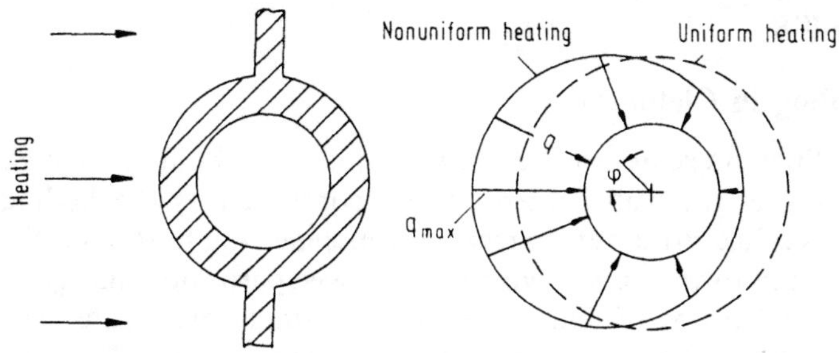

Fig. 13.39. Nonuniform heating of a tube. Distribution of the heat fluxes over the inner circumference

where $\bar{q}$ is the mean heat flux computed from the supplied heat flow rate and the surface, and q_{max} is its maximum value. This profile is so chosen that it fulfills the conditions

$$q = q_{max} \quad \text{for} \quad \varphi = 0$$

and

$$\bar{q} = \frac{2}{\pi} \int_{\varphi=0}^{\pi/2} q \, d\varphi .$$

One introduces now a *degree of irregularity* or *flux tilt* $f = q_{max}/\bar{q}$, which indicates by how much the maximum heat flux deviates from the mean heat flux. It is

$$\frac{q}{\bar{q}} = \frac{1-f}{\frac{2}{\pi} - 1} (\cos\varphi - 1) + f. \tag{13.155}$$

The profile of the heat flux is completely characterized by the degree of irregularity, if one assumes a cosine-shaped distribution. With a large degree of irregularity, and, thus, with a large maximum heat flux in comparison to the mean value, the boiling crisis can begin at the location of the maximum heat flux, although the mean heat flux is moderate, as the data of Lee [13.69] and Alekseev [13.70] showed. Reference values for the critical heat flux are obtained on the basis of these data by first computing the critical heat flux from one of the empirical equations for the uniformly heated tube, Eqs. (13.150) to (13.152). From this one obtains the mean critical heat flux, by multiplying these calculated values according to the suggestion of Collier [13.71] by the factor 0.9, if the degree of irregularity is below 1.5 and by the factor 0.8, if the degree of irregularity lies between 1.5 and 2.0.

13.10 Spray Cooling

13.10.1 Spray Cooling in Channels

Whereas at low qualities nucleate boiling converts into film boiling, at high qualities the liquid film at the wall is torn up and consumed by the boiling process. The heating surface dries out. Farther downstream the heat from the wall is transferred mainly to the vapor, which becomes superheated and gives off heat to the entrained drops of liquid, which increasingly evaporate. The vapor forms the continuous phase, while the entrained drops of liquid are the dispersed phase. The heat transfer to such a flow is generally called spray cooling. One speaks also of heat transfer in the post-dryout region.

During spray cooling, drops of liquid deposit upon the wall and destroy there the vapor boundary layer while simultaneously absorbing heat. The heat transfer is consequently considerably better than during film boiling. Heat transfer coefficients for water lie during spray cooling between $5 \cdot 10^3$ and $5 \cdot 10^4$ W/m²K, depending upon the mass flux and the quality, whereas during film boiling, values are achieved between roughly 300 and 1500 W/m²K.

In once-through vapor generators in power plants, in refrigeration plants, or in chemical plants, heat is almost always transferred over a large part of the heating surface by spray cooling, because pure liquid enters the vapor generator and superheated vapor leaves it. Occasionally, the spray cooling is intentionally produced by injecting drops of liquid into a flow of vapor or gas, in order to augment the cooling effect.

Figure 13.40 shows schematically the course of the wall, vapor, and liquid temperatures, for constant heat flux, as a function of the quality x^* according to Eq. (13.18)

$$x^* = \frac{1}{\Delta h_v}\left(\frac{\Phi}{\dot{M}} + h_1 - h'\right),$$

which can be ascribed to every point z on the basis of the energy balance. Until the boiling crisis, the vapor and liquid are in thermal equilibrium. After the inception of the boiling crisis, the wall is cooled mainly by vapor and scarcely at all by liquid. The heat transfer between wall and fluid becomes poorer. In

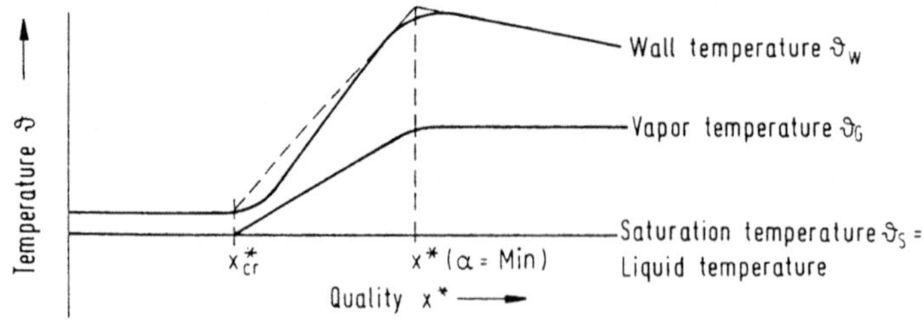

Fig. 13.40. Behavior of the wall, vapor, and liquid temperatures after dryout

the case of imposed heat flux, the wall temperature rises. Because heat is transferred to the vapor, its temperature also rises. The drops of liquid remain at the saturation temperature, if the superheating of the very small drops is disregarded. A nonequilibrium condition between vapor and liquid is thus established. Because the heat transfer coefficient at first decreases as a result of the dryout of the wall and because the heat is transferred mainly at the wall, the wall temperature rises more sharply than the vapor temperature because $q \approx \alpha(\vartheta_w - \vartheta_G)$ (α decreases, ϑ_G increases). The heat transfer coefficient falls and reaches a minimum at the point $x^*(\alpha = \text{Min})$ in Fig. 13.40.

As a result of the vapor superheating, farther downstream the heat is increasingly transferred from the vapor to the drops, which evaporate more and more. Thereby, the quality and the velocity of the vapor flow increase. Downstream, the heat transfer between wall and fluid improves, so that the wall temperature then rises only slowly or, as shown in Fig. 13.40, even falls. If the droplets have evaporated to a great extent, then the relations for single-phase heat transfer apply. Vapor and wall temperature rise steadily for this case of imposed heat flux.

Because there is thermal nonequilibrium between vapor and liquid after the dryout, the actual quality x^*_{real} is different from the quality x^*_{th} computed assuming thermal equilibrium and which is also called the thermodynamic quality. The actual quality is obtained from an energy balance, Fig. 13.41, between inlet cross section and any random cross section after the dryout

$$\dot{M}h_1 + \Phi = \dot{M}x^*_{\text{real}}h_G + \dot{M}(1 - x^*_{\text{real}})h',$$

from which follows

$$x^*_{\text{real}} = \frac{1}{h_G - h'}\left(\frac{\Phi}{\dot{M}} + h_1 - h'\right). \tag{13.156}$$

The actual quality coincides with the thermodynamic quality, if the vapor and liquid are in thermal equilibrium. Then, $h_G = h''$ in the above equation, and one obtains $x^*_{\text{th}} = x^*_{\text{real}}$. Because part of the supplied heat is consumed for the superheating of the vapor, less vapor is formed than in equilibrium in which only saturated vapor is produced. This yields

$$x^*_{\text{real}} < x^*_{\text{th}},$$

which is also shown by comparison of Eq. (13.156) with Eq. (13.18).

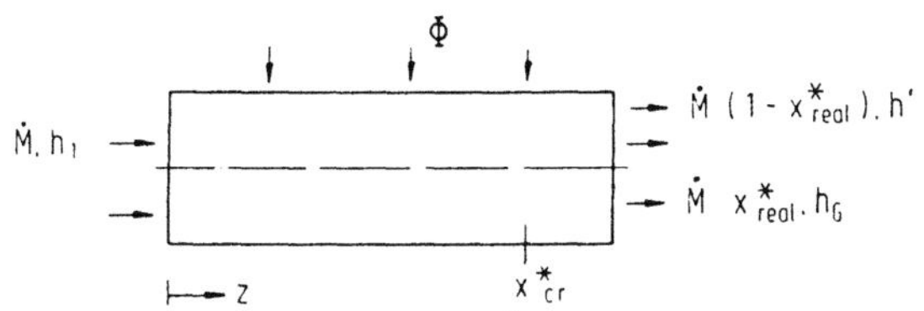

Fig. 13.41. Determination of the actual quality x^*_{real}

Because the vapor temperature is not constant over a cross section but rather rises toward the wall, the drops of liquid in the wall-adjacent boundary layer evaporate before those in the core flow. At very high wall temperatures, the heat transfer by radiation from the wall to the drops can no longer be disregarded.

The heat is transferred from the wall to the fluid in different ways:

—by convection from the wall to the vapor;
—by convection from the superheated vapor to the drops in the vapor core;
—by convection and conduction to the drops, which penetrate into the thermal boundary layer without wetting the wall; one speaks of dry collisions;
—by convection and conduction from the wall to drops, which touch the wall and briefly roll or glide along it; one speaks of wet collisions;
—by radiation from the heating surface to the vapor and to the drops.

Of these components, the heat exchange between the wall and the drops, which penetrate into the boundary layer or touch the wall, is small in comparison to the heat flow rate transferred convectively to the vapor. This is due to the great difference between the wall and saturation temperatures, according to the data of Iloeje et al. [13.73]. The radiation component is also very small even at 1000 °C. Heat transfer coefficients for radiation are then in the order of magnitude of 100 W/m^{2}K, whereas the heat transfer coefficients for convection lie between $5 \cdot 10^3$ and $5 \cdot 10^4$ W/m^{2}K. The controlling mechanisms for the heat transfer are, accordingly, the heat transfer by convection from the wall to the vapor and by convection from the superheated vapor to the drops in the vapor core. Almost all models of computation are based upon this concept.

How strongly the vapor is superheated depends for a given fluid upon the mass flux and the pressure. If mass flux and pressure are low, then the heat transfer between vapor and drops is slight, the drops are large, and the vapor, therefore, is strongly superheated. In the limiting case, the supplied heat serves exclusively for the superheating of the vapor; the actual quality remains constant after the critical heat flux. For imposed heat flux, the vapor temperature and the wall temperature rise at first linearly with the flow path or, as the case may be, with the thermodynamic quality, Fig. 13.42. The heat transfer between vapor

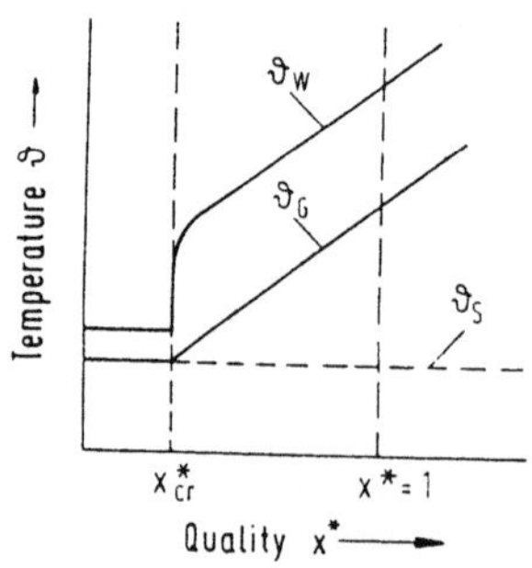

Fig. 13.42. Behavior of wall, vapor, and liquid temperatures illustrating thermal nonequilibrium between vapor and liquid drops

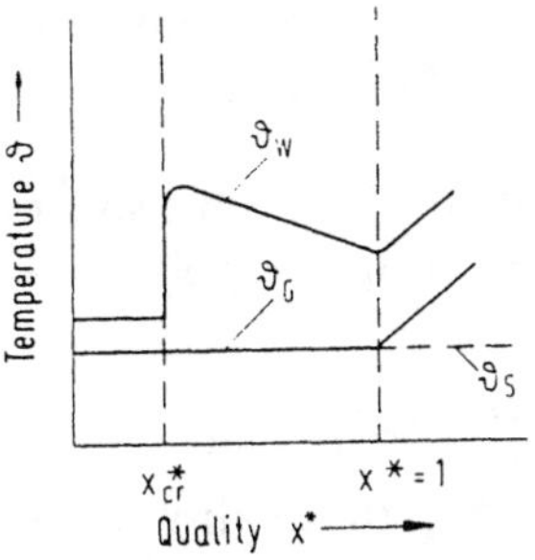

Fig. 13.43. Behavior of wall, vapor, and liquid temperatures with thermal equilibrium between vapor and liquid drops

and drops is completely eliminated, and there prevails complete thermodynamic nonequilibrium. Conversely, if mass flux and pressure are large, then the heat transfer rate between vapor and drops is large, the drops are small, and the vapor is, therefore, only slightly superheated. In the limiting case, the heat transfer rate between vapor and drops is infinitely large. The supplied heat serves solely for the evaporation of the drops. Vapor and liquid both have saturation temperature and are, therefore, in thermal equilibrium. The thermodynamic quality is then identical with the actual quality. The rise in the actual quality causes an increase in the vapor velocity, which removes more heat from the heating surface, with the result that the wall temperature is lowered, Fig. 13.43.

The actual wall and fluid temperatures lie between the two extreme cases of thermal equilibrium and of complete nonequilibrium. Actually, after the inception of the boiling crisis, the vapor and liquid temperatures increasingly separate. At first a thermal imbalance is formed. This is then eliminated further downstream as the liquid drops evaporate, depending upon the extent of the heat transfer between vapor and drops.

13.10.1.1 A Calculation Model

Numerous approximation procedures have been developed for the calculation of heat transfer during spray cooling. A survey is found in the works of Collier [13.74] and of Mayinger and Langner [13.75]. Of the many approximate procedures, we shall discuss here only the one suggested by Bennett and Hewitt [13.76], because it is plausible and has been well tested by experiments. According to it, one can imagine the heat transfer as taking place in three stages in series.

In a first stage after the dryout, the heat is transferred mainly to the vapor. Drops do not yet evaporate because for this to happen the vapor must first be sufficiently superheated. In Fig. 13.40, this is the stage between x^*_{cr} and $x^*(\alpha = \text{Min})$. The heat transfer between vapor and drops is poor. The actual quality remains at first unchanged and equal to that at the point of dryout.

In a second stage—in Fig. 13.40 $x^* > x^*(\alpha = \text{Min})$—the superheating of the vapor is large enough so that drop evaporation sets in. The vapor temperature no longer rises appreciably. According to the suggestion of Köhler [13.72], this process can be simulated mathematically by assuming the enthalpy of the vapor to be constant. All of the heat supplied to the wall serves for the evaporation of part of the drops and for the heating of the newly formed vapor to the temperature of the vapor already present.

There follows a third stage in which the vapor content has become so large that the relations for single-phase convective heat transfer again apply.

The energy balance for the first stage reads

$$\dot{M}x^*_{\rm cr}h''_{\rm cr} + \dot{M}(1 - x^*_{\rm cr})h'_{\rm cr} + qA_1 = \dot{M}x^*_{1\,\rm real}h_{\rm G1} + \dot{M}(1 - x^*_{1\,\rm real})h_{\rm L1},$$

if A_1 is the heat transfer surface area in the first stage, $x^*_{1\,\rm real}$ the quality, $h_{\rm G1}$ the specific enthalpy of the vapor, and $h_{\rm L1}$ the enthalpy of the liquid at the end of the first stage. Now, however, the quality in the first stage is constant at $x^*_{\rm cr} = x^*_{1\,\rm real}$ and, because, as assumed, heat is supplied only to the vapor, the specific enthalpy of the liquid is constant at $h'_{\rm cr} = h_{\rm L1}$. The energy balance can thus be rearranged into

$$h_{\rm G1} = h''_{\rm cr} + \frac{qA_1}{\dot{M}x^*_{\rm cr}}. \tag{13.157}$$

From this is obtained the vapor temperature $\vartheta_{\rm G1}$ at the end of the first stage, if the surface area A_1 is known. Assuming that all the supplied heat serves to produce vapor, one finds the thermodynamic quality $x^*_{1\rm th}$ at the end of the first stage

$$qA_1 = \dot{M}(x^*_{1\rm th} - x^*_{\rm cr})\Delta h_{\rm v}.$$

Here a constant evaporation enthalpy is assumed. Together with Eq. (13.157), this yields for the thermodynamic quality at the end of the first stage

$$x^*_{1\rm th} = x^*_{\rm cr}\left(1 + \frac{h_{\rm G1} - h''_{\rm cr}}{\Delta h_{\rm v}}\right). \tag{13.158}$$

In order to calculate from Eq. (13.157) the temperature $\vartheta_{\rm G}$ at any random point of the first stage, one must substitute in place of A_1 the heat transfer surface area up to this point. Thereby, the wall temperature in the first stage is also established, because

$$q = \alpha_{\rm 2Ph}(\vartheta_{\rm w} - \vartheta_{\rm G})$$

where $\alpha_{\rm 2Ph}$ is the heat transfer coefficient of the two-phase flow, which can be calculated according to the application equations of Sect. 13.8. Because of the high quality and especially because of the large void fraction after the inception of spray cooling, one can also calculate, as shown by Köhler [13.72], the heat transfer coefficient by using empirical correlations for the heat transfer of single-phase flows and by forming there the Reynolds number with the vapor velocity, assuming that there is no slip between vapor and liquid. Then, according to Eq. (13.17), the void fraction is

$$\varepsilon = \frac{x^*}{x^* + (1 - x^*)\varrho_{\rm G}/\varrho_{\rm L}},$$

and the vapor velocity, according to Eq. (13.14), is

$$w_G = \frac{\dot{m}}{\varrho_G}[x^* + (1 - x^*)\varrho_G/\varrho_L],$$

where $x^* = x^*_{real}$ is the actual quality. This yields the Reynolds number

$$Re = \frac{w_G \varrho_G d}{\eta_G} = \frac{\dot{m} d}{\eta_G}[x^*_{real} + (1 - x^*_{real})\varrho_G/\varrho_L]. \qquad (13.159)$$

For the calculation of the heat transfer coefficient α_{2Ph} one can apply the simple McAdams equation or the more accurate empirical equation formulated by Gnielinski [13.77]

$$Nu = \frac{\alpha_{2Ph} d}{\lambda_G} = \frac{(\zeta/8)(Re - 1000)Pr_G}{1 + 12.7\sqrt{\zeta/8}(Pr_G^{2/3} - 1)} \qquad (13.160)$$

with the friction factor

$$\zeta = (0.79 \ln Re - 1.64)^{-2}. \qquad (13.161)$$

In order to determine the as yet unknown heat transfer surface area A_1 of the first stage or the enthalpy h_{G1} in Eqs. (13.157) and (13.158), we formulate the energy balance of the second stage. In this stage, a liquid mass flow rate $\Delta\dot{M}$ is evaporated and subsequently superheated to the temperature of the vapor. The energy balance now reads

$$\dot{M}_G h_{G1} + \dot{M}_L h_{L1} + qA_2 = (\dot{M}_G + \Delta\dot{M})h_{G2} + (\dot{M}_L - \Delta\dot{M})h_{L2},$$

where A_2 is the heat transfer surface area of the second stage, h_{G2} the specific enthalpy of the vapor, and h_{L2} the specific enthalpy of the liquid at the end of the second stage. Because the vapor temperature no longer rises, the specific enthalpy of the vapor remains constant at $h_{G1} = h_{G2}$. In addition, the liquid drops are in the saturated state, and thus by good approximation $h_{L1} = h_{L2} = h'$. Using this, we can simplify the energy balance to

$$qA_2 = \Delta\dot{M}(h_{G1} - h') = \Delta\dot{M}(h_{G1} - h'') + \Delta\dot{M}\Delta h_v.$$

The liquid mass evaporated per unit time is determined from the heat flow rate, which is supplied convectively to the liquid drops across their surface area A_T

$$\alpha_G A_T(\vartheta_G - \vartheta_s) = \Delta\dot{M}\Delta h_v$$

or with $h_G - h'' = h_{G1} - h'' = \bar{c}_{pG}(\vartheta_G - \vartheta_s)$

$$\Delta\dot{M} = \frac{\alpha_G A_T(h_{G1} - h'')}{\bar{c}_{pG}\Delta h_v}.$$

Thus, the energy balance now reads

$$qA_2 = \frac{\alpha_G A_T}{\bar{c}_{pG}\Delta h_v}(h_{G1} - h'')^2 + \frac{\alpha_G A_T}{\bar{c}_{pG}}(h_{G1} - h'').$$

From this is obtained the enthalpy difference $h_{G1} - h''$ as

$$h_{G1} - h'' = \frac{\Delta h_v}{2}\left(\sqrt{1 + \frac{4qA_2\bar{c}_{pG}}{\alpha_G A_T \Delta h_v}} - 1\right). \tag{13.162}$$

This equation presupposes that one knows the wall surface area A_2, the heat transfer coefficient α_G between vapor and liquid drops, and the representative drop surface area A_T. For this, one must know how the drop spectrum is dependent upon the relative velocity between gas and liquid and upon the pertinent property values. Such measurements are very difficult and have not as yet been carried out with sufficient accuracy. On the other hand, the unknowns $\alpha_G A_T/A_2$ from Eq. (13.162) can be calculated from the measured variables and correlations stated for these. Based upon experiments with water in the regions

$$50\,\text{bar} \leqq p \leqq 200\,\text{bar}$$

$$300\,\text{kg/m}^2\text{s} \leqq \dot{m} \leqq 2500\,\text{kg/m}^2\text{s}$$

$$100\,\text{kW/m}^2 \leqq \dot{q} \leqq 600\,\text{kW/m}^2$$

Köhler [13.72] calculated the unknowns $\alpha_G A_T/A_2$ from Eq. (13.162) and has correlated the values empirically. Accordingly,

$$\alpha_G \frac{A_T}{A_2} = 2.336\left(\frac{\dot{m}}{b}\right)^{1.33} \quad \text{for} \quad \frac{\dot{m}}{b} \leqq 1249 \frac{\text{kg}}{\text{m}^2\text{s}}\frac{1}{\text{mm}}$$

and

$$\alpha_G \frac{A_T}{A_2} = 1.231 \cdot 10^{-8}\left(\frac{\dot{m}}{b}\right)^4 \quad \text{for} \quad \frac{\dot{m}}{b} > 1249 \frac{\text{kg}}{\text{m}^2\text{s}}\frac{1}{\text{mm}}. \tag{13.163}$$

Here, the quantity b is the Laplace constant, which governs the drop size,

$$b = \sqrt{\frac{2\sigma}{g(\varrho' - \varrho'')}}.$$

One obtains $\alpha_G A_T/A_2$ in $\text{W/m}^2\text{K}$.

In such a purely empirical correlation, the physical processes, which have not yet been sufficiently researched, are hidden. The empirical constants correct also the approximate assumptions of the model, allowing, for example, for the fact that, contrary to the assumption, heat from the wall is transferred to the drops striking it.

According to Köhler, the second stage ends when the thermodynamic quality reaches the value

$$x^*_{2\,\text{th}} = 0.7 + 0.02p \tag{13.164}$$

where the pressure p is to be entered in bar. According to Eq. (13.16), void fractions ε of 0.99 correspond to this value, if one assumes equal velocities of vapor and liquid. On the other hand, the thermodynamic quality at the end of the second stage is obtained also from an energy balance for this stage, which reads, when assuming a constant evaporation enthalpy

$$\frac{qA_2}{\dot{M}} = (x^*_{2\text{th}} - x^*_{1\text{th}})\Delta h_v.$$

Entering the quality according to Eq. (13.164), yields the surface area A_2 of the second stage. For the determination of the wall temperature from

$$\vartheta_w = \vartheta_G + q/\alpha_{2\text{Ph}}$$

one can calculate the heat transfer coefficient of the two-phase flow $\alpha_{2\text{Ph}}$ from Eq. (13.160).

For the *practical calculation* of the heat transfer, the quality, and the wall temperature according to the described model, it is expedient to proceed in the following manner:

—One calculates first the enthalpy difference $h_{G1} - h''$ of the second stage from Eq. (13.162) while allowing for Eq. (13.163). Because the specific enthalpy of the vapor in the second stage is assumed to be constant, one then also knows the enthalpy h_{G1} or, as the case may be, the vapor temperature at the end of the first stage.

—With the specific enthalpy h_{G1}, the surface area A_1 of the first stage is also established according to Eq. (13.157)

—The quality at the end of the first stage is obtained from Eq. (13.158), and the wall temperature from

$$\vartheta_w = \vartheta_G + q/\alpha_{2\text{Ph}}$$

with the heat transfer coefficient of the two-phase flow according to Eq. (13.160).

—The quality at the end of the second stage is obtained from Eq. (13.164). The wall temperature is obtained from the same equations as for the first stage.

—In the third stage $x^* > x^*_{2\text{th}}$ ($x^*_{2\text{th}}$ according to Eq. (13.164)), the relations for the single-phase heat transfer apply.

13.10.1.2 Application Equations

Besides the just discussed model, which is supported by numerous data, there are many application equations for the calculation of heat transfer coefficients, which reproduce data in rather narrowly limited ranges of conditions. A survey is found in the mentioned work by Mayinger and Langner [13.75]. It shows that the heat transfer coefficients calculated according to the application equations deviate strongly from one another, with the result that one also obtains very divergent wall temperatures, depending upon which equation is used. A much-used correlation, which was also recommended by official authorities for the calculation of heat transfer during emergency cooling of nuclear reactors, has been published by Groeneveld [13.78]. It reads

$$Nu = aRe^{n_1} Pr_G^{n_2} Y^{n_3} \tag{13.165}$$

with

$$Nu = \frac{\alpha_{2Ph} d}{\lambda_G}, \quad Re = \frac{\dot{m} d}{\eta_G}[x^* + (1 - x^*)\varrho_G/\varrho_L],$$

the Prandtl number Pr_G of the vapor at wall temperature and the parameter

$$Y = 1 - 0.1\left[\left(\frac{\varrho_L}{\varrho_G} - 1\right)(1 - x^*)\right]^{0.4}.$$

In Eq. (13.165),

$$a = 3.27 \cdot 10^{-3}, \quad n_1 = 0.901, \quad n_2 = 1.32, \quad n_3 = -1.50.$$

The equation reproduces measurements on vertical tubes and annuli. In the case of annuli, the hydraulic diameter

$$d = 4A_0/U$$

is to be entered, where A_0 is the heated surface area and U its circumference.

For additional equations, one should refer to the survey of Mayinger and Langner [13.75].

13.10.2 Spray Cooling of Hot Surfaces

One can effectively cool hot surfaces by spraying a liquid on them. Examples are the spray cooling during continuous casting of steels and nonferrous metals directly after the emergence of the molten metal from the pouring ladle, the spray cooling for the protection of fire-endangered objects during fires, or the coating of surfaces with liquids, which upon cooling solidify and form a protective coating. The heat fluxes, especially during spray cooling of hot surfaces with water, are substantial and of the order of magnitude of 10^5 to 10^6 W/m^2.

In order to achieve a sufficient spray cooling, the coolant—in most cases water—is applied by spray nozzles. For this purpose, various types of nozzles come into question. In nozzles, the pressure difference across the nozzle serves to spray the liquid. Rotating disks shoot out a centrally supplied liquid as drops. Finally, in pneumatic sprayers, the liquid is sprayed with the aid of a gas, usually as very fine drops. Each nozzle is characterized by a definite drop spectrum and a definite distribution of drop velocities for each set of operating conditions. Figure 13.44 shows as an example the water dispersion 200 mm downstream behind a fan spray nozzle.

Depending upon the type of water dispersion, very different values of the local heat transfer coefficient can be expected. In order to investigate the heat transfer coefficient, it is expedient to first become familiar with several fundamentals of spray dispersion. Figure 13.45 shows as an example the histogram

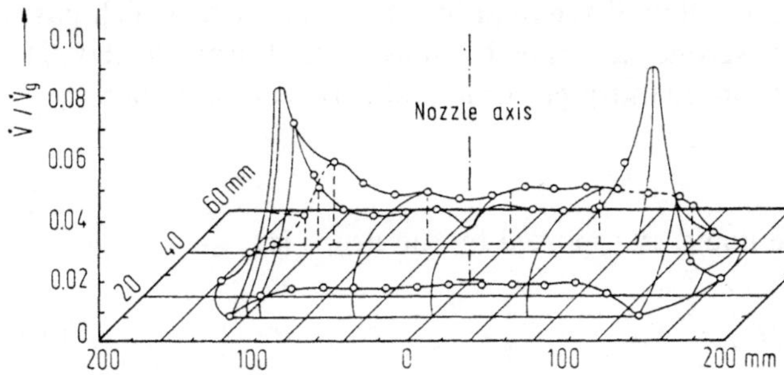

Fig. 13.44. Water dispersion 200 mm behind a fan spray nozzle, according to [13.79]. $\dot{V}$ local volumetric flow, $\dot{V}_g$ entire volumetric flow, water velocity at the end of nozzle 15 m/s, nozzle F 11/5

of the distribution of drop diameters into classes of size d_i downstream behind a nozzle. The quantity of drops n_i of a class i can be summed to yield the total N of all drops

$$N = \sum_i n_i .$$

Such a histogram can be described by a distribution curve. The logarithmic normal distribution plotted in Fig. 13.45 reproduces satisfactorily—except in the vicinity of the maximum—the measured results. Similarly, the drop velocities are not distributed uniformly, as Fig. 13.46 shows for the same nozzle. In general, large drops have a greater velocity than small ones. A value for the

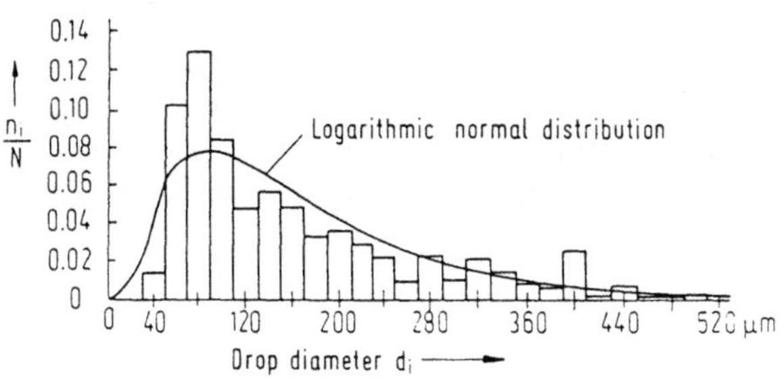

Fig. 13.45. Histogram of a drop size distribution, according to [13.80]. Nozzle F 12/60

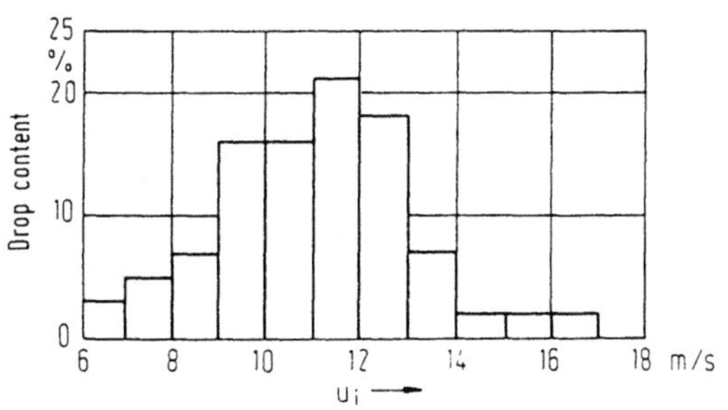

Fig. 13.46. Distribution of drop velocities into classes, according to [13.80]. Nozzle F 12/60

characterization of the spray flow is the quantity X of the drops, which pass a cross section of $1\,\text{cm}^2$ per second. Let x_i be the mass flux of drops of class i (SI unit $1/\text{m}^2\text{s}$). The entire drop quantity per time and surface unit is then

$$X = \sum_i x_i$$

and the mass flux of the liquid transported by the drops is

$$\dot{m}_L = \varrho_L \sum_i x_i V_i = \varrho_L X \sum_i \frac{x_i}{X} V_i, \tag{13.166}$$

if $V_i = \pi d_i^3/6$ is the volume of a drop of class i. We assume that the mass flow rate $\dot{M}_L$ of the liquid exiting from the nozzle and also the mass flux $\dot{m}_L = \dot{M}_L/A$, impacting the area A, are known. Because the drop flux x_i is proportional to the drop population n_i of the respective class i, and $X \sim N$, there holds

$$\frac{x_i}{X} = \frac{n_i}{N},$$

where the ratio n_i/N is known by means of the histogram. Thus, Eq. (13.166) can be rearranged into the equation

$$X = \frac{\dot{m}_L}{\varrho_L \sum_i \frac{n_i}{N} V_i}, \tag{13.167}$$

from which one can calculate the value X. Because of

$$x_i = \frac{n_i}{N} X,$$

the drop flux x_i of class i is also established. The "mean impact velocity" w_L of the drops on a plate is effectively defined by a momentum balance

$$\dot{m}_L w_L = \varrho_L \sum_i x_i w_i V_i.$$

When allowing for $x_i = n_i X/N$, one obtains from this

$$w_L = \frac{\varrho_L X \sum_i \frac{n_i}{N} w_i V_i}{\dot{m}_L}. \tag{13.168}$$

We will consider in the following a single drop of a definite class and investigate its *hydrodynamic behavior* upon impact on a plate. Whether a drop bursts into smaller drops during impact depends upon how large its kinetic energy is in comparison with the surface energy which holds the drop together. Both energies can be combined into the Weber number

$$We = (w_{Ln}^2 \varrho_L d)/\sigma,$$

which represents the ratio of the kinetic energy to the surface energy. The velocity vertical to the plate is designated by w_{Ln}, and the surface tension between drop and surrounding gas is given by σ. If the Weber number is small, then the surface energy dominates. The drop does not fragment, but rather is held together by the surface forces. A large Weber number signifies a kinetic energy that is large in comparison with the surface energy. The drop bursts into many smaller single drops, of which each has a lesser kinetic and higher surface energy than the original drop. As experiments [13.81–13.84] showed, one can undertake approximately the following classification:

—As long as $We < 30$, the drop bounces up and away from the plate without noticeably changing its shape.
—In the region $30 \leqq We \leqq 80$, the drops are strongly deformed during impact on the plate and again assume spherical shape when reflected by the plate. Especially with Weber numbers near 80, one or two small drops can separate from a larger drop.
—In the region $We > 80$, the drop spreads out over the plate and forms a liquid film, which finally breaks up into many small single drops.

Weber numbers for drops produced by nozzles are almost always over 80. Thus, one calculates a Weber number of over 80, if a water drop of 0.5 mm diameter has a velocity of more than 3.41 m/s; a drop of 2 mm diameter at equal temperature would have to have a velocity of more than 1.71 m/s.

If one considers drops with $We \geqq 80$, then several stages of drop disintegration can be distinguished during impact on a wall, Fig. 13.47.

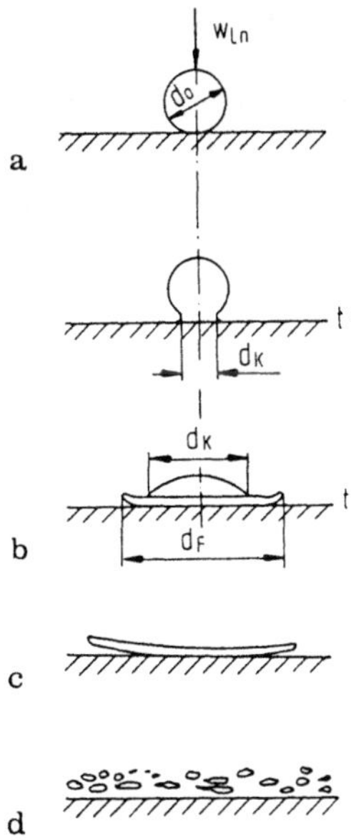

Fig. 13.47 a–d. Stages of drop disintegration [13.80] with $We > 80$ and $t^+ = d_0/w_{Ln}$

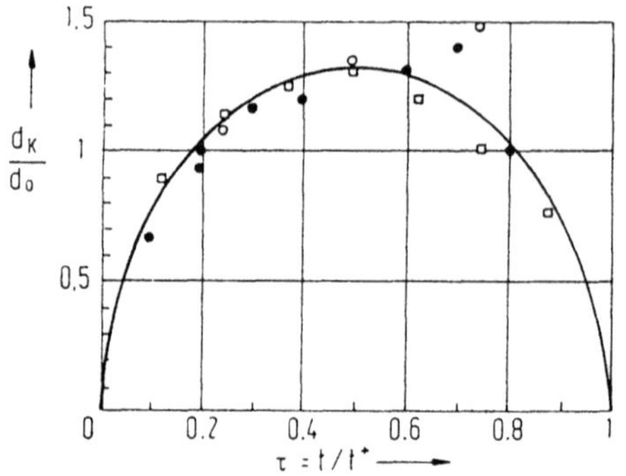

Fig. 13.48. Dome diameter as a function of time, according to [13.81–13.84]. $184 \leqq We \leqq 1500$; $2.3\,\text{mm} \leqq d_0 \leqq 6\,\text{mm}$; $400\,°\text{C} \leqq \vartheta_w \leqq 900\,°\text{C}$

—The almost spherically shaped drop of diameter d_0 impacts upon the plate with the velocity w_{Ln} (Fig. 3.47a).
—A short time after the impact, the drop begins to be deformed. At the base, a film is formed over which the remainder of the drop arches in spherical shape. The film spreads out farther at the expense of the dome (Fig. 13.47b).
—The dome has disappeared. There still exists at the wall only a film-shaped liquid disk slightly arched at the edge (Fig. 13.47c).
—The film disintegrates into a large number of small drops, of which each has a kinetic energy smaller than that of the original drop (Fig. 13.47d).

On very hot plates with temperatures far above the Leidenfrost temperature, a vapor film is formed under the drop some time after impact. Because of the insulating effect of the vapor film, the heat transfer from the wall to the drop is poor and, as Bolle and Moureau [13.80] proved, is negligibly small in comparison with the heat transfer during the time of direct contact between drop and wall.

The diameter d_K at the foot of the dome grows at first after impact, then becomes smaller to the degree that the dome disappears with the growth of the film. Figure 13.48 shows the dome diameter d_K as a function of time. The dimensionless time $\tau = t/t^+$ is formed with

$$t^+ = d_0/w_{Ln},$$

which denotes the time for the dome to disappear. The curve plotted in Fig. 13.48 can be described by [13.80]

$$\left(\frac{d_K}{d_0}\right)^2 = 6.97\tau(1-\tau). \tag{13.169}$$

13.10.2.1 A Calculation Model

Starting from the idea that the heat transfer is determined almost exclusively by the drops that are in contact with the wall, Bolle and Moureau [13.80] developed a model for the calculation of the heat transfer. The basic features of this model will be explained in the following, because the model reproduces clearly the essential limiting values.

We will consider a single drop that strikes a hot surface. The critical Weber number at which the drop breaks up into many smaller drops is $We \geqq 80$. Heat is transferred mainly by the wall as long as the drop spreads out along the wall and does not break up. During this time, there is a strong convection in the drop, so that one can assume a uniform liquid temperature. Characteristic times for the direct contact are of the order of magnitude 10^{-5} s. One can thus think of the process as the sudden lowering of the wall temperature of the plate to the temperature of the drop by contact of the drop with the plate. Because the plate during the short time of contact can be regarded as a semi-infinite body, one can reduce the problem to the known solution of the heat conduction equation for a semi-infinite body, whose wall temperature is suddenly lowered from ϑ_w to ϑ_L.

We will consider a liquid circular element with a radius of r' and a surface area $2\pi r' \mathrm{d}r'$. It has been spread across the heating surface during the time t_0, so that at this time the heating surface gives off over this elemental area a heat transfer rate approaching infinity, which, however, dissipates very quickly. According to the solution of the heat conduction equation for the semi-infinite body [13.85], the heat transfer rate given off by the plate is

$$\mathrm{d}\Phi = 2\pi r' \mathrm{d}r' \frac{b}{\sqrt{\pi(t - t_0)}} (\vartheta_w - \vartheta_L) \tag{13.170}$$

with the thermal penetration coefficient $b = \sqrt{\lambda \varrho c}$ (SI unit $\mathrm{Ws^{1/2}/Km^2}$), in which λ is the thermal conductivity (SI unit W/Km), ϱ the density (SI unit $\mathrm{kg/m^3}$), and c the specific heat of the wall (SI unit J/kgK). The heat flow rate is variable with time, and the plate can transfer heat only as long as the liquid exists. According to Fig. 13.49, the liquid starts growing from $d' = 2r'$, reaches a maximum, then recedes until the drop disappears. The liquid front thus runs twice across the point r', first at the time t_0 when it reaches the point, and then at t_1

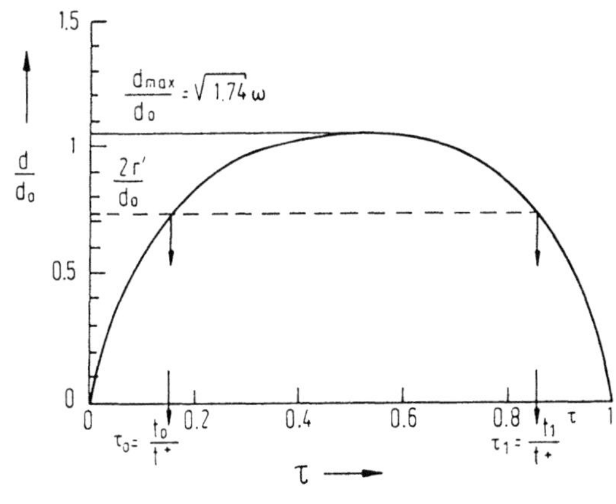

Fig. 13.49. The determination of the times t_0 and t_1

when it is disappearing. The mean heat flow rate, that is transferred between the two times, is

$$d\bar{\Phi} = \frac{1}{t_1 - t_0} \int_{t_0}^{t_1} 2\pi r' \, dr' \frac{b}{\sqrt{\pi(t - t_0)}} \vartheta_w - \vartheta_L) \, dt. \tag{13.171}$$

The integration yields

$$d\bar{\Phi} = 4\pi r' dr' \frac{b}{\sqrt{\pi(t_1 - t_0)}} (\vartheta_w - \vartheta_L). \tag{13.172}$$

In order to determine the times t_0 and t_1, during which the ring surface gives off heat to the drop, Bolle and Moureau [13.80] assumed that the diameter of any wetted circular surface is proportional to the dome diameter d_K, according to Fig. 13.49,

$$d = \omega d_K,$$

in which ω is an experimentally determined parameter. With Eq. (13.169), one then obtains

$$d^2 = \omega^2 d_K^2 = 6.97 \omega^2 d_0^2 (\tau - \tau^2). \tag{13.173}$$

In addition, the maximum diameter of the wetted surface, as can be determined by differentiation, is

$$d_{max} = \sqrt{1.74} \omega d_0. \tag{13.174}$$

Figure 13.49 presents the time dependence of d/d_0, as calculated from Eq. (13.173). At time $\tau_0 = t_0/t^+$, the drop has spread out up to the radius r'; at time $\tau_1 = t_1/t^+$, it recedes to a smaller radius r'. The times t_0 and t_1 are thus obtained as roots of Eq. (13.173), if one assumes $d = 2r'$. One finds

$$(t_1 - t_0)^{1/2} = t^{+1/2} \left(1 - \frac{4r'^2}{1.74 \omega^2 d_0^2} \right)^{1/4}. \tag{13.175}$$

According to Eq. (13.172), the mean heat flow rate across the elemental surface area $2\pi r' dr'$, is, therefore,

$$d\bar{\Phi} = 4\pi r' dr' \frac{b}{\sqrt{\pi}} (\vartheta_w - \vartheta_L) \frac{1}{\sqrt{t^+}} \left(1 - \frac{4r'^2}{1.74 \omega^2 d_0^2} \right)^{-1/4}. \tag{13.176}$$

The heat given off during the time interval $t_1 - t_0$ is

$$dQ = d\bar{\Phi}(t_1 - t_0) = 4\sqrt{\pi} r' dr' b (\vartheta_w - \vartheta_L) \sqrt{t^+} \left(1 - \frac{4r'^2}{1.74 \omega^2 d_0^2} \right)^{1/4}. \tag{13.177}$$

Integration gives the total heat transfer:

$$Q = 4\sqrt{\pi} b (\vartheta_w - \vartheta_L) \sqrt{t^+} \int_{r'=0}^{d_{max}/2} \left(1 - \frac{4r'^2}{1.74 \omega^2 d_0^2} \right)^{1/4} r' dr'.$$

With d_{max} according to Eq. (13.174) and $t^+ = d_0/w_{Ln}$, one obtains

$$Q = 1.23\omega^2 b(\vartheta_w - \vartheta_L)\frac{d_0^{2.5}}{\sqrt{w_{Ln}}}. \tag{13.178}$$

According to the investigations of Bolle and Moureau [13.80], data and experiments coincide best, if one assumes $\omega^2 = 2/3$.

The heat given off is accordingly proportional to the product of the thermal penetration coefficient $b = \sqrt{\lambda \varrho c}$ of the wall, the difference between wall and liquid temperatures, the 2.5th power of the drop diameter and the inverse of the square root of the impact velocity of the drop. The drops of a class i, whose mass flux is given by $x_i = n_i X/N$ with X according to Eq. (13.167), withdraw from the heating surface a heat flux

$$q_i = Q_i x_i.$$

Here Q_i is obtained from Eq. (13.178), if one enters there the diameter d_{0i} and the impact velocity $w_{Ln,i}$ of the drop class i, and additionally $\omega^2 = 2/3$. Addition of all the individual heat fluxes yields the total heat flux given off

$$q = \sum_i Q_i x_i = 0.82 b(\vartheta_w - \vartheta_L)\sum_i \frac{d_{0i}^{2.5}}{\sqrt{w_{Ln,i}}} x_i. \tag{13.179}$$

The unwetted heating surface gives off additional heat by radiation. In general, the wetted surface is very small in comparison to the unwetted surface. According to the investigations of Bolle and Moureau [13.80], the portion of the wetted surface is less than 1% of the entire surface. It is, therefore, legitimate to enter the entire surface as a radiating surface. If one assumes that the heated surface gives off radiant heat to an environment at the temperature ϑ_L, and one designates the emissivity of the heated surface by ε, then the total heat flux given off by radiation and convection is

$$q = \varepsilon\sigma_S(T_w^4 - T_L^4) + 0.82\sqrt{\lambda \varrho c}(\vartheta_w - \vartheta_L)\sum_i \frac{d_{0i}^{2.5}}{\sqrt{w_{Ln,i}}} x_i. \tag{13.180}$$

As is recognized from this, the distribution of the drop sizes and of the drop velocities, as depicted for a definite nozzle in Figs. 13.45 and 13.46, has a decisive influence on the transferred heat flux. As one further recognizes from the equation, the smaller drops contribute only slightly to the heat flux, although their number is generally much larger than that of the large drops. It is, therefore, to be recommended that one uses nozzles with a narrow drop spectrum, which encompasses mainly large drops. An evaluation of Eq. (13.180) for two nozzles with different drop spectrums and velocity distributions [13.80] further showed that, independent of the nozzle, the heat flux given off to water drops by a heated surface with a temperature of 727 °C had a maximum, if the drop size lay between 0.3 and 0.4 mm, Fig. 13.50. The contribution of small water drops under 0.16 mm to the heat flux was very small. These values can serve as

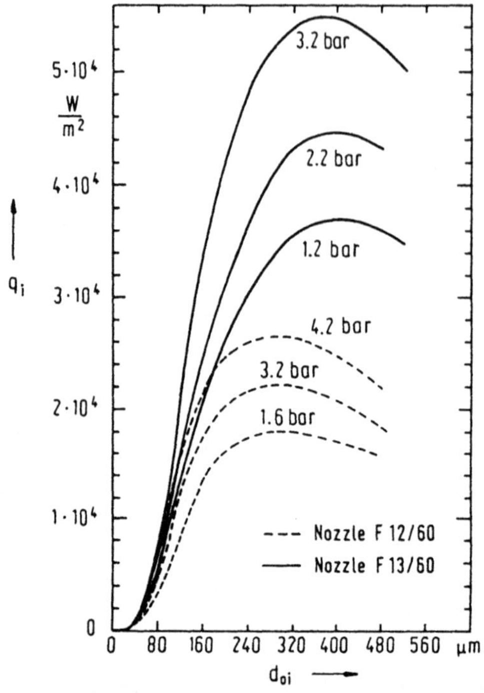

Fig. 13.50. Calculated heat flux q_i as a function of the drop diameter d_{0i} of a class i, according to [13.80]

approximate determining factors for a favorable drop size. To generalize these findings based on two nozzles is, however, difficult, because it is not easy to measure drop spectra and velocity distributions.

13.10.2.2 Application Equations

The practical computation of heat fluxes according to Eq. (13.180) is time-consuming and usually fails because one does not know the drop spectrum, velocity distribution, or mass flux of drops. One has, therefore, reproduced data on heat transfer during spray cooling of hot surfaces by using simple empirical correlations of the form

$$q = \alpha(\dot{m}_L)(\vartheta_w - \vartheta_L)$$

in the form of exponential functions. The radiation component of the heat flux given off is not contained in these. According to Bolle and Moureau [13.80],

$$\alpha = 423\dot{m}_L^{0.556} \tag{13.181}$$

with α in W/m²K and $\dot{m}_L$ in kg/m²s. The equation reproduces measurements made on steel surfaces cooled with water in the region

$$627\,°C < \vartheta_W < 927\,°C$$

and

$$1\,\mathrm{kg/m^2s} < \dot{m}_L < 7\,\mathrm{kg/m^2s}.$$

The cooling water temperature was 20 °C.

In addition, the heat transfer coefficient on plates, which are cooled from below by sprayed water, is

$$\alpha = 360\dot{m}_L^{0.556} \tag{13.182}$$

with α in $\mathrm{W/m^2K}$ and $\dot{m}_L$ in $\mathrm{kg/m^2s}$ in the region

$$727\,°\mathrm{C} < \vartheta_w < 1027\,°\mathrm{C}$$

and

$$0.8\,\mathrm{kg/m^2s} < \dot{m}_L < 2.5\,\mathrm{kg/m^2s}.$$

According to data of Müller [13.79] we have

$$\alpha = 10 w_{Ln} + (107 + 0.688 w_{Ln})\dot{V}_s \tag{13.183}$$

with α in $\mathrm{W/m^2K}$; the impact velocity w_{Ln} in m/s, is assumed to be equal to the exit velocity from the nozzle, and the volume flux $\dot{V}_s$ of the water is to be entered in $\mathrm{dm^3/m^2s}$. The equation is valid for the cooling of hot steel surfaces with temperatures up to 1250 °C, by water from fan-spray nozzles in the region

$$0.2\,\mathrm{dm^3/m^2s} \leqq \dot{V}_s \leqq 10\,\mathrm{dm^3/m^2s}$$

with impact velocities

$$10\,\mathrm{m/s} \leqq w_{Ln} \leqq 35\,\mathrm{m/s}.$$

The values according to Eq. (13.183) lie some 20% lower than those according to Eq. (13.181). Such deviations, however, are to be expected because of measurement uncertainty.

If surfaces other than those of steel are to be cooled, until more precise findings are available, it is recommended in accordance with Eq. (13.182) that the heat transfer coefficients calculated from the previous equations be multiplied by the ratio of the thermal penetration coefficients

$$b/b_{St}.$$

The thermal penetration coefficient of rust-free steel is $b_{St} = 9410\,\mathrm{Ws^{1/2}/Km^2}$ ($\lambda = 23\,\mathrm{W/Km}, \varrho = 7700\,\mathrm{kg/m^3}, c = 500\,\mathrm{J/kgK}$) and b is the thermal penetration coefficient of the cooled surface.

14 Heat Transfer During Boiling of Mixtures in Free Convection

Frequently, in industrial processing procedures, liquid mixtures of two or more components have to be evaporated in order to separate them from one another. Examples are the concentrating of solutions, the thickening of lyes, the recovery of solvents, the distillation of sea water for the production of drinking water, or the separation of substances by evaporation during distillation. Heat and mass transfer are closely connected with one another during the evaporation of multi-component mixtures, and the quantity of vapor produced usually is determined primarily by the mass transfer in contrast to the evaporation of pure substances. It is known from previous experiments that heat transfer coefficients during evaporation of mixtures can be substantially smaller than those of the pure components of the mixture. On the other hand, marked improvements of the heat transfer have been noted, if one of the components of the mixture is surface-active. This leads to a reduction of the surface tension and, thus, as was presented in Chap. 10, to a greater bubble density. Simultaneously, the bubble frequency and, in turn, also the heat transfer increases. Mixtures of organic or inorganic liquids contain, however, surface-active components only in certain cases (soaps, addition of wetting agents), so that one usually must reckon with a decrease of the heat transfer coefficient in comparison with that of the pure component.

14.1 Binary Mixtures. Physical Fundamentals

Bonilla and Perry [14.1] were the first to investigate a rather large number of binary mixtures of organic liquids and of water with organic liquids. The heat transfer coefficients obtained from their data for ethanol-water mixtures, together with the findings of Preusser [14.2], are plotted in Fig. 14.1 as an example. Heat flux and pressure are given in the diagram. As is recognized, the heat transfer coefficients α of the mixture are clearly smaller than the values α_{id} that would be obtained if one were to interpolate linearly between the heat transfer coefficients of the pure components. One recognizes also a clear decrease of the heat transfer coefficient in the region, in which vapor and liquid compositions $y-x$ are strongly different, as is seen by a comparison with the upper curve in Fig. 14.1.

Such a decrease of the heat transfer in the region of great differences in vapor and liquid composition is observed in many mixtures. It is explained by

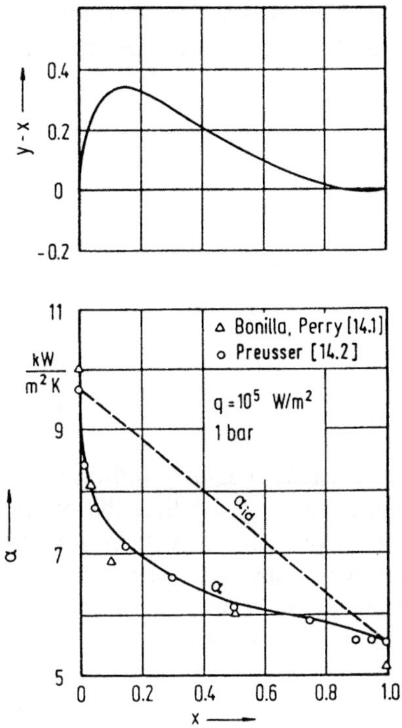

Fig. 14.1. Heat transfer coefficients for boiling ethanol-water mixtures. y is the mole fraction of ethanol in the vapor and x is the mole fraction of ethanol in the liquid

the fact that during vapor bubble formation, the individual components are converted from liquid to vapor in varying amounts. The more the more easily volatile component is enriched with vapor, the poorer in this component the liquid becomes in the vicinity of the bubble. One must distinguish between the processes in the vicinity of the wall and those in the core of the liquid. On the heating surface, vapor bubbles are formed at preferred points that favor nucleation. The bubbles grow by addition of heat, until the buoyant forces are large enough for the bubbles to depart. Because the departure diameter is relatively small, only a very small amount of the added heat flux is required for bubble formation at the wall. The greatest part of the heat flux is transferred at first from the heating surface to the liquid, which, in turn, supplies the energy to the rising bubbles in the core of the liquid so that new vapor is formed. Thus, the heating wall produces first of all the bubble nuclei required for nucleate boiling, and, secondly, transfers to the liquid boundary layer a heat flux that is conveyed by convection or conduction across the column of liquid to the rising bubbles. This process for mixtures is not different from that for pure substances. There are, however, fundamental differences with respect to nucleation and also with respect to the heat transfer to the rising bubbles, because the heat transfer of both processes is still determined by the mass transfer.

We shall first consider nucleation at the wall. As the earlier investigations, Sect. 10.1, concerning bubble formation of pure substances showed, a liquid, in which vapor bubbles form, is always superheated in comparison to a liquid at equal pressure without bubble formation. In a liquid superheated by $\Delta\vartheta$, according to Eq. (10.12), a bubble of radius

$$r = \frac{2\sigma T_s}{\varrho'' \Delta h_v \Delta\vartheta}\left(1 + \frac{\Delta\vartheta}{T_s}\omega\right)$$

with

$$\omega = \frac{\Delta\varrho T_s}{\varrho'\sigma}\frac{d}{d\vartheta}\left(\frac{\varrho'\sigma}{\Delta\varrho}\right)$$

or, approximately, because generally $\Delta\vartheta\omega/T_s \ll 1$

$$r = \frac{2\sigma T_s}{\varrho'' \Delta h_v \Delta\vartheta} \tag{14.1}$$

is just viable. Smaller bubbles condense again, because the surrounding liquid is too cold for them; larger bubbles can continue to grow. So that the bubble can be viable at all, bubbles of radius r must be present or must be formed. The gas and vapor residue locked in fine cavities in the wall serve as nuclei. Nuclei of a definite critical size must be present, and the wall must exhibit a definite superheating, [13.56], in order that nuclei be activated.

14.1.1 Bubble Formation and Wall Superheat

We calculate the critical nucleus size and the necessary superheating in binary mixtures from the Thomson equation for binary mixtures. Like the Thomson equation for pure substances, Eqs. (10.7) and (10.8), this one also follows from the conditions of phase equilibrium. For our purpose, we shall consider a spherical bubble of radius r in a large volume of liquid of the composition x (x is the mole fraction of the more easily volatile component in the liquid, and y that in the vapor of a binary mixture). Both bubble and liquid have the equilibrium temperature ϑ. There continues to be valid the condition for mechanical equilibrium

$$p_G - p_L = \frac{2\sigma}{r} \tag{14.2}$$

and the conditions for equilibrium with respect to mass transfer

$$\mu_{1L}(p_L, \vartheta, x) = \mu_{1G}(p_G, \vartheta, y) \tag{14.3}$$

$$\mu_{2L}(p_L, \vartheta, x) = \mu_{2G}(p_G, \vartheta, y) \tag{14.4}$$

with the chemical potential μ (SI unit J/kg or J/mole, here J/mole). From these equations, one obtains by series expansion as in the case of pure substances, Sect. 10.1, the Thomson equation

$$p_G = p_0 - \frac{yV_1' + (1-y)V_2'}{\bar{V}'' - [yV_1' + (1-y)V_2']}\frac{2\sigma}{r} \tag{14.5}$$

and

$$p_L = p_0 - \frac{\bar{V}''}{\bar{V}'' - [yV'_1 + (1-y)V'_2]} \frac{2\sigma}{r}. \tag{14.6}$$

Here, p_0 is the saturation pressure at the temperature ϑ and the liquid composition x, if the phase interface between vapor and liquid is flat. The molar volume of the vapor (SI unit m³/mole) is indicated by $\bar{V}'' = yV''_1 + (1-y)V''_2$, and the partial molar volumes (SI unit m³/mole) of the components 1 and 2 in the liquid by V'_1 and V'_2 and in the vapor phase by V''_1 and V''_2, respectively. The value occurring in the denominator of the two Eqs. (14.5) and (14.6)

$$\bar{V}'' - [yV'_1 + (1-y)V'_2] = y(V''_1 - V'_1) + (1-y)(V''_2 - V'_2)$$

is also known as the molar conversion volume. It indicates by how much the molar volume is changed if a liquid is converted at constant pressure and temperature into a vapor, and if liquid and vapor have the same composition y. The liquid is, therefore, completely evaporated, or, conversely, the vapor of composition y is completely condensed into a liquid of composition y.

The saturation pressure p_0 at the plane phase interface in the Thomson Eqs. (14.5) and (14.6) is dependent in binary mixtures upon the temperature ϑ and the liquid composition x, $p_0 = p_0(\vartheta, x)$. In Fig. 14.2, the interrelations are graphically depicted. At prescribed values of temperature and of liquid composition x, in accordance with Eq. (14.5), the vapor pressure p_G is smaller by

$$p_0 - p_G = \Delta p_G = \frac{yV'_1 + (1-y)V'_2}{\bar{V}'' - [yV'_1 + (1-y)V'_2]} \frac{2\sigma}{r}$$

than the vapor pressure $p_0(\vartheta, x)$ at the plane phase interface, and the liquid pressure p_L in keeping with Eq. (14.6) is smaller by

$$p_0 - p_L = \Delta p_L = \frac{\bar{V}''}{\bar{V}'' - [yV'_1 + (1-y)V'_2)} \frac{2\sigma}{r}$$

than the vapor pressure $p_0(\vartheta, x)$ of the plane phase interface. Because the surface tension σ of the mixture with the prescribed composition x is temperature-dependent, the curves for the vapor pressure p_G and the liquid pressure p_L do not run parallel to the vapor pressure curve p_0, given at a plane phase interface.

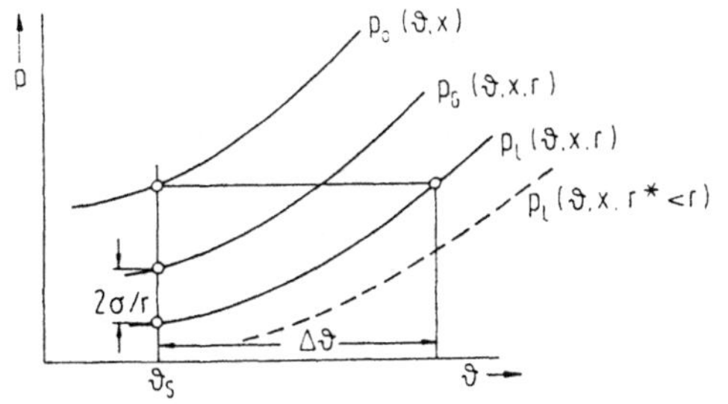

Fig. 14.2. Vapor and liquid pressures with equilibrium between liquid and a spherical vapor bubble in a binary mixture

If one does not assume the boiling temperature ϑ of the liquid-vapor bubble system, but rather the pressure p_0, then, corresponding to Fig. 14.2, the liquid must be superheated by $\Delta\vartheta$ in comparison with the system with a plane phase interface. Just as during bubble formation of pure substances, the necessary superheating increases with decreasing radius r of the vapor bubble.

For the calculation of the required superheating, one differentiates, as with pure substances, Sect. 10.1, Eq. (14.6), with respect to temperature

$$\left(\frac{\partial p_L}{\partial \vartheta}\right)_x = \left(\frac{\partial p_0}{\partial \vartheta}\right)_x - \frac{\partial}{\partial \vartheta}\left[\frac{\bar{V}''}{\bar{V}'' - [yV'_1 + (1-y)V'_2]}\frac{2\sigma}{r}\right]_x. \tag{14.7}$$

The derivation $(\partial p_0/\partial\vartheta)_x$ is the slope of the vapor pressure curve $p_0(\vartheta)$, given the assumed liquid composition. It results from the differential equations of the phase interface of a binary mixture [14.3], if $x = \text{const.}$ is assumed there. One obtains an equation related to the equation of Clausius–Clapeyron, in which H'_i and $H''_i, i = 1, 2$, are the partial molar enthalpies (SI unit J/mole) of the liquid and the vapor:

$$\left(\frac{\partial p_0}{\partial \vartheta}\right)_x = \frac{y(H''_1 - H'_1) + (1-y)(H''_2 - H'_2)}{T[y(V''_1 - V'_1) + (1-y)(V''_2 - V'_2)]}. \tag{14.8}$$

The term in the numerator of the right side states by how much the enthalpy changes, if a liquid of the composition y is completely evaporated at constant pressure and temperature. This value is known as the molar enthalpy of transfer $\Delta\bar{H}_u$. Because of

$$\begin{aligned} &y(H''_1 - H'_1) + (1-y)(H''_2 - H'_2) \\ &\quad = yH''_1 + (1-y)H''_2 - [xH'_1 + (1-x)H'_2] - (y-x)(H'_1 - H'_2), \end{aligned}$$

in which there is [14.3, p. 114]

$$H'_1 - H'_2 = \left(\frac{\partial \bar{H}'}{\partial x}\right)_{T,p}$$

and with the molar enthalpy of evaporation

$$\Delta\bar{H}_v = yH''_1 + (1-y)H''_2 - [xH'_1 + (1-x)H'_2]$$

the molar enthalpy of transfer can also be reduced to the molar enthalpy of evaporation

$$\Delta\bar{H}_u = y(H''_1 - H'_1) + (1-y)(H''_2 - H'_2) = \Delta\bar{H}_v - (y-x)\left(\frac{\partial \bar{H}'}{\partial x}\right)_{T,p}.$$

It represents the heat that must be added to a large quantity of liquid in order to produce vapor. The composition of the liquid is very little changed by the evaporation process, because the quantity of liquid may be quite large. The molar transfer enthalpy can thus be depicted as the differential molar heat of evaporation of the isobarically produced vapor. The differential quotient $(\partial\bar{H}'/\partial x)_{T,p}$ represents the slope of the tangent to the boiling line for the liquid

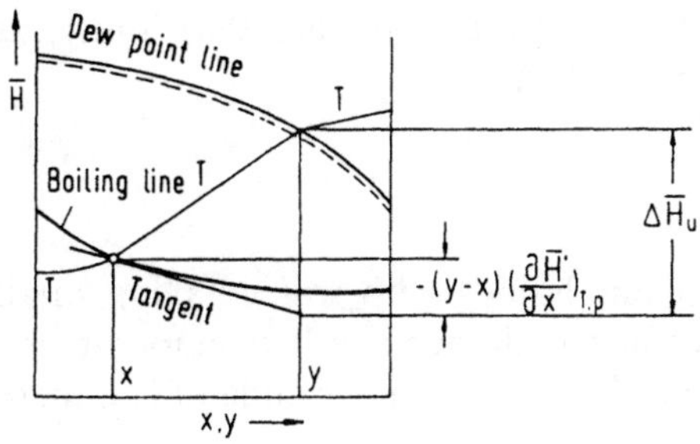

Fig. 14.3. The differential evaporation process showing the molar transfer enthalpy

with the composition x. Thereby is obtained the molar enthalpy of transfer from the $\bar{H}, x$ diagram according to Fig. 14.3. In an analogous manner, the molar volume of transfer is obtained in the $\bar{V}, x$ diagram.

The expression in the numerator on the right side of Eq. (14.8) takes the place of the enthalpy of evaporation for pure substances, whereas the molar volume of transfer in the square brackets of the denominator takes the place of the volume increase during evaporation. Substitution of Eq. (14.8) in Eq. (14.7) and integration between the boundaries ϑ_s and $\vartheta_s + \Delta\vartheta$, while allowing for $p_L(\vartheta_s + \Delta\vartheta, x, r) = p_0(\vartheta, x)$, Fig. 14.2, yields the expression

$$\left[\frac{\bar{V}''}{\bar{V}'' - [yV'_1 + (1-y)V'_2]}\frac{2\sigma}{r}\right]_{\vartheta_s + \Delta\vartheta} = \int_{\vartheta_s}^{\vartheta_s+\Delta\vartheta} \frac{\Delta\bar{H}_v - (y-x)\left(\dfrac{\partial \bar{H}'}{\partial x}\right)_{T,p}}{T_s[y(V''_1 - V'_1) + (1-y)(V''_2 - V'_2)]}\,d\vartheta. \tag{14.9}$$

Because the liquid superheating $\Delta\vartheta$ is small, one can develop the expressions on both sides into a Taylor series and truncate the series after the term that is linear in $\Delta\vartheta$. One then finds the relation between the bubble radius r and the superheating $\Delta\vartheta$:

$$r = \frac{2\sigma\bar{V}''T_s}{\Delta\bar{H}_v\left[1 - (y-x)\dfrac{1}{\Delta\bar{H}_v}\left(\dfrac{\partial\bar{H}'}{\partial x}\right)_{T,p}\right]\Delta\vartheta}\left(1 + \frac{\Delta\vartheta}{T_s}\omega\right) \tag{14.10}$$

with

$$\omega = \frac{y(V''_1 - V'_1) + (1-y)(V''_2 - V'_2)}{\sigma\bar{V}''}T_s \times \frac{\partial}{\partial\vartheta}\left[\frac{\sigma\bar{V}''}{y(V''_1 - V'_1) + (1-y)(V''_2 - V'_2)}\right]_x.$$

It converts into the earlier Eq. (10.12) for pure substances, if one assumes $y = x = 1$.

The term $(\Delta\vartheta/T_s)|\omega|$ is almost always very much smaller than unity, as was explained in Sect. 10.1, and also almost always

$$\left|(y-x)\frac{1}{\Delta\bar{H}_v}\left(\frac{\partial\bar{H}'}{\partial x}\right)_{T,p}\right| \ll 1.$$

If one further disregards mixtures of organic liquids with water having small molar fractions x of organic components and, simultaneously, low temperatures, where the mentioned expression can reach the order of magnitude of 0.2, there results to a rough approximation

$$r = \frac{2\sigma\bar{V}''T_s}{\Delta\bar{H}_v\Delta\vartheta} \tag{14.11}$$

which coincides with the approximation expression for pure substances, Eq. (14.1).

The liquid superheating necessary for the existence of a viable bubble in a binary mixture is approximately of the same magnitude as for pure substances and is mainly determined by the properties of the mixture to be found in Eq. (14.11). The formation of viable bubbles depends then, first of all, upon the properties of the mixture and obeys approximately the same laws that apply for pure substances.

Nucleation is, of course, necessary for further bubble growth, but the heat flow rate transferred for nucleation is insignificant, if compared with the total heat transfer. The latter is determined above all by the further bubble growth after nucleation. The rate of growth depends upon the pressure difference between bubble and liquid, the forces of inertia and friction in the liquid, the surface tension, the transport of heat and the transport of the lower boiling component by the liquid to the bubble surface. On the other hand, the forces of inertia and friction, the gradients of pressure, temperature, and condensation, as well as the compressibility can usually be disregarded in vapor, [14.4]. As

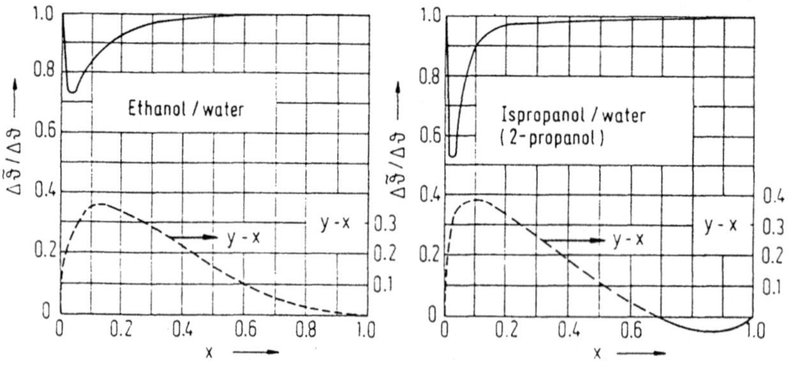

Fig. 14.4. Influence of diffusion on liquid superheating and bubble growth, Eq. (14.12), $p = 1$ bar. Dashed curve: $y - x$

a result of the transport of the more easily volatile component by the liquid to the bubble surface, the liquid in the vicinity of the bubble surface becomes poor in this component. Because of this mass transfer, the saturation temperature at the surface of a bubble growing in the liquid is greater than the saturation temperature of the original liquid.

Scriven [14.5] first investigated the influence of mass transfer on the growth of a spherical bubble in a binary mixture of initially equal temperature and composition, by solving the mass, momentum, and energy balances for a spherically symmetrical bubble while assuming negligible friction, inertia, and surface forces. There is obtained the ratio of the liquid superheating $\Delta\tilde{\vartheta}$ in a binary mixture and of the liquid superheating $\Delta\vartheta$ that would occur in a binary mixture, if the diffusion were negligible, i.e., if there were no concentration difference present in the liquid. To a good approximation we have

$$\frac{\Delta\tilde{\vartheta}}{\Delta\vartheta} = \frac{1}{1-(\xi_G-\xi_L)\frac{c_{pL}}{\Delta h_v}\left(\frac{\partial\vartheta}{\partial\xi_L}\right)_p}, \tag{14.12}$$

where ξ_L is the mass fraction of the lower boiling component in the liquid and ξ_G is the accompanying equilibrium value in the vapor. Because the rate of growth of vapor bubbles is proportional to the superheating, this equation reproduces the influence of the diffusion on the velocity of growth. Figure 14.4 shows the dependence of $\Delta\tilde{\vartheta}/\Delta\vartheta$ on the molar fraction x in the liquid of ethanol-water and isopropanol-water mixtures at a pressure of 1 bar. In the figure, the curves $y-x$ are also shown as dashed lines. It is clearly recognized that the diffusion there leads to a strong reduction of the superheating, and, thus, leads to a reduction of the rate of growth where the composition of vapor and liquid $y-x$ in equilibrium are different by the greatest amount. Superheating and, thus, the rate of growth are the greatest, $\Delta\tilde{\vartheta}/\Delta\vartheta = 1$, when the vapor and liquid composition can no longer be distinguished because of the equalizing effect of the diffusion, $y-x=0$.

14.2 Heat Transfer in Binary Mixtures

As the previous discussion showed, the bubble growth and, thus, also the heat transfer are strongly influenced by the difference $y-x$ of the molar fractions in the vapor and liquid phases. One may expect, therefore, that the more the composition of liquid and vapor will differ, the more the heat transfer coefficients will decrease. In this regard, it is a question of the amount of the difference $y-x$, and not of whether the sign is plus or minus. These phenomena are supported by many experiments, several examples of which will be presented. They are taken from papers by Stephan and Körner [14.6, 14.7] and by Happel and Stephan [14.8].

Figure 14.5 shows measured heat transfer coefficients for the mixture n-heptane/methylcyclohexane ($C_7H_{16}/CH_3C_6H_{11}$) during evaporation in a

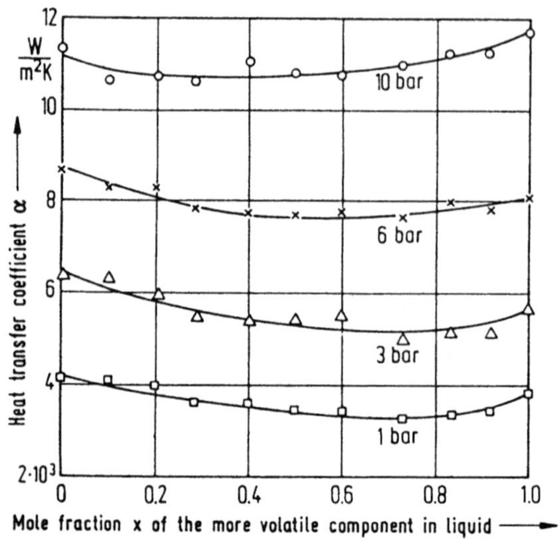

Fig. 14.5. Heat transfer coefficient α for the system n-heptane/methylcyclohexane. Heat flux $q = 10^5$ W/m^2

horizontal tube. They change only slightly with the composition. The boiling point (bubble point) and dew point lines practically coincide, with the result that bubble growth is not prevented by a difference in the composition of vapor and liquid. The slight concentration dependence of the heat transfer coefficients can be attributed to the change in property values along with the composition. One is not committing a large error, if one interpolates the heat transfer coefficient of the mixture linearly in conformity with the molar fractions between the heat transfer coefficients of the pure components.

If one investigates a mixture, which exhibits in certain regions quite large differences in the composition of the vapor and the liquid, one finds that the heat transfer is reduced to a rather large degree in these regions, as is the case in the boiling diagram of the mixture acetone/n-butanol ($(CH_3)_2CO/C_4H_9OH$), Fig. 14.6. This can be seen in Fig. 14.7, in which the heat transfer coefficient is plotted as a function of the composition. Where the difference $y - x$ is the greatest, the heat transfer coefficient decreases the most.

This phenomenon is especially noticeable during the boiling of mixtures that have an azeotropic point. To illustrate this, we will consider the mixture methanol/benzene (CH_3OH/C_6H_6), Figs. 14.8 and 14.9. The mixture shows a strong reduction of the heat transfer in the region of great differences $y - x$. Approaching the azeotropic point, the heat transfer increases again. Thus, it is confirmed that in the vicinity of an azeotropic point the heat transfer coefficient decreases, not because of the plus or minus sign of the difference $y - x$, but rather because its absolute value is decisive.

As is recognized from Figs. 14.7 and 14.9, the heat transfer decreases with the difference in concentration more strongly at high pressures than at low pressures. This can be explained by the fact that the number of vapor bubbles being formed per surface unit increases with the pressure. The vapor bubbles are more densely packed, and there is less surface available for the mass flow

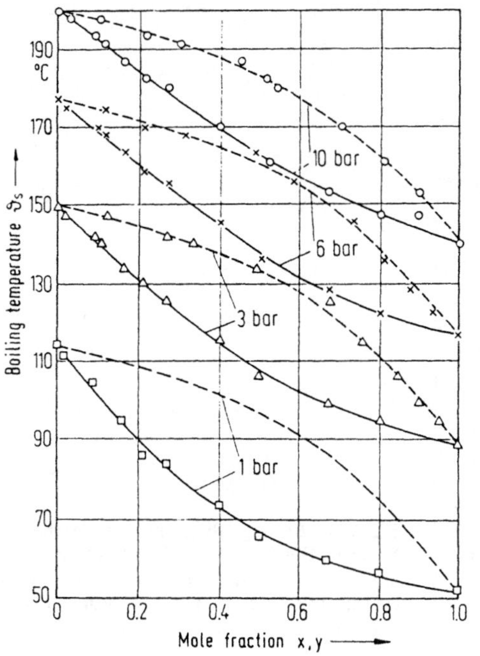

Fig. 14.6. Boiling diagram of acetone/n-butanol. Solid curve: mole fraction x of the more volatile substance in the liquid; dashed line: mole fraction y of the more volatile substance in the vapor

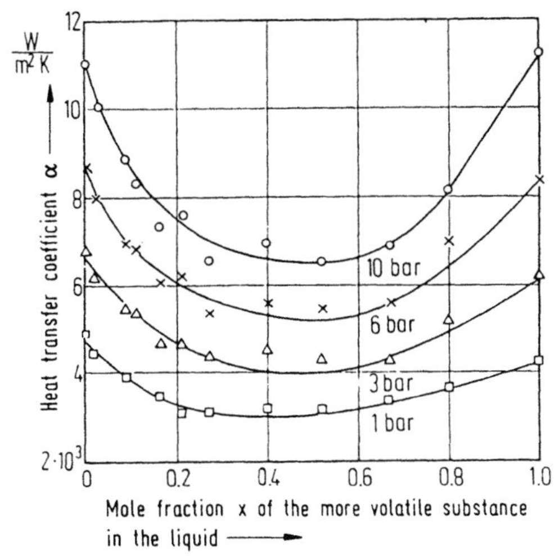

Fig. 14.7. Heat transfer coefficient α for the system acetone/n-butanol. Heat flux $q = 10^5\ \mathrm{W/m^2}$

rate in the liquid space. The delivery of low boiling components is thus more strongly impeded, which leads to a further reduction in the heat transfer. Similar results have been found for many other liquid mixtures. One will find a survey of the previous data on heat transfer in a review by Stephan [14.9].

One notices several peculiarities during the boiling of *oil-refrigerant mixtures*, which can be encountered frequently in the evaporators of refrigeration

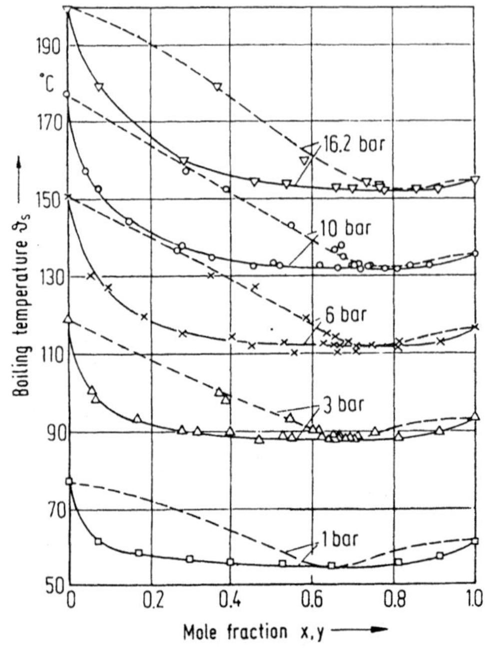

Fig. 14.8. Boiling diagram for methanol/benzene. Solid curve: mole fraction x of the more volatile substance in the liquid; dashed curve: mole fraction y of the more volatile substance in the vapor

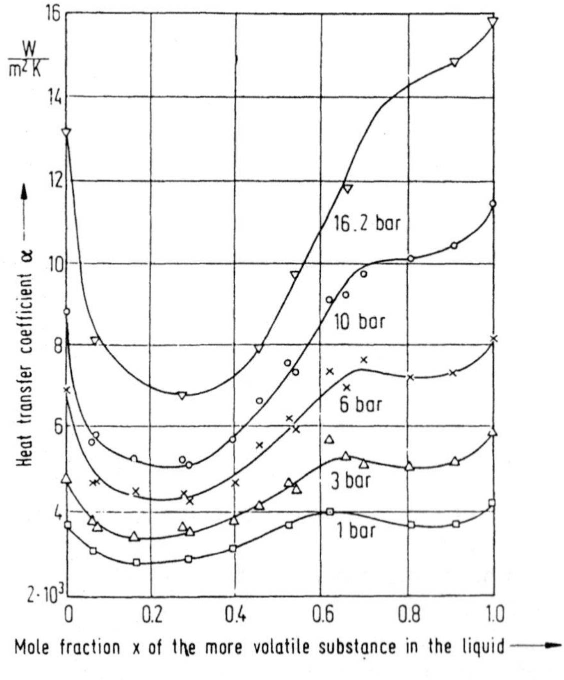

Fig. 14.9. Heat transfer coefficient α for the system methanol/benzene. Heat flux $q = 10^5\ \mathrm{W/m^2}$

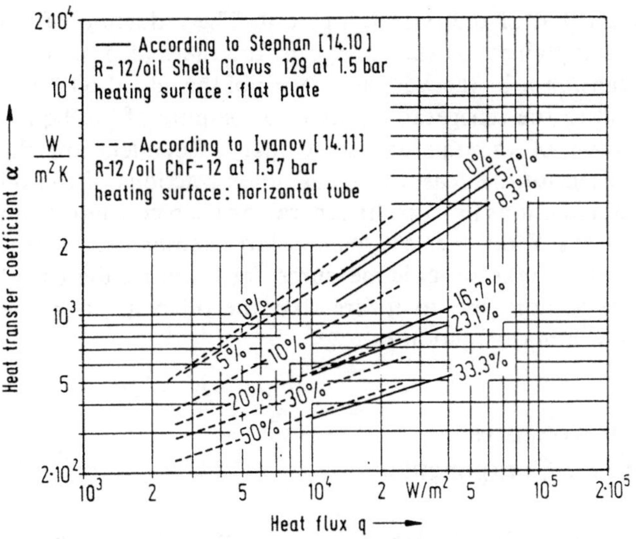

Fig. 14.10. Heat transfer coefficient of R-12/oil mixtures. Parameter: mass fraction of the oil

plants, because the refrigerant always drags along some lubricating oil from the compressor into the evaporator. As a characteristic example, Fig. 14.10 shows that oil in the refrigerant reduces the heat transfer. Slight amounts of oil have only a slight influence on the heat transfer coefficient. According to Fig. 14.11, depending upon the type of oil and the heat flux, small amounts of oil can lead to a slight decrease of the heat transfer coefficient, and amounts up to 3% by mass fraction can even lead to an increase, with the result that the heat

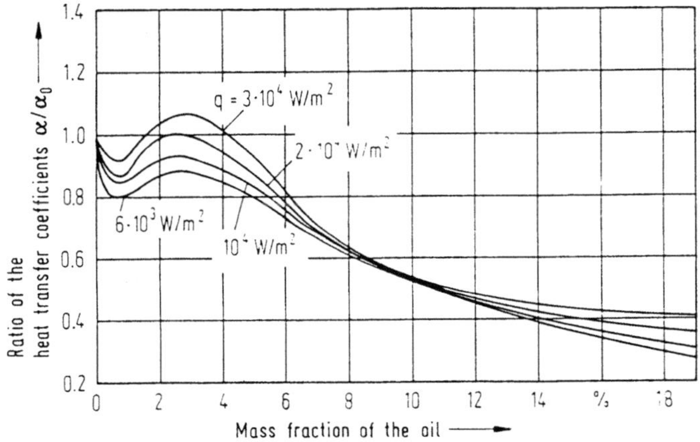

Fig. 14.11. Ratio of the heat transfer coefficients α/α_0, according to Henrici and Hesse [14.12]. α is the heat transfer coefficient of the oil-refrigerant mixture, α_0 that of the oil-free refrigerant R-114 at 1.285 bar

transfer can be better than that of the pure refrigerant. The reduction of the heat transfer coefficient is explained by the change in the thermal properties, especially of the surface tension near the wall, as a result of the added oil. Even very small additions of oil to the refrigerant leads to a foaming of the liquid, because the surface tension of the mixture is strongly dependent on the composition. The surface tension decreases at small mass fractions of the oil but increases at larger mass fractions [14.12]. In general, one can conclude from Fig. 14.11 that mass fractions of oil over 5% lead to large reductions in heat transfer. One should, therefore, take suitable measures, for example, the use of an oil separator after the compressor, to insure that the oil content in the evaporator is always below a mass fraction of 5%.

14.3 Heat Transfer in Mixtures with more than two Components

Investigations of the heat transfer in mixtures with more than three components are as yet unknown, and until now the heat transfer of only a few ternary mixtures has been measured. Preusser [14.13] has investigated in detail the heat transfer in mixtures of acetone/methanol/water and of methanol/ethanol/water during boiling on horizontal tubes, whereas Grigorjev [14.14] has conducted measurements of the heat transfer with the same mixtures on a flat plate. De Dood [14.15] reports on measurements of heat transfer during boiling of mixtures of n-hexane/ethanol/n-heptane and of n-hexane/ethanol/2-butanol, also on a flat plate.

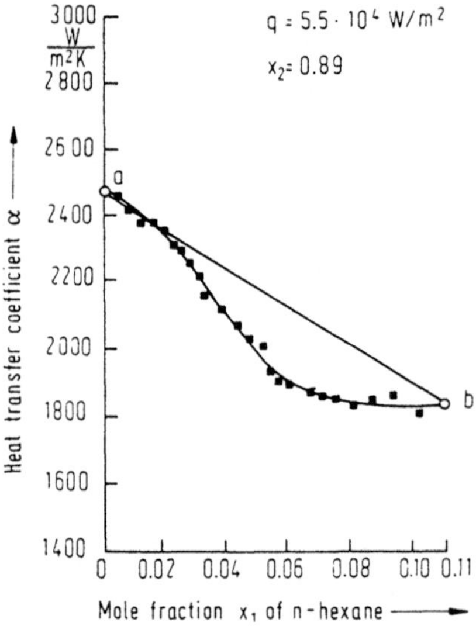

Fig. 14.12. Heat transfer to the three-component mixture n-hexane/ethanol/2-butanol, according to de Dood [14.15]. Mole fraction x_2 of ethanol is constant at $x_2 = 0.89$. Point a: binary mixture of ethanol/2-butanol; point b: binary mixture of n-hexane/ethanol

Basically, the findings for binary mixtures were confirmed. As an example, Fig. 14.12 shows the results of the measurements by de Dood for the ternary mixture n-hexane/ethanol/2-butanol. These mixtures were produced from the two binary mixtures ethanol/2-butanol and n-hexane/ethanol in such a manner that in each of the formed ternary mixtures the mole fraction x_2 of the ethanol remained constant, $x_2 = 0.89$. Therefore, the mole fraction of the n-hexane in the ternary mixture can change between the limiting value $x_1 = 0$ (binary mixture ethanol/2-butanol) and $x_1 = 1 - x_2 = 0.11$ (binary mixture n-hexane/ethanol). As is recognized, the heat transfer coefficient decreases disproportionally when the two binary mixtures ethanol/2-butanol and n-hexane/ethanol are mixed. The few other measurements of de Dood led to the same result, provided that the mixture did not foam as a consequence of the change in surface tension because of the composition. In the case of strong foam formation near the wall, the heat transfer could be substantially smaller. According to the data of Preusser [14.13], however, depending upon the type of mixture, the heat transfer coefficient of the ternary mixture can also lie above that of the two associated binary mixtures.

14.4 Empirical Correlations and Application Equations

14.4.1 Binary Mixtures of Liquids

We assume complete miscibility of the liquids. Essentially two methods have proven reliable for the reproduction of heat transfer data. One starts from empirical correlations for pure substances. Such correlations usually contain dimensionless parameters, which can now be formed with the properties of the binary mixtures. By use of an additional term, allowance is made for the decrease of heat transfer resulting from the obstruction of bubble growth by diffusion. The equation stated by Preusser [14.13] is of this type

$$\alpha = \alpha_0 \left[1 + \left| (y - x) \left(\frac{\partial y}{\partial x} \right)_p \right| \right]^{-0.0733} . \tag{14.13}$$

In it the heat transfer coefficient α_0 can be calculated from the equation established for pure substances, Eq. (11.9). One must, however, enter the properties of the mixture. Equation (14.13) is based upon data at atmospheric pressure and is valid only for this pressure.

As shown by Preusser, a considerable amount of the reduction of heat transfer in mixtures, compared to pure substances, is conditioned by the change in the thermal properties, whereas the additional term in the square brackets makes a comparatively small contribution. It lies between 0.8 and close to unity for most hydrocarbon mixtures and the mixtures of hydrocarbons with water. All other methods of calculation and models described in the following serve, in the final analysis, to avoid the lengthy computation of the property values of the mixture.

Such a correlation, which avoids the difficult and time-consuming calculation of the property values of the mixture, has been suggested by Stephan and Körner [14.6]. It starts from the fact that, for the transfer of a definite heat flux to the mixture, a larger wall superheat $\Delta\vartheta = \vartheta_w - \vartheta_s$ is required than for the evaporation of pure substances. The saturation temperature ϑ_s in this case is the boiling temperature of the mixture with the mean composition x of the liquid. In order to calculate the superheat, an "ideal" wall superheat $\Delta\vartheta_{id}$ is defined by

$$\Delta\vartheta_{id} = x_1\Delta\vartheta_1 + x_2\Delta\vartheta_2, \tag{14.14}$$

in which the temperature differences $\Delta\vartheta_1$ and $\Delta\vartheta_2$ between the wall and saturation temperatures result from the heat transfer coefficients α_1 and α_2 of the pure substances 1 and 2 with the heat flux q of the mixture according to

$$\Delta\vartheta_1 = q/\alpha_1 \quad \text{and} \quad \Delta\vartheta_2 = q/\alpha_2$$

and can be calculated by way of example with the aid of Eq. (11.9). According to Eq. (14.14), one adds the driving temperature differences, which would occur during evaporation of the pure substances, in conformity with the molar fractions of the two components. The actual driving temperature difference $\Delta\vartheta$ is different from the ideal one

$$\Delta\vartheta = \Delta\vartheta_{id} + \Delta\vartheta_E \quad \text{or} \quad \Delta\vartheta = \Delta\vartheta_{id}(1+\theta) \tag{14.15}$$

with

$$\theta = \Delta\vartheta_E/\Delta\vartheta_{id}.$$

The supplementary term depends mainly on the difference between vapor and liquid composition, and is always positive because of the reduction of the heat transfer in the mixture. Experiments with many mixtures yielded the simple linear relation

$$\theta = K_{12}|y - x|, \tag{14.16}$$

where K_{12} is a positive number, which to a good approximation is independent of the composition. One can interpret K_{12} as a binary interaction parameter that must be determined for a given mixture and pressure. In the pressure range between 1 and 10 bar, the pressure dependence of K_{12} could be reproduced approximately by the empirical equation

$$K_{12} = K_{12}^0(0.88 + 0.12p/p_0) \tag{14.17}$$

with $p_0 = 1$ bar. The value K_{12}^0 is different for every mixture, but is independent of pressure. Table 14.1 reproduces the value K_{12}^0 for different binary mixtures. A mean value for K_{12}^0 for all these mixtures is approximately 1.4.

Equation (14.17) is valid only in the limited pressure region between 1 and 10 bar. At higher pressures, the value of K_{12} grows more than linearly with the pressure. For a more precise investigation of the influence of pressure upon the heat transfer during boiling of mixtures, especially at pressures above 10 bar, further experiments are required.

Table 14.1 Values K_{12}^0 for several binary mixtures

Mixture	K_{12}^0
Acetone/Ethanol	0.75
Acetone/Butanol	1.18
Acetone/Methanol	1.19
Acetone/Water (copper heating surface)	1.40
(nickel heating surface)	0.81
Acetone/Benzene	0.42
Ethanol/Cylohexane	1.31
Ethanol/Water (copper heating surface)	1.21
(nickel heating surface)	0.71
Benzene/Toluene	1.44
n-Heptane/Methylcyclohexane	1.95
i-Propanol/Water	2.04
Methanol/Ethanol	1.39
Methanol/Benzene	1.08
Methanol/Amylalcohol	0.80
Methanol/Water	0.56
Methylethylketone/Water	1.21
Propanol/Water	3.29
Water/Glycol	1.47
Water/Glycerol	1.50
Water/Pyridine	3.56

Measurements by Bier et al. [14.16] on the binary mixture of sulphurhexafluoride (SF_6) and the refrigerant R-13B1 (CF_3Br) in the region $0.56 \leqq p/p_{cr} \leqq 0.97$ extending almost up to the critical pressure could be correlated by an equation of the form

$$\frac{\alpha}{\hat{\alpha}} = 1 - K(p^*)(y - x)^{1/2} \tag{14.18}$$

with

$$p^* = p/p_{cr}, K(p^*) = -0.50 + 3.65p^* + \frac{0.08}{1 - p^*}$$

and

$$\hat{\alpha} = \alpha_1 x_1 + \alpha_2 x_2.$$

The quantity x_1 is the molar fraction of the sulphurhexafluoride, α_1 and α_2 are the heat transfer coefficients of the pure substances using the heat flux of the mixture and at the existing pressure.

Finally, mention should be made of an equation by Schlünder [14.17], which starts from the assumptions of the film theory, according to which one imagines the phase interface between vapor and liquid as being plane and assumes that the liquid mass flux to be evaporated is flowing vertically to the phase interface. After several simplifications, one obtains for the quantities defined in

Eq. (14.15)

$$\theta = 1 + \frac{1}{\Delta\vartheta_{id}}(\vartheta_{s2} - \vartheta_{s1})(y - x)\left[1 - \exp\left(-\frac{q}{\varrho_L \beta_L \Delta h_v}\right)\right], \tag{14.19}$$

where the mass transfer coefficient $\beta_L = 2 \cdot 10^{-4}$ m/s was determined by adjustment to measurements and is of the same order of magnitude as the values, that one knows from the falling film evaporation. The temperatures ϑ_{s1} and ϑ_{s2} are the boiling temperatures of the pure substances. The equation also reproduces well the pressure dependence of the data of Bier et al. [14.16].

14.4.2 Liquid Mixtures with more than two Components

Equation (14.16) for binary mixtures was extended by Stephan and Preusser [14.18] to mixtures with more than two components. Equation (14.15) continues to be valid. According to it, the driving temperature difference $\Delta\vartheta$ is split up into an "ideal amount" $\Delta\vartheta_{id}$ and a surplus or excess amount $\Delta\vartheta_E$. For a mixture of K components, the amount $\Delta\vartheta_{id}$ is defined by

$$\Delta\vartheta_{id} = \sum_{i=1}^{K} x_i \Delta\vartheta_i, \tag{14.20}$$

and

$$\frac{\Delta\vartheta_E}{\Delta\vartheta_{id}} = \theta = \left|\sum_{i=1}^{K-1} K_{iK}(y_i - x_i)\right|. \tag{14.21}$$

Here K_{iK} are interaction parameters that must be ordered in a sequence according to increasing boiling points of the pure substances. A detailed explanation for

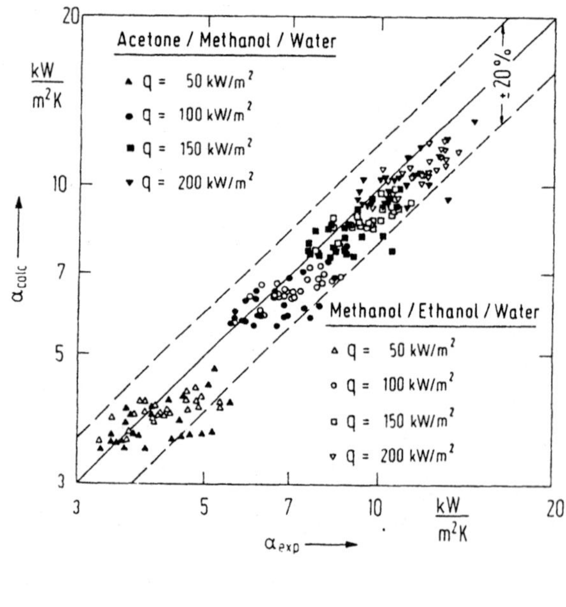

Fig. 14.13. Comparison of calculated (α_{calc}) and measured heat transfer coefficients (α_{exp}) for three-component mixtures, according to [14.18]

this can be found in a work by Stephan [14.9]. For a ternary mixture, Eq. (14.21) is reduced to

$$\theta = |K_{13}(y_1 - x_1) + K_{23}(y_2 - x_2)|, \tag{14.22}$$

which contains the associated binary mixtures as limiting cases. If one assumes $y_1 = x_1 = 0$, then one reduces the ternary to the binary mixture of components 2 and 3, whereas for $y_2 = x_2 = 0$, the associated binary mixture consists of components 1 and 3. One recognizes from this that the coefficients K_{iK} in Eq. (14.21) are identical with the values for binary mixtures given in Table 14.1. Heat transfer coefficients calculated according to this method reproduce quite well the previously known test data for ternary mixtures, as Fig. 14.13 shows in an exemplary manner.

14.4.3 Solutions of Solids in Liquids

The boiling point rises in an isobaric solution of solids in liquids. The driving temperature difference between wall and saturation temperature thus decreases with the amount of the dissolved solids. The vapor consists of the pure solvent. In the vicinity of the wall, the solvent must overcome a mass transfer resistance in order to get from the liquid to the bubble surface. This impedes bubble growth, and, in comparison with the pure solvent, reduces the heat transfer, exactly as in the case of the binary and multi-component mixtures of liquids discussed earlier. This is confirmed by numerous data [14.19–14.24].

As an example, Fig. 14.14 shows the results of the measurements of Feldkamp [14.24]. Depicted is the ratio α/α_w of the heat transfer coefficient α of a sodium chloride solution to that of pure water α_w at the same heat flux. The parameter is the mass fraction ξ of the dissolved salt. The heat transfer coefficient is decreased considerably by the addition of salt to the water, and a weak decrease in the coefficient with increasing heat flux is also recognizable. This is explained by the higher concentration of salt in the layers adjacent to the wall, which results in a decrease of the driving temperature gradient directly at the wall with higher heat fluxes.

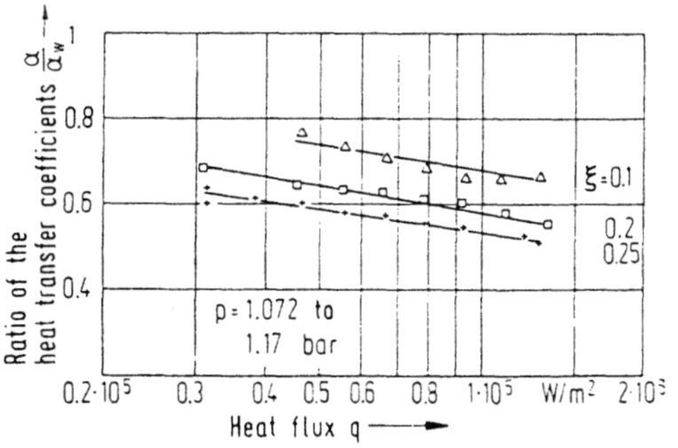

Fig. 14.14. Heat transfer coefficients α/α_w of sodium chloride solutions, according to [14.24]

Experiments with aqueous solutions of sucrose, sodium chloride, sodium hydroxide, and ammonium nitrate in the pressure range between 1 and 16 bar were correlated by Feldkamp [14.24] who expanded an empirical equation suggested by Sagan [14.23], with a standard deviation of 12.1% and a mean error of 15.36%. This equation reads

$$\frac{\alpha}{\alpha_w} = \left(\frac{\varrho_L}{\varrho_{Lw}}\right)^{0.816} \left(\frac{\varrho_G \Delta h_v c_{pw}}{\varrho_{Gw} \Delta h_{v,w} c_p}\right)^{-0.716} \left(\frac{\varrho_L - \varrho_G}{\varrho_{Lw} - \varrho_{Gw}}\right)^{-1.0795} \cdot \left(\frac{\eta}{\eta_w}\right)^{-0.1} \left(\frac{\sigma}{\sigma_w}\right)^{-1.2135} \left(\frac{\lambda}{\lambda_w}\right)^{0.284} . \tag{14.23}$$

The index L signifies liquid, G gas, and w water. The property values $\varrho_L, c_p, \eta, \lambda$ of the solution and the surface tension σ are to be evaluated at the mean temperature $\bar{\vartheta} = (\vartheta_w + \vartheta_s)/2$ and the mass fraction ξ of the solution.

14.5 Maximum Heat Flux, Transition Boiling, and Film Boiling

Measurements of the maximum heat flux of aqueous solutions and of hydrocarbon mixtures on horizontal wires [14.25–14.29] yielded values, which, depending upon the composition, were larger by a factor of up to two than the maximum heat fluxes of the pure components. This is understandable, if one considers that strong differences in the liquid composition are formed around thin wires. The bubble growth is substantially impeded by this, and larger heat fluxes are required until the heating surface is largely covered by vapor and the maximum heat flux is achieved. Maxima and minima were established as functions of the composition [14.28], the maxima at large differences $y - x$ in the composition of vapor and liquid, and the minima in the vicinity of azeotropic points.

The results are completely different, if the liquid mixture boils outside a horizontal tube, whose diameter in contrast to that of a wire is larger by several orders of magnitude than the departure diameter of the formed vapor bubbles. The maximum heat flux then always lies between the values for the pure components. Figure 14.15 shows as an example maximum heat fluxes of mixtures of acetone/water, methanol/ethanol, and acetone/methanol. The values are reproduced by a correlation of Krushilin and Subbotin [14.30]

$$q_{max} = 1830 \frac{T_s^{0.32} \sigma^{0.21} (\varrho' - \varrho'')^{0.48} (\varrho'' \Delta h_v)^{0.36} \lambda'^{0.4}}{\varrho'^{0.31} c_p'^{0.08} M_L^{0.14}} \tag{14.24}$$

in which all property values are to be entered in SI units; M_L is the molar mass of the liquid mixture in kg/kmol. One obtains q_{max} in W/m^2. The influence of the mixture composition and of the pressure are unsatisfactorily reproduced by this equation because of the lack of data. As the data show, the maximum heat flux as a function of pressure passes through a maximum just as in the case of

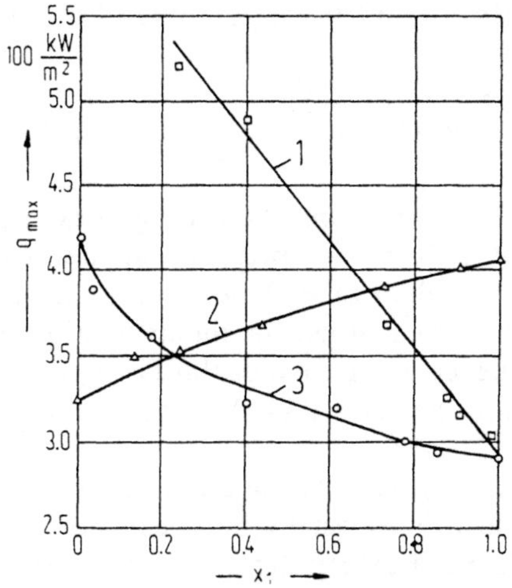

Fig. 14.15. Maximum heat flux of *1* acetone/water, *2* methanol/ethanol, *3* acetone/methanol at 1 bar, according to [14.13]

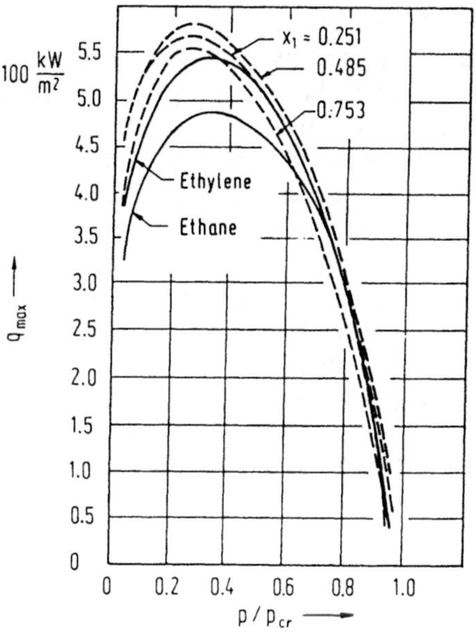

Fig. 14.16. Maximum heat flux of ethane, ethylene, and their mixtures. x_1 = mole fraction of ethylene, according to [14.31]

pure substances, Fig. 11.16. As an example, Fig. 14.16 shows the results of Wright and Colver [14.31]. Their data could be correlated well by an equation of Noyes [14.32]

$$q_{max} = 1.19\varrho''\Delta h_v(ga')^{1/3}\left(\frac{\varrho' - \varrho''}{\varrho''}\right)^{0.56}(Pr')^{1/12}, \qquad (14.25)$$

which is valid for $p/p_{cr} \leqq 0.3$. In the pressure range $p/p_{cr} > 0.3$, the test data were between the values computed from Eq. (14.25) and those computed from an another equation of Noyes

$$q_{\max} = 0.144\varrho''\Delta h_v\left(\frac{g\sigma}{\varrho'}\right)^{1/4}\left(\frac{\varrho' - \varrho''}{\varrho''}\right)^{1/2}(Pr')^{-0.245}. \tag{14.26}$$

Only a few investigations deal with *transition boiling*. According to experiments of Happel and Stephan [14.8], the influence of pressure on the heat transfer in the transition region is slight. The data of Preusser [14.2] showed a linear dependence of the heat transfer coefficient upon the molar fraction in binary mixtures. The very pronounced influence of a second or third component on the heat transfer in the transition region indicates that the mass exchange plays an appreciable role.

In comparison with pure substances, only a few investigations are known that deal with *film boiling* of mixtures, and it is especially not clear, how the heat transfer to the boiling mixture is influenced by a rough film surface or by the emission of heat by radiation, which is no longer negligible during film boiling. A critical analysis of the laminar film evaporation of mixtures with allowance for diffusion [14.33] showed that the phase interface temperature and thus also the vapor composition are determined by the mass transport to the phase interface. Because the liquid during film boiling does not touch the heating surface, one should not expect the deposition of solids on the heating surface. If on the other hand the vapor velocities are very small, solid particles separated at the phase interface between vapor and liquid can diffuse to the wall and precipitate there as a coating. In order to avoid this phenomenon, the temperature difference $\vartheta_w - \vartheta_s$ must be so large that the vapor velocities are sufficient to prevent the diffusion of solid particles to the wall. On the other hand, the temperature difference $\vartheta_w - \vartheta_s$ must not reach such high values that a direct contact between liquid and wall is made as a result of the fluctuating motion of the vapor-liquid phase interface. The favorable temperature differences, at which no deposition occurs, are very high, and, according to experiments of Farrar and Marschall [14.33], are between 600 and 800 °C in aqueous solutions of calcium sulfate.

14.6 Boiling of Immiscible Liquids

In a binary mixture consisting of two immiscible liquid phases and the vapor phase, the boiling temperature is clearly determined, according to the Gibbs phase rule, by prescribing the pressure. Reference should be made to the explanations in Chap. 7. The boiling temperature of the mixture is less than that of the pure components. If, for example, one mixes water with perchloroethylene, two immiscible liquid phases are formed, whose boiling temperature at a pressure of 1 bar lies at 87.8 °C, whereas pure water boils at about 100 °C and pure perchloroethylene at about 121 °C. If, therefore, one wishes to evaporate a liquid that would decompose at its boiling temperature, one can add an immiscible liquid or its vapor and thus lower the boiling temperature.

The heat transfer to such immiscible liquids is determined to a great extent by which of the two liquid phases touches the heating surface. If, for example, a mixture of water/perchloroethylene fills a container with a horizontal heating surface at the bottom, then the lower phase contains mainly the heavier and less volatile perchloroethylene, whereas the upper phase consists primarily of water. At sufficiently large heat fluxes, either nucleate or film boiling occurs, depending upon the heat flux.

As experiments of Bragg and Westwater [14.34] showed, the rising bubbles merge, partially at the phase interface between the two liquids, and then diffuse as large bubbles through the upper phase. The perchloroethylene-rich, lower phase had to be 6 mm high in order that the water-rich, upper phase did not touch the wall, whereas the water-rich, upper phase had to be 45 mm high. Only then was there thermal equilibrium in the mixture. Lower levels of the liquid layers led to noticeably lower boiling temperatures than in the case of pure substances. The boiling temperature of the mixture was then above the equilibrium temperature, because the time of contact between rising bubbles and liquid was too short for the vapor to achieve its equilibrium composition.

If water is mixed into boiling perchloroethylene, the boiling temperature falls. At a constant wall temperature, the driving temperature difference $\Delta\vartheta = \vartheta_w - \vartheta_s$ and, thus, also the heat flux increase. The curve of the heat flux $q(\Delta\vartheta)$ lies above that for pure perchloroethylene, so that up to 30% greater heat fluxes can be transferred at the same temperature difference.

In the likewise immiscible liquid mixture of n-hexane/water, the water-rich phase is heavier and is, therefore, at the bottom of the container. The boiling temperature of the pure n-hexane is 68.3 °C at 1 bar, that of the n-hexane/water mixture 61.7 °C. During sufficient superheating, film boiling prevailed at the bottom of the container. The rising vapor condensed again at the colder phase interface between the liquids and produced vapor bubbles in the upper phase. The heat transfer at the bottom was poor because of the film boiling, whereas at the phase interface between the liquids, there was a far better heat transfer because of nucleate boiling. Because upon inception of boiling more heat was removed by nucleate boiling than was added at the bottom of the container, the liquid cooled down strongly in the vicinity of the phase interface and became up to 33 °C subcooled [14.34].

In spite of this very complicated phenomenon, the heat transfer during the boiling of immiscible liquids is determined to a great extent by the properties of the liquid at the heating surface. Bragg and Westwater could reproduce their measurements, as long as there was film boiling at the wall, quite well with an equation of Hamill and Baumeister [14.35], which starts with the known equations for film boiling, see Sect. 11.8, and allows for the influence of radiation and the convective heat transport in the case of a subcooled liquid. Their equation is given by

$$\alpha = \alpha_f + 0.88\alpha_r + 0.12\alpha_c\theta. \tag{14.27}$$

α_f is the heat transfer coefficient for film boiling, which is calculated from

$$\alpha_f = 0.41 \left[\frac{\lambda_G^3 \varrho_G (\varrho_L - \varrho_G) g \Delta h}{\eta_G (\vartheta_w - \vartheta_s) \sqrt{\dfrac{\sigma}{g(\varrho_L - \varrho_G)}}} \right]^{1/4} \tag{14.28}$$

with $\Delta h = \Delta h_v + 0.95 c_{vG} (\vartheta_w - \vartheta_s)$. The quantity α_r is the heat transfer coefficient for radiation, Eq. (11.39). The heat transfer coefficient α_c is obtained from one of the known equations for heat transfer during turbulent free convection in the vapor film [for example, 14.36]. A subcooling factor

$$\theta = \frac{\vartheta_s - \vartheta_b}{\vartheta_w - \vartheta_s}$$

plays a role, if the temperature ϑ_b in the core of the liquid is below the boiling temperature. The properties are those of the phase at the heated surface. A condition for the use of these equations is that the fluid layers are thicker than 4.5 cm. The above equations are not valid if there is nucleate boiling at the wall. For this case, the heat transfer with boiling immiscible liquids has not yet been sufficiently researched.

15 Heat Transfer During Boiling of Mixtures in Forced Flow

As in the case of pure substances, during the boiling of mixtures in forced flow in tubes or channels, there is a decrease in saturation temperature because of the pressure drop along the path of flow. In addition, because the more volatile component is converted into vapor first, the liquid becomes richer in the less volatile boiling component. In many industrial processes where the tube diameter is not very small, and, thus, the pressure drop not very large, the rise in boiling temperature, because of the increase in the less volatile component in the liquid, compensates the decrease in boiling temperature as a consequence of the pressure drop. It is thus possible that the saturation temperature even rises downstream. In a two-fluid heat exchanger, the driving temperature gradient decreases. In general, the heat transfer coefficient also decreases.

Figure 15.1 shows the course of the vapor temperature ϑ_G in the core of an annular flow of an evaporating liquid film which is directed upward in a vertical evaporator tube. The course of the phase interface temperature ϑ_I on the vapor side of the liquid film is also plotted in the figure. The mixture evaporated was ethanol/water. The concentration of the higher boiling component in the liquid causes in most cases an increase in the viscosity, which likewise contributes to a reduction in the heat transfer coefficient.

As in the case of boiling of pure substances in forced flow, one has to distinguish between saturated and convective boiling.

15.1 Saturated Boiling

In the region of saturated boiling, the qualities are still so small that the heat transfer is determined by the formation of vapor bubbles. The heat transfer coefficient $\alpha_{2\,Ph}$ of the two-phase flow is calculated, therefore, as was described in Sect. 13.7. One forms it from two quantities, following the simple example of Eq. (13.111):

$$\alpha_{2\,Ph} = \alpha_B + \alpha_K,$$

where the heat transfer coefficient α_B during boiling of mixtures in free convection is to be computed according to the equations of Sect. 11.4, whereas one finds the convective contribution α_K to the heat transfer coefficient from the equation

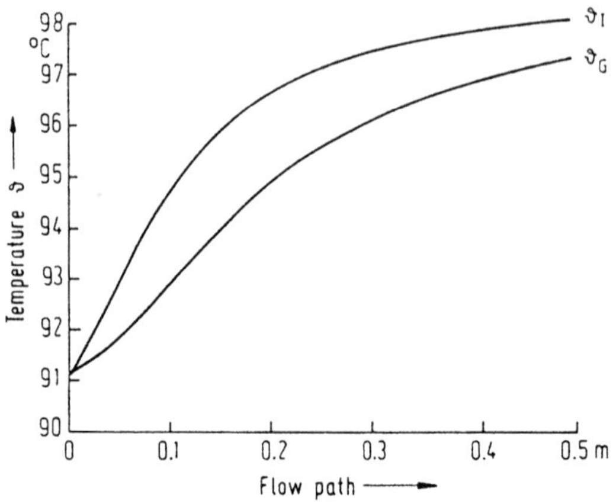

Fig. 15.1. Temperature ϑ_G of the vapor and ϑ_I at the phase interface of a boiling ethanol/water mixture with upwards directed annular flow in a vertical tube of 37 mm inner diameter, according to Shock [15.1]. At inlet: mass flow rate $\dot{M} = 0.1$ kg/s, quality $x^* = 0.05$, mole fraction of ethanol $x = 0.041$. Heat flux $q = 2 \cdot 10^6$ W/m².

for single-phase heat transfer during turbulent tube flow

$$Nu = \frac{\alpha_K d}{\lambda_L} = 0.023\, Re^{0.8} Pr^{0.4}$$

with the Reynolds number

$$Re = \dot{m}_L d/\eta_L = [\dot{m}(1 - x^*)d/\eta_L]$$

and the Prandtl number $Pr = \nu_L/a_L$ of the liquid.

15.2 Convective Boiling

For the region of convective boiling, Bennett and Chen [15.2] have modified their Eq. (13.117), which is valid for the convective boiling of pure substances,

$$\alpha_{2Ph} = S\alpha_B + F\alpha_K$$

In this equation, the factor $S \leqq 1$ allows for the fact that in a forced flow more heat is transferred convectively through the temperature boundary layer and that, therefore, for the same heat flux, fewer bubbles are formed than in free convection. The factor $F \geqq 1$ allows for the augmentation of the convective heat transfer in comparison to the single-phase flow, because the vapor in two-phase flows exerts a shear stress on the liquid. In mixtures, further modifications are necessary. Because the lower boiling component is converted into the vapor phase first, the wall-adjacent liquid becomes enriched with the higher boiling component. As a result, the temperature profile is changed and thus also the factor S. For binary mixtures the factor S should be corrected by the temperature ratio $\Delta\tilde{\vartheta}/\Delta\vartheta$, Eq. (14.12), according to Scriven [14.5], i.e., instead of entering the term $S\alpha_B$ one should enter the expression $S\alpha_B\Delta\tilde{\vartheta}/\Delta\vartheta = S^*\alpha_B$, in which $\Delta\tilde{\vartheta}$

is the driving temperature difference with allowance for the mass transfer and $\Delta\vartheta$ is the liquid superheating that would set in, if the diffusion were negligible. The heat transfer coefficient α_B during nucleate boiling is, therefore, not related to the liquid superheating $\Delta\vartheta$, but rather to the effective liquid superheating $\Delta\tilde{\vartheta}$.

Because the driving temperature difference is changed, the definition of the heat transfer coefficient α_K or the factor F in the equation for pure substances must be changed. For the heat transfer coefficient one now sets α_K^*, and for mixtures one makes the assumption

$$\alpha_{2\,Ph} = S^*\alpha_B + F\alpha_K^*. \tag{15.1}$$

with

$$S^* = S\frac{\Delta\tilde{\vartheta}}{\Delta\vartheta} = \frac{1}{1 + 2.53\cdot 10^{-6} Re_{2\Phi}^{1.17}} \frac{1}{1 - (\xi_G - \xi_L)\dfrac{c_{pL}}{\Delta h_v}\left(\dfrac{\partial\vartheta}{\partial\xi_L}\right)_p}. \tag{15.2}$$

The factor S for pure substances is given here by Eq. (13.119) and the quotient $\Delta\tilde{\vartheta}/\Delta\vartheta$ by Eq. (14.12). As was already explained in connection with Eq. (13.119),

$$Re_{2\Phi} = ReF^{1.25} = \frac{\dot{m}(1 - x^*)d}{\eta_L} F^{1.25},$$

in which $F(X_{tt})$ is either to be taken from Fig. 13.27 or to be calculated from Eq. (13.121a). The quantity α_B is the heat transfer coefficient during nucleate boiling without allowance for the mass transfer. It would be obtained, if one were to bring a homogeneous liquid mixture to a boiling condition and the produced vapor were to be in equilibrium with the liquid. Relying on their work on the convective boiling of pure substances, Bennett and Chen [15.2] computed the heat transfer coefficient α_B from Eq. (13.118) of Forster and Zuber [13.38]. However, as was explained in Sect. 13.8, this equation was formulated in 1955 and does not reproduce more recent data satisfactorily. For mixtures, then, it also seems more expedient to calculate the heat transfer coefficient α_B according to one of the application equations of Sect. 11.4, for example, Eq. (11.9), while using the property values of the mixture.

In order to calculate the heat transfer coefficient α_K^* in Eq. (15.1), one must take into consideration the fact that the driving temperature difference for the convective portion of the heat transfer is reduced in comparison with that for pure substances. According to Bennett and Chen [15.2], one obtains the heat transfer coefficient α_K^*, assuming that the mass transfer influences, by means of convection, only the driving temperature difference, but not the heat transfer coefficient. There then holds for the convective heat transfer in the single-component system

$$q_{K,1\,Ph} = \alpha_K(\vartheta_w - \vartheta_B),$$

where ϑ_w is the wall temperature and ϑ_B the mean fluid temperature (adiabatic mixture temperature). Correspondingly, there holds in a binary mixture

$$q_{K,2Ph} = \alpha_K(\vartheta_w - \vartheta_I),$$

where the equilibrium temperature at the phase interface between the vapor core and the wall-adjacent liquid is entered for the temperature ϑ_I in the driving temperature difference $\vartheta_w - \vartheta_I$. If one defines a heat transfer coefficient α_K^* in the mixture with the aid of the temperature difference $\vartheta_w - \vartheta_B$, then

$$q_{K,2Ph} = \alpha_K^*(\vartheta_w - \vartheta_B)$$

or

$$\alpha_K^* = \alpha_K \frac{\vartheta_w - \vartheta_I}{\vartheta_w - \vartheta_B}. \tag{15.3}$$

The temperature ϑ_I is dependent upon the composition at the phase interface. The mass balance for the lower boiling component in the liquid is

$$j_{L,I} = \varrho_L \beta_L (\xi_{L,I} - \xi_{L,B}),$$

where $j_{L,I}$ signifies the mass flux (SI unit kg/m^{2}s) at the phase interface. The mass transfer coefficient β_L is defined by this equation. The energy balance is formulated with the limiting assumption that all the added heat serves for the formation of vapor. The portion $q\xi_{G,B}$ of the heat flux is used for the lower boiling component and the portion $q(1 - \xi_{G,B})$ for the higher boiling one. The energy balance for the lower boiling component thus yields

$$q\xi_{G,B} = j_{L,I} \Delta h_v, \tag{15.4}$$

if Δh_v is the evaporation enthalpy of the lower boiling component. From these two equations, one obtains

$$\xi_{L,I} - \xi_{L,B} = \frac{q}{\Delta h_v \varrho_L \beta_L} \xi_{G,I}. \tag{15.5}$$

If, on the other hand, we linearize the boiling line in the region between $\xi_{L,I}$ and $\xi_{L,B}$, we get

$$\left(\frac{\partial \vartheta}{\partial \xi_L}\right)_p = \frac{\vartheta_I - \vartheta_B}{\xi_{L,I} - \xi_{L,B}}.$$

One can also write for this

$$\frac{\vartheta_w - \vartheta_I}{\vartheta_w - \vartheta_B} = 1 - \frac{\xi_{L,I} - \xi_{L,B}}{\vartheta_w - \vartheta_B} \left(\frac{\partial \vartheta}{\partial \xi_L}\right)_p. \tag{15.6}$$

If the difference $\xi_{L,I} - \xi_{L,B}$ is then eliminated with the aid of Eq. (15.5), there follows

$$\frac{\vartheta_w - \vartheta_I}{\vartheta_w - \vartheta_B} = 1 - \frac{q}{\Delta h_v \varrho_L \beta_L} \left(\frac{\partial \vartheta}{\partial \xi_L}\right)_p \frac{\xi_{G,I}}{\vartheta_w - \vartheta_B}. \tag{15.7}$$

One obtains the mass transfer coefficient β_L occurring here in the liquid phase from the equation for the mass transfer of the turbulent flow, into which must be entered according to Bennett and Chen [15.2] the Reynolds number $Re_{2\Phi}$ of the two-phase flow. The Sherwood number is then given by

$$Sh = \frac{\beta_L d}{D} = 0.023 Re_{2\Phi}^{0.8} Sc^{0.4}. \tag{15.8}$$

with

$$Re_{2\Phi} = Re F^{1.25} = \frac{\dot{m}(1 - x^*)d}{\eta_L} F^{1.25}.$$

When using the procedure of Bennett and Chen some precaution is advised, because until now there are only a few known experiments on the testing of the assumptions and, thus, of the accuracy of the procedure. Bennett and Chen used only data on aqueous solutions of ethylene glycol. These could be reproduced by that method of calculation with an average error of $\pm$ 14.9%. The most far reaching of the many assumptions is surely the one which assumes that all of the added heat serves in the mixture for vapor formation. Actually, a significant amount of the added heat is needed to heat the liquid and the vapor to the boiling temperature, which is constantly increasing along the path of flow.

This amount can be approximately calculated according to a suggestion of Butterworth [15.3], by allowing for the fact that the added heat flow rate effects an enthalpy increase

$$d\Phi = \dot{M} dh$$

and that a part of this heat flow rate

$$d\Phi_G = \dot{M}_G c_{pG} d\vartheta_G$$

is supplied to the vapor and raises its enthalpy. Thus,

$$\frac{d\Phi_G}{d\Phi} = \frac{\dot{M}_G c_{pG} d\vartheta_G}{\dot{M} dh} = x^* c_{pG} \frac{d\vartheta_G}{dh}$$

resulting in

$$d\Phi_G = x^* c_{pG} \frac{d\vartheta_G}{dh} d\Phi = \alpha_G(\vartheta_I - \vartheta_B),$$

in which α_G is the heat transfer coefficient on the side of the gas. Thus, there is

$$\vartheta_I - \vartheta_B = \frac{1}{\alpha_G} x^* c_{pG} \frac{d\vartheta_G}{dh} q.$$

On the other hand, the driving temperature difference for the heat transfer of the two-phase flow is

$$\vartheta_w - \vartheta_I = q/\alpha_{2Ph}.$$

Addition of these two equations yields

$$\vartheta_w - \vartheta_B = q\left(\frac{1}{\alpha_{2Ph}} + \frac{1}{\alpha_G} x^* c_{pG} \frac{d\vartheta_G}{dh}\right).$$

If one now defines an effective heat transfer coefficient $(\alpha_{2Ph})_{eff}$ of the two-phase flow by $(\alpha_{2Ph})_{eff} = q/(\vartheta_w - \vartheta_B)$, then this coefficient is given by

$$\frac{1}{(\alpha_{2Ph})_{eff}} = \frac{1}{\alpha_{2Ph}} + \frac{1}{\alpha_G} x^* c_{pG} \frac{d\vartheta_G}{dh}. \tag{15.9}$$

15.3 The Practical Calculation of the Heat Transfer Coefficient During Convective Boiling

In order to calculate the heat transfer coefficient α_{2Ph} according to Eq. (15.1) during convective boiling of mixtures and, subsequently, the effective heat transfer coefficient $(\alpha_{2Ph})_{eff}$ according to Eq. (15.9), it is expedient to proceed according to the following plan.

—One calculates first the Martinelli parameter with the properties of the mixture

$$X_{tt} = \left(\frac{\varrho_G}{\varrho_L}\right)^{0.5} \left(\frac{\eta_L}{\eta_G}\right)^{0.1} \left(\frac{1 - x^*}{x^*}\right)^{0.9}.$$

—With the aid of X_{tt}, one obtains the factor F from Fig. 13.27 or from Eq. (13.121a), and, thus, also

$$Re_{2\Phi} = ReF^{1.25} = \frac{\dot{m}(1 - x^*)d}{\eta_L} F^{1.25}.$$

—With the Reynolds number $Re_{2\Phi}$, one finds the factor S from Fig. 13.26 or from Eq. (13.119a).
—The factor S^* thus results from Eq. (15.2).
—The heat transfer coefficient α_B is calculated from one of the application equations given in Sect. 11.4, for example, from Eq. (11.9), into which one must enter as property values those of the mixture.
—Because of Eq. (15.3), the heat transfer coefficient α_K^* is obtained from the value α_K and the temperature ratio $(\vartheta_w - \vartheta_I)/(\vartheta_w - \vartheta_B)$. One calculates α_K from Eq. (13.120) for the convective heat transfer during turbulent tube flow. The temperature ratio follows from Eq. (15.7), after the mass transfer coefficient was determined from Eq. (15.8).
—After α_{2Ph} has been calculated in this manner, the effective heat transfer coefficient must be determined. One obtains it from Eq. (15.9), in which the heat transfer coefficient on the gas side is evaluated from Eq. (13.120) for the heat transfer during turbulent tube flow. One enters as property values those of the gas. The Reynolds number of the gas is $Re_G = \dot{m}x^*d/\eta_G$. In order to

constitute the expression $x^* \mathrm{d}\vartheta_G/\mathrm{d}h$, one will assume, just as in the case of the condensation of vapor mixtures, Sect. 6.4, that in every cross section, thermodynamic equilibrium prevails between vapor and liquid; then, $h(\vartheta_G)$, $x^*(\vartheta_G)$, and thus also $x^* \mathrm{d}\vartheta_G/\mathrm{d}h$ are formed.

A simpler, empirical equation has been suggested by Varma and Sharma [15.4]. It is based upon data for the refrigerant mixture of R-12 and R-22 (boiling point at 1 bar of R-12: $\vartheta_s = -30\,°\mathrm{C}$, of R-22: $\vartheta_s = -41\,°\mathrm{C}$) during forced flow in a horizontal tube. The mass fluxes were between 120 and 300 $\mathrm{kg/m^2s}$, the heat fluxes between 3250 and 15200 $\mathrm{W/m^2}$. The data could be described by the following empirical equation:

$$\frac{\alpha_{2\mathrm{Ph}}}{\alpha_K} = 8.275 X_{tt}^{-0.253} \left(\frac{q}{\dot{m} \Delta h_v}\right)^{0.120} (1 - |y - x|)^{-0.9} \tag{15.10}$$

Here the Martinelli parameter X_{tt} given earlier and the entire mass flux $\dot{m}$ are used. The heat transfer coefficient α_K is given by Eq. (13.120) for the heat transfer during turbulent tube flow, where the property values of the mixture must be entered. The mean deviation from the experimental results amounted to about $\pm 30\%$, if the mass fraction of R-22 was small, and about $\pm 15\%$ at higher mass fractions of R-22.

16 Enhancement of Heat Transfer During Boiling

16.1 General Observations

Heat transfer coefficients of boiling water in free convection lie in installed plants between $2 \cdot 10^3$ W/m^{2}K and are rarely larger than $5 \cdot 10^4$ W/m^{2}K, whereas those of ammonia, alcohols, paraffins, and other organic substances seldom reach values over 1500 W/m^{2}K. In flooded refrigerant evaporators, which are usually operated in the transition region between single-phase free convection and nucleate boiling, to avoid too large temperature differences heat transfer coefficients may even lie only between 500 and 800 W/m^{2}K. On the other hand, the heat transfer coefficients on the side of the heating agent can be significantly larger, especially if heat is supplied there by condensation of a superheated vapor or convectively by water. The decisive thermal resistance is then on the boiling side, and one should look for measures to reduce this thermal resistance.

It has long been known, that heat transfer can be enhanced by many techniques. Jakob and Fritz [16.1] were the first to demonstrate that heat transfer during the boiling of water on rough heating surfaces is better than on smooth heating surfaces. As they showed, Fig. 16.1, the heat transfer to boiling water can be augmented by a factor of 1.3, if one sand blasts the heating surface, and even up to a factor of 5, if one machines in a roughness pattern. Since then it has been pointed out repeatedly in the literature that heat transfer during evaporation can be augmented by use of roughened or knurled heating surfaces. One must, however, take into consideration the fact that, with most industrial fluids, roughened surfaces tend to become dirty faster than smooth ones; therefore, the heat transfer coefficient decreases with time. Applying special profiles to heating surfaces has also proven successful for the augmentation of heat transfer. The following measures are to be considered for augmentation of heat transfer during evaporation:

—roughening of the heating surface
—structuring or coating the surface
—production of artificial nucleation sites by sintering and
—special shaping of the heating surface
—addition of gases or liquids
—addition of solids

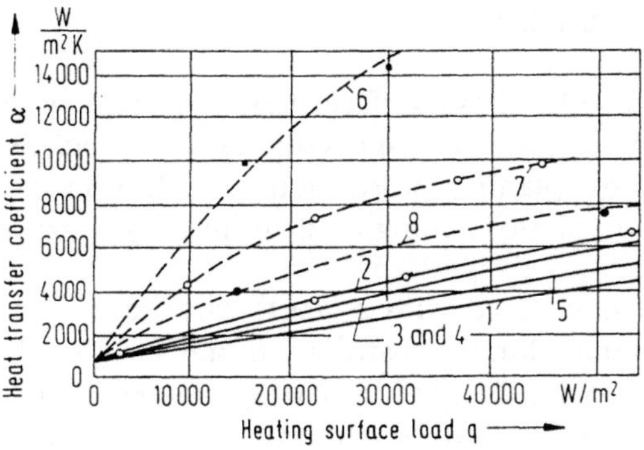

Fig. 16.1. Heat transfer from rough heating surfaces to boiling water, according to [16.1]. *1* smooth tube; *2–5* heating surfaces roughened by sand-blasting; *6–8* heating surfaces with a roughness grating

What enhancement of heat transfer can be achieved by these measures will be discussed in the following sections.

16.2 Roughening of Heating Surfaces

Data on heating surfaces, that were roughened by rolling, turning, grinding, or filing, showed that the heat transfer coefficient for evaporation can be described to a good approximation by the depth of surface smoothness R_p. The latter is defined in DIN 4762 and is a measure for the mean roughness of a heating surface. At pressures close to atmospheric pressure, the heat transfer coefficient evaluated at a constant heat flux is proportional to $R_p^{0.133}$. This interrelation discovered by Stephan [11.14] has been confirmed by other authors [16.2–16.4]. However, it holds in this form only if the roughnesses are distributed according to a Gauss error function, as is generally the case for industrial surfaces, and fails, for example, in the case of corroded surfaces, because then the heating surface is roughened primarily at the grain boundaries of the surface crystals, thus resulting in a more or less uniform surface pattern.

According to investigations of Fedders [16.2], the influence of roughness decreases somewhat with increasing heat flux. For example, the heat transfer coefficient of boiling water at the pressure of 19.6 bar, and a heat flux of 10^6 W/m², is proportional to $R_p^{0.1}$. The augmentation of the heat transfer by roughening is relatively small because of the weak dependence on the depth of surface smoothness. Industrial surfaces have a depth of surface smoothness of approximately 1 μm. If the depth of surface smoothness were raised by roughening up to 10 μm, which is quite possible, the heat transfer coefficient would climb only by a factor of around 1.36. Data of Berenson [16.5] indicating that the heat transfer coefficient could be raised by roughening by up to a factor of 6, seem to be exaggerated and have not withstood a more precise testing by Fedders [16.2].

16.3 Structuring or Coating of Surfaces

Structured surfaces are understood to be those surfaces that have received a definite structure by processing. As processes for manufacture can be considered the following: turning or planing, cross-planing, production of nucleation sites by fine planing, chemical milling or photo etching, and drilling by laser or electron beam.

By photo etching, the surface is covered with a photoactive layer, that contains open areas acting as a pattern for the distribution and form of the cavities to be produced. After developing, the photoactive layer is resistant to etching fluids, but the area for nucleation sites can be removed by etching agents. With this process, which was first suggested by Schimmelpfennig [16.6] for the production of evaporator heating surfaces, one obtains obtuse cone shaped recesses of 0.1 mm depth and 0.01 to 0.15 mm diameter at the base. Even smaller nucleation sites of roughly 0.1 mm depth and 0.05 mm hole diameter can be produced by electron beam drilling. The nucleation site density lies at $10^4\ \mathrm{cm}^{-2}$. Figure 16.2 shows several structured surfaces.

If one assumes, just as in the case of finned tubes, that the heat transfer grows at least proportionally with the growth of surface area, then a doubling of the outer surface would lead to at least a doubling of the heat transfer coefficient at an equal temperature difference. If, for example, one wanted to enlarge by a factor of two the surface of a tube by the pyramids shown in Fig. 16.2b, one

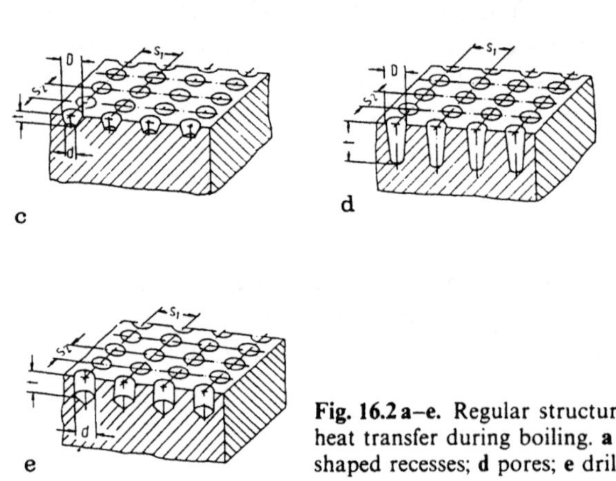

Fig. 16.2 a–e. Regular structured surfaces for the improvement of heat transfer during boiling. **a** grooves; **b** pyramids; **c** obtuse cone shaped recesses; **d** pores; **e** drilled holes

could achieve this by making the pyramids 0.15 mm high and their bases 0.2 mm long. Presumably a tube with such slight roughening could be drawn and profiled at the same time during production without substantial additional costs. Reliable data on the achievable improvements of heat transfer with this type of surface are not yet available.

The few known experiments show that the roughness and thus the heat transfer during boiling can be augmented by etching the heating surface, but the heating surface then becomes more subject to corrosion and other damage. According to experiments of Nix et al. [16.7], the shape of the produced surface depends strongly upon the time of etching. After etching of short duration, there arose on the surface irregularities whose peaks were removed by a rather long treatment with acid while the roughness continued to increase. The data showed a clear interrelationship between the roughness produced by etching and the heat transfer coefficient during boiling.

By etching of heated surfaces, fine pores connected by tiny channels can also be formed. Almgren and Smith [16.8] have used the possibility of producing pores by etching in order to augment the heat transfer during boiling. By heating a copper heating surface and then treating it with hydrochloric acid, they produced a rough surface with numerous pores. For boiling water, the heat transfer coefficient was raised by approximately 30%. Whether better results can be produced by other procedures has not yet been systematically investigated.

A doubling of the heat transfer coefficient seems possible by a coating with a thin teflon layer of 0.01 mm thickness [16.7–16.10]. The cause for this is the inability of water to wet teflon, which facilitates the formation of large vapor bubbles. At the same time, the teflon bond with the original surface favors the formation of many tiny channels, which further favor bubble formation.

According to a suggestion of Huelle [16.11], the heat transfer coefficient in an evaporator tube can be raised by a factor of 1.5 to 1.8 by the installation of a netlike fabric on the tube wall. However, the net installation leads also to a pressure drop that is higher by a factor of 1.5 to 3.5. Presumably, it causes less an intensification of vapor bubble formation than a thinning of the wall-adjacent liquid layer.

16.4 Production of Artificial Nucleation Sites by Sintering

Far greater elevations in heat transfer have been achieved by the installation of heating surfaces with a sprayed on porous surface layer. The diameter of the sprayed on solid particles is between 0.1 and 0.5 mm, so that a sufficient number of tiny hollow spaces are formed on the surface. The layer thickness is about 1 mm. As a result of the production process described in the patent literature [16.12], the surface is extraordinarily rough and covered with a maze of fine capillaries. As a result of the capillary effect, liquid enters into the cavities. Vapor is generated internally by thin film evaporation. Also playing a role is the fact that the bubbles in the narrow capillaries must be brought to a sufficient overpressure, so that they can overcome the capillary forces and leave the sintered layer. After leaving this layer, they quickly expand, and there is strong liquid displacement, which leads to a further augmentation of the heat transfer.

The increase of the heat transfer depends upon the boiling liquid and upon the pore radius of the sinter layer. This radius is characterized by an equivalent pore radius. One measures it, by dipping the porous area at one end into a liquid and then measuring the height of rise z of the liquid in the porous layer. The equivalent pore radius r_0 is obtained from the "capillary rise formula"

$$r_0 = \frac{2\sigma}{\varrho_L z g},$$

where σ is the surface tension, ϱ_L the density of the liquid, and g the gravitational constant.

The equivalent pore radius with respect to the optimum heat transfer during boiling of water at 1 bar ranges between 0.0381 and 0.127 mm [16.12]. By optimum equivalent pore radius is meant that equivalent pore radius requiring the smallest superheating $\vartheta_w - \vartheta_s$ for the transfer of a particular heat flux; for it the heat transfer coefficient is a maximum.

Data up to now suggest that there exists such an optimum pore radius, because, with increasing superheating, smaller pores also contribute to bubble formation, curve a in Fig. 16.3. On the other hand, with increasing superheating, the pores fill more readily with vapor, which because of its poor heat conduction impedes heat transfer. More large pores are therefore needed so that the vapor generation can increase with increasing superheating, curve b in Fig. 16.3. Superimposing curves a and b yields curve c in Fig. 16.3, which shows that the necessary superheating has a relative minimum at a definite pore radius.

Because the value r of a viable bubble belonging to a definite superheating is given approximately according to Eq. (10.12) by

$$r = \frac{2\sigma}{\varrho'' \Delta h_v} \frac{T_s}{\Delta\vartheta},$$

which contains as determining property value the parameter $2\sigma/\varrho''\Delta h_v$ (SI unit m), it was recommended in the mentioned application for patent [16.12] that

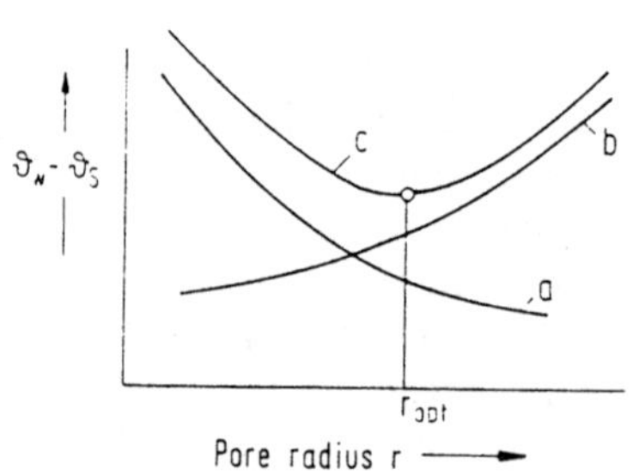

Fig. 16.3. Most favorable pore radius for heat transfer during boiling on sintered heating surfaces

sintering material of the same courseness be used for substances with values equal to those of this parameter. Given the present state of knowledge, one can scarcely say more than that for every liquid and heat flux there is an optimum granulation, whose value must be sought by several carefully designed experiments.

The augmentation of heat transfer achieved in some cases with sintered surfaces is remarkable. For example, a heat transfer coefficient of 2270 W/m^{2}K is obtained according to Fig. 16.4 during the boiling of refrigerant R-11 on a smooth tube at a heat flux of 31540 W/m^2. With a sintered tube for the same heat flux, a 16 times larger heat transfer coefficient of 38 000 W/m^{2}K was achieved. Such a high heat transfer coefficient is usually advantageous only if the heat transfer is also augmented on the heating agent side.

For practical applications, one must allow for the fact that bubble formation in sintered tubes begins only after a sufficient wall superheat has been exceeded. This superheat is low for water, but liquids with a good wetting capacity, such as the halogenated hydrocarbons used as refrigerants, penetrate into the larger pores and make these ineffective as nucleation sites. In order that the vapor be driven out of the small pores and the bubble formation started, a higher initial superheat is needed for such substances than is the case in smooth tubes with a normal technically rough surface. This hysteresis effect can be considerable during the transition from free convection to nucleate boiling. Once the bubble formation has started, however, the wall superheat for the same heat flux is smaller than in smooth tubes.

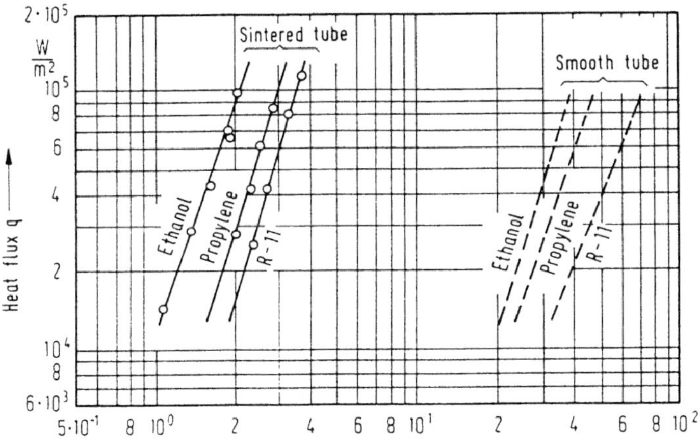

Fig. 16.4. Heat transfer during boiling (wall superheat $\Delta\vartheta = \vartheta_w - \vartheta_s$) on tubes with sintered surfaces and on smooth tubes, according to [16.14]

16.5 Tubes with Formed Surfaces, Tubes with T-shaped Fins

According to a procedure of Union Carbide [16.15], a tube wall is grooved by a noncutting grooving process in such a way that the groove that has just been formed is narrowed down by the formation of the next groove. There results a tube surface like the one in Fig. 16.5. The distance between the grooves is 0.1 to 0.9 mm, with a depth of approximately 0.2 to 0.6 mm. The liquid evaporates only in the indentations. The better heat transfer might result chiefly from the fact that the vapor produced in the indentations experiences a rise in pressure because of the confinement and, when forced through the openings into the liquid, causes very high velocities of vapor flow with good heat transfer. Sufficient liquid for the production of new vapor can flow laterally from the circumferential grooves, thereby eliminating to a great extent the usually rather long waiting period for the formation of bubbles. Optimal groove intervals depend, as in the case of tubes with sintered surfaces, upon the surface tension of the fluid used. Liquids with a greater surface tension require a greater groove interval than

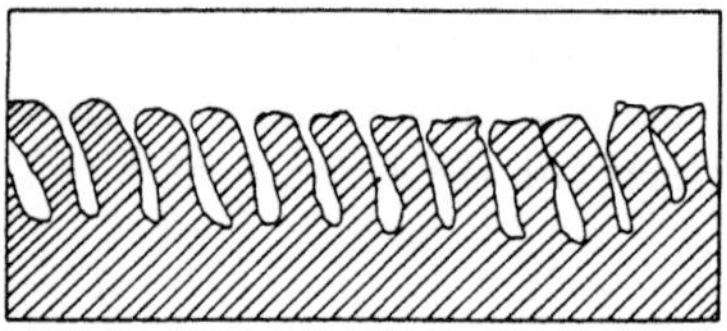

Fig. 16.5. Formed surface

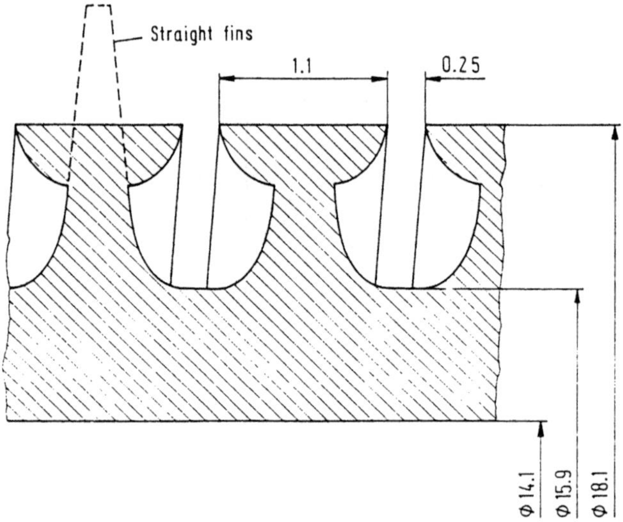

Fig. 16.6. Tube with T-shaped fins (GEWA-T-Tube, manufacturer Wieland Werke, Ulm, Germany). Tube with straight fins: surface area $A = 0.159\,m^2/m$, smooth tube surface area $0.050\,m^2/m$, area ratio $\varphi = 3.18$

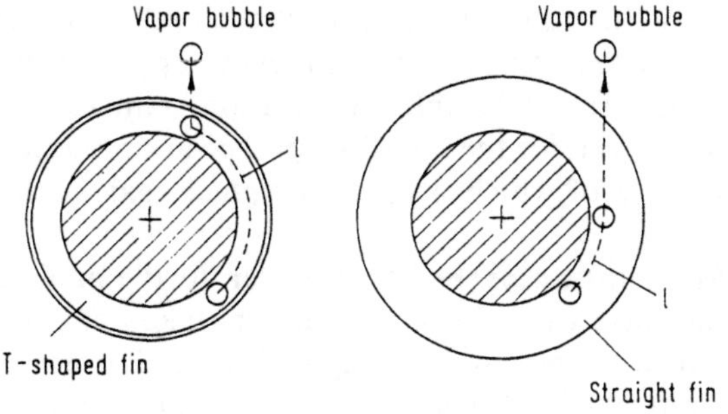

Fig. 16.7. Path l of a vapor bubble in a T-shaped fin passage

do those with a smaller surface tension. There are as yet no experiments available from which to judge which optimal groove interval should be ascribed to a definite liquid at a specified heat flux.

Tubes with T-shaped fins are produced by cutting into the tip of a fin and then bending it to both sides, Fig. 16.6. There results around the tube a helical channel that is connected with the environment by a narrow slit. Heat transfer coefficients are approximately 1.2 to 1.4 times larger than in the original tube with straight fins [16.16] and are to a great extent independent of the position of a tube in the horizontal bundle of a flooded evaporator. The vapor bubbles formed in the narrow helical channel might be expected, at lower heat fluxes, to leave the channel only upon reaching its upper side, and, therefore, have contact with the tube surface over a longer distance l than do the bubbles of a tube with straight fins, Fig. 16.7. On their way to the upper side of the channel they sweep along other bubbles. The convective portion of the heat transfer in the channels as well as the bubble frequency and, thus, also the heat transfer coefficient become larger than in tubes with straight fins.

16.6 Addition of Gases or Liquids

The heat transfer during boiling is generally raised by the addition of noncondensable gases. For example, air dissolved in refrigerants can lead to a temporarily better heat transfer, which decreases again in the course of time, however, because the air collects first at the colder and lower sites in the evaporator. In order to maintain reproducible results, one usually makes sure that the liquids used are degassed, because otherwise the heat transfer is not constant with time, but instead varies in an uncontrollable way with the changes in heat duty. The presence of air can lead to the corrosion of the heating surface because of the oxygen component in the air. Until now, there have been almost no systematic experiments on the influence of inert gases on heat transfer during evaporation.

More expedient than the use of inert gases is an augmentation of the heat transfer by feeding in a part of the produced vapor. Because the vapor has practically the same temperature as the liquid, no heat exchange takes place, to be sure, but nevertheless the flow in the evaporator is influenced appreciably, and, if the vapor is supplied in the form of bubbles, nucleation is increased and thus the heat transfer raised. However, for the recycling of the vapor, pumps and regulators are necessary, the costs of which must be weighed against the gains through the savings in area.

By the addition of surface-active substances the heat transfer during boiling can be augmented appreciably, because these lower the energy necessary for bubble formation. By the addition of 1 percent by weight of a wetting agent (ercantol BDX) to water, Kirschbaum [16.17] has measured an augmentation of the heat transfer in flowed through evaporator tubes by more than a factor of 2.5. Ten years earlier, Brooks and Badger [16.18] had already indicated the strong influence of surface tension on heat transfer during evaporation. As follows from Eq. (11.9),

$$\alpha \sim \sigma^{-0.32}.$$

The surface tension of water is approximately $70 \cdot 10^{-3}$ N/m and is clearly larger than that of the refrigerants and hydrocarbons of $8 \cdot 10^{-3}$ to $25 \cdot 10^{-3}$ N/m, so that by the addition of surface-active substances to water, the surface tension can be more strongly reduced than is the case for refrigerants and hydrocarbons. Until now no systematic experiments have been conducted on how much the heat transfer during boiling can be augmented by the addition of slight amounts of surface-active substances. Especially, there remains unclarified which surface-active substances come into question as additives for any given purpose.

16.7 Addition of Solids

Solids can serve in boiling liquids as nuclei for bubble formation. One observes, for example, that individual vapor bubbles in evaporators do not always form directly on the heating surface, but rather can form in a thin wall-adjacent superheated liquid layer. This can be explained among other things by the presence of foreign substances which favor nucleation. According to data of Elrod et al. [16.19], even the addition of 40 ppb of finely dispersed solid particles to boiling water leads to a noticeable augmentation of heat transfer. In the experiments mentioned, the size of the particles was smaller than 1 to 3 μm. For a solid matter content of 20 to 25 ppb there resulted a heat transfer coefficient of 6590 W/m^{2}K for $\Delta\vartheta = 5.56$ K and a pressure of 37 bar. If the solid matter content was raised to values of between 40 and 150 ppb, then the heat transfer coefficient was raised by a factor of 1.22 to the value 8039 W/m^{2}K. It has not yet been clarified whether the heat transfer can be further improved by addition of particles of other sizes or in other amounts. It can be surmised, however, that there is a critical concentration below which the nucleation and, thus, the heat transfer are promoted and above which surface deposits of the solid occur, which causes a reduction of the heat transfer coefficient.

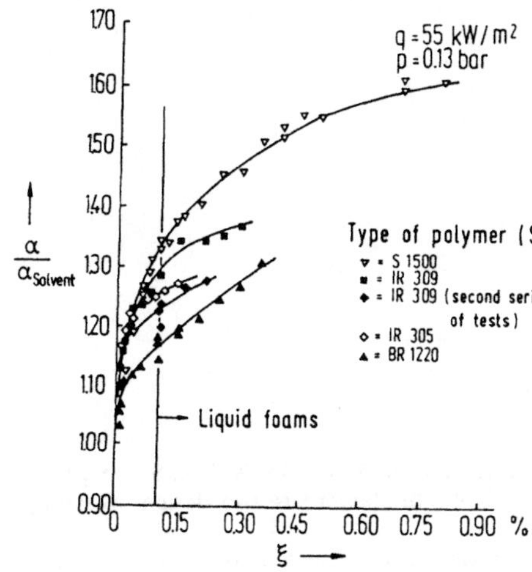

Fig. 16.8. Augmentation of heat transfer during boiling by addition of polymers to n-heptane, according to [16.20]

The addition of polymers also resulted in an augmentation of the heat transfer, depending upon the type and quantity of the polymer. Figure 16.8 shows the results of de Dood [16.20], in which the heat transfer coefficient α of the polymer solution relative to the value $\alpha_{solvent}$ of the pure solvent is plotted as a function of the mass fraction ξ of the added polymer. Above 0.1 percent of weight of the added polymer, the liquid begins to foam. The polymers are partially absorbed at the heating surface and favor there nucleation and the growth of vapor bubbles. The bubble frequency therefore increases [16.21]. The polymers can form at the phase interface a netlike structure [16.22], by which the movement of the liquid on the bubble surface is suppressed and the viscosity there is much greater than in the core of the liquid. A sufficiently large molar mass ($\geqq 10^5$ to 10^6 kg/kmol) of the polymers is important, because molecules with too small molar masses can diffuse away from the surface of the bubble because of their better mobility.

Bibliography

Chapter 1

1.1 Butterworth, D.: Condensers: Basic heat transfer and fluid flow. In: Kakaç, S.; Bergles, A. E; Mayinger, F.: Heat exchangers, thermal-hydraulic fundamentals and design. Washington: Hemisphere 1981, p.289–313

1.2 Berman, L. D.: On the effect of molecular-kinetic resistance upon heat transfer with condensation. Int. J. Heat Mass Transfer 10(1967) 1463

1.3 Nusselt, W.: Die Oberflächenkondensation des Wasserdampfes. VDI Z. 60(1916) 541–546, 569–575

Chapter 2

2.1 Chen, M. M.: An analytical study of laminar film condensation. Part 1, Flat Plates. Part 2, Single and multiple horizontal tubes. Trans. Am. Soc. Mech. Eng., Ser. C. J. Heat Transfer 83(1961) 48–60

2.2 Rohsenow, W. M.: Heat transfer and temperature distribution in laminar film condensation. Trans. Am. Soc. Mech. Eng., Ser. C. J. Heat Transfer 78(1956) 1645–1648

2.3 Koh, Y. C. H.; Sparrow, E. M.; Hartnett, J. P.: The two-phase boundary layer in laminar film condensation. Int. J. Heat Mass Transfer 2(1961) 69–82

2.4 Sparrow, E. M.; Eckert, E. R. G.: Effects of superheated vapor and non-condensable gases on laminar film condensation. Am. Inst. Chem. Eng. J., 7(1961) 473–477

2.5 Stephan, K.; Laesecke, A.: The influence of suction in condensation of mixed vapors. Wärme Stoffübertrag. 13(1980) 115–123

2.6 Sparrow, E. M.; Minkowycz, W. J.; Saddy, M.: Forced convection condensation in the presence of noncondensables and interfacial resistance. Int. J. Heat Mass Transfer 10(1967) 1829–1845

2.7 Dukler, H. E.; Bergelin, O. P.: Characteristics of flow in falling liquid films. Chem. Eng. Progr. 48(1952) 557–563

2.8 Kirschbaum, E.: Neues zum Wärmeübergang mit und ohne Änderung des Aggregatzustandes. Chem. Ing. Tech. 24(1952) 393–402

2.9 Brauer, H.: Strömung und Wärmeübergang bei Rieselfilmen. VDI-Forschungsh. 457(1956)

2.10 Grimley, S. S.: Liquid flow conditions in packed towers. Trans. Inst. Chem. Eng. (London) 23(1945) 228–235

2.11 Van der Walt, J.; Kröger, D. G.: Heat transfer resistances during film condensation. Proc. Vth Int. Heat Transfer Conf., Tokyo 1974, Vol. III, p. 284–288

2.12 Voskresenskij, K. D.: Calculation of heat transfer during film condensation with condensate properties as a function of temperature (in Russian). Izv. Akad. Nauk SSSR (1948) 1023–1028

2.13 Grigull, U.: Wärmeübergang bei der Kondensation mit turbulenter Wasserhaut. Forsch. Ingenieurwes. 13(1942) 49–57 und VDI Z. 86(1942) 444–445

2.14 Mostofizadeh, Ch.; Stephan, K.: Strömung und Wärmeübergang bei der Oberflächenverdampfung und Filmkondensation. Wärme Stoffübertrag. 15(1981) 93–115

2.15 Henstock, W. H.; Hanratty, Th. J.: The interfacial drag and the height of the wall layer in annular flows. Am. Inst. Chem. Eng. J. 6(1976) 990–1000

2.16 Isashenko, V. P.: Heat transfer during condensation (in Russian). Moscow: Energia 1977

2.17 Labunzov, D. A.: Heat transfer during film condensation of pure vapors on vertical cooled surfaces and horizontal tubes (in Russian). Teploenergetika 7(1957) 72–79

2.18 Blangetti, F.: Lokaler Wärmeübergang bei der Kondensation mit überlagerter Konvektion im vertikalen Rohr. Diss. Univ. Karlsruhe 1979
2.19 Kern, D. Q.: Mathematical development of tube loading in horizontal condensers. Am. Inst. Chem. Eng. J. 4(1958) 157–160
2.20 Short, B. E.; Brown, H. E.: Condensation of vapours in vertical banks of horizontal tubes. Proc. Inst. Mech. Eng., General Discussion Heat Transfer, 1951, p. 27–31
2.21 Shekriladze, I. G.; Gomelauri, V. I.: Theoretical study of laminar film condensation of flowing vapor. Int. J. Heat Mass Transfer 9(1966) 581–591

Chapter 3

3.1 Watson, R. G. H.; Birt, D. C. P.; Honour, G. W.; Ash, B. W.: The promotion of dropwise condensation by Montan wax. Part I. Heat transfer measurements. J. Appl. Chem. 12(1962) 539–552
3.2 Erb, R. A.; Thelen, E.: Promoting permanent dropwise condensation. Ind. Eng. Chem. 57(1965) 49–52
3.3 Woodruff, D. W.; Westwater, J. W.: Steam condensation on electroplated gold: Effect of plating thickness. Int. J. Heat Mass Transfer 22(1979) 629–632
3.4 Krischer, S.; Grigull, U.: Mikroskopische Untersuchung der Tropfenkondensation. Wärme Stoffübertrag. 4(1971) 48–59
3.5 Hampson, H.; Özisik, N.: An investigation into the condensation of steam. Proc. Inst. Mech. Eng. 1, Ser. B, 1952, p. 282–293
3.6 Wenzel, H.: Versuche über Tropfenkondensation. Allg. Wärmetech. 8(1957) 33–39
3.7 Welch, J. F.; Westwater, J. W.: Microscopic study of dropwise condensation. Int. Dev. Heat Transfer 2(1961) 302–309
3.8 Kast, W.: Theoretische und experimentelle Untersuchung der Wärmeübertragung bei Tropfenkondensation. Fortschrittsber. VDI-Z., Reihe 3, Nr. 6, Düsseldorf 1965
3.9 Le Fèvre, E. J.; Rose, J. W.: An experimental study of heat transfer by dropwise condensation. Int. J. Heat Mass Transfer 8(1965) 1117–1133
3.10 Tanner, D. W.; Pope, D.; Potter, C. J.; West, D.: Heat transfer in dropwise condensation at low pressures in the absence and presence of non-condensable gas. Int. J. Heat Mass Transfer 11(1968) 181–190
3.11 Griffith, P.; Lee, M. S.: The effect of surface thermal properties and finish on dropwise condensation. Int. J. Heat Mass Transfer 10(1967) 697–707
3.12 Brown, A. R.; Thomas, M. A.: Filmwise and dropwise condensation of steam at low pressures. Proc. IIIrd Int. Heat Transf. Conf., Chicago 1966, Vol. II, p. 300–305
3.13 Eucken, A.: Energie- und Stoffaustausch an Grenzflächen. Naturwissenschaften 25(1937) 209–218
3.14 Wicke, E.: Einige Probleme des Stoff- und Wärmeaustausches an Grenzflächen. Chem. Ing. Tech. 23(1951) 5–12
3.15 Kast, W.: Wärmeübergang bei Tropfenkondensation. Chem. Ing. Tech. 35(1963) 163–168
3.16 Jakob, M.: Heat Transfer. Vol. I. New York: Wiley 1949, p.694
3.17 Umur, A.; Griffith, P.: Mechanism of dropwise condensation. Trans. Am. Soc. Mech. Eng., Ser. C. J. Heat Transfer 87(1965) 275–282
3.18 McCormick, J. L.; Baer, E.: On the mechanism of heat transfer in dropwise condensation. J. Colloid Sci. 18(1963) 208–216

Chapter 4

4.1 Jakob, M.; Erk, S.; Eck, H.: Verbesserte Messungen und Berechnungen des Wärmeübergangs beim Kondensieren strömenden Dampfes in einem vertikalen Rohr. Phys. Z. 36(1935) 73–84
4.2 Rohsenow, W. M.; Webber, J. H.; Ling, A. T.: Effect of vapor velocity on laminar and turbulent film condensation. Trans. Am. Soc. Mech. Eng., Ser. C. J. Heat Transf. 78(1956) 1637–1643
4.3 Jones, W. P.; Renz, U.: Condensation from a turbulent stream into a vertical surface. Int. J. Heat Mass Transfer 17(1974) 1009–1014
4.4 Carpenter, E. F.; Colburn, A. P.: The effect of vapor velocity on condensation inside tubes. Proc. Inst. Mech. Eng., General Discussion Heat Transfer, 1951, p. 20–26
4.5 Mostofizadeh, Ch.; Stephan, K.: Strömung und Wärmeübergang bei der Oberflächenverdampfung und Filmkondensation. Wärme Stoffübertrag. 15(1981) 93–115

4.6 Dukler, A. E.: Fluid mechanics and heat transfer in vertical falling film systems. Chem. Eng. Prog. Symp. Ser. 56(1960) 1–10

4.7 Deissler, R. G.: Analytical and experimental investigation of adiabatic turbulent flow in smooth tubes. Nat. Advis. Comm. Aeronaut. Tech. Notes 2138(1950) 24–25

4.8 Deissler, R. G.: Analysis of turbulent heat transfer, mass transfer and friction in smooth tubes at high Prandtl and Schmidt numbers. Nat. Advis. Comm. Aeronaut. Tech. Notes 3145(1954) 8

4.9 von Kàrmàn, Th: Mechanische Ähnlichkeit und Turbulenz. Nachr. Ges. Wiss. Göttingen, Math. Phys. Kl. 58(1930) und Verhandlg. d. III. Int. Kongr. Tech. Mechanik, Stockholm, Teil I, 85(1930). Nat. Advis. Comm. Aeronaut. Tech. Notes 611(1931)

4.10 Chun, K. R.; Seban, R. A.: Heat transfer to evaporating liquid films. Trans. Am. Soc. Mech. Eng., Ser. C. J. Heat Transfer 93(1971) 391–396

4.11 Domanski, I. V.; Solokov, V.I.: Heat transfer from a heated tube wall to boiling, downward flowing liquid films (in Russian). Zh. Prikl. Khim. 40(1967) 66–71

4.12 Hewitt, G. F.: Analysis of annular two-phase flow, application of the Dukler analysis to vertical upward flow in a tube. AERE-R 3680, H.M.S.O.

4.13 Lee, J.: Turbulent velocity profile of a vertical film flow. Chem. Eng. Sci. 20(1965) 533–536

4.14 Levich, V. G.: Physiochemical Hydrodynamics. Englewood Cliffs, New Jersey: Prentice Hall 1962

4.15 Lamourelle, A. P.; Sandall, O. C.: Gas absorption into turbulent liquid. Chem. Eng. Sci. 27(1972) 1035–1043

4.16 Mills, A. F.; Chung, D. K.: Heat transfer across turbulent falling films. Int. J. Heat Mass Transfer 16(1973) 694–696

4.17 Brötz, W.: Über die Vorausberechnung der Absorptionsgeschwindigkeit von Gasen in strömenden Flüssigkeiten. Chem. Ing. Tech. 26(1954) 470–478

4.18 Blangetti, F.: Lokaler Wärmeübergang bei der Kondensation mit überlagerter Konvektion im vertikalen Rohr. Diss. Univ. Karlsruhe 1979

4.19 Van Driest, E. R.: On turbulent flow near a wall. J. Aerosol Sci. 23(1956) 1007–1010

4.20 Pantankar, S. V.; Spalding, D. B.: Heat and mass transfer in boundary layers. 2nd ed. London: Intertext Books 1970, p. 21

4.21 Bergelin, O. P.; Kegel, P. K.; Carpenter, F. G.; Gazley, C.: Co-current gas liquid flow. Part II. Flow in vertical tubes. Am. Soc. Mech. Eng., Heat Transfer Fluid Mech. Inst. (1949) 19–28

4.22 Shah, M. M.: A general correlation for heat transfer during film condensation inside pipes. Int. J. Heat Mass Transfer 22(1979) 547–556

4.23 Wallis, G. B.: One dimensional two phase flow. New York: McGraw-Hill 1969

4.24 Tandon, T. N.; Varma, H. K.; Gupta, C. P.: A new flow regime map for condensation inside horizontal tubes. Trans. Am. Soc. Mech. Eng., Ser. C. J. Heat Transfer 104(1982) 763–768

4.25 Jaster, H.; Kosky, P. G.: Condensation heat transfer in a mixed flow regime. Int. J. Heat Mass Transfer 19(1976) 95–99

4.26 Zivi, S. M.: Estimation of steady state void fraction by means of the principle of minimum entropy production. Trans. Am. Soc. Mech. Eng., Ser. C. J. Heat Transfer 86(1964) 247–252

4.27 Collier, J. G.: Convective boiling and condensation. New York: McGraw-Hill 1972

4.28 Mayinger, F.: Strömung und Wärmeübergang in Gas-Flüssigkeits-Gemischen. Wien: Springer 1982

4.29 Ahmad, S. Y.: Axial distribution of bulk temperature and void fraction in a heated channel with inlet subcooling. Trans. Am. Soc. Mech. Eng., Ser. C. J. Heat Transfer 92(1970) 595–609

4.30 Butterworth, D.: Developments in the design of shell-and-tube condensers. Am. Soc. Mech. Eng., preprint 77-WA/HT-24

4.31 Lockhart, R. W.; Martinelli, R. C.: Proposed correlation data for isothermal two-phase, two-component flow in pipes. Chem. Eng. Prog. 45(1949) 39–48

4.32 Palen, J. W.; Breber, G.; Taborek, J.: Prediction of flow regimes in horizontal tube condensation. Heat Transfer Eng. 1(1979) 47–57

Chapter 5

5.1 Sukhatme, S. P.; Rohsenow, W. M.: Heat transfer during film condensation of a liquid metal vapour. Report MIT 2995–1,1964

5.2 Chen, M. M.: An analytic study of laminar film condensation. Part I: Flat plates. Part II: Single and multiple horizontal tubes. Trans. Am. Soc. Mech. Eng., Ser. C. J. Heat Transfer 81(1959) 48–60

5.3 Koh, J. C. Y.; Sparrow, E. M.; Hartnett, J. P.: The two-phase boundary layer in laminar film condensation. Int. J. Heat Mass Transfer 2(1961) 69–82
5.4 Subbotin, V. I.; Ivanovskii, M. N.; et al.: Heat transfer during the condensation of potassium vapor (in Russian). Teplofiz. Vys. Temp. 2 (1964) 612–622
5.5 Labunzov, D. A.; Smirnow, S. I.: Heat transfer in condensation of liquid metal vapours. IIIrd Int. Heat Transf. Conf., Chicago 1966, Vol. II, p. 328–336
5.6 Aldyev, I. T.; Kondratyev, N. S.; Mukkin, V. A.; Kpshidze, M. E.; Porfentyeva, I.; Kisselev, V. V.: Thermal resistance of phase transition with condensation of potassium vapour. IIIrd Int. Heat Transf. Conf., Chicago 1966, Vol. II, p. 313–317
5.7 Barry, R. E.; Balzhizer, R. E.: Condensation of sodium at high heat fluxes. IIIrd Int. Heat Transf. Conf., Chicago 1966, Vol. II, p. 318–328
5.8 Kröger, D. G.; Rohensow, W. M.: Film condensation of saturated potassium vapour. Int. J. Heat Mass Transfer 10(1967) 1891–1894
5.9 Wilcox, S. J.; Rohsenow, W. M.: Film condensation of liquid metals—precision of measurement. Report MIT 7145-62, 1969
5.10 Lee, J.: Turbulent film condensation. Am. Inst. Chem. Eng. J. 10(1964) 540–544
5.11 Deissler, R. G.: Analysis of turbulent heat transfer, mass transfer and friction in smooth tubes at high Prandtl and Schmidt numbers. Nat. Advis. Comm. Aeronaut. Tech. Notes 3145 (1954)
5.12 von Kàrmàn, Th.: Mechanische Ähnlichkeit und Turbulenz. Nachr. Ges. Wiss. Göttingen, Math. Phys. Kl. 58(1930) und Verhandlg. d. III. Int. Kongr. Tech. Mechanik, Stockholm, Teil I, (1930) 85; Nat. Advis. Comm. Aeronaut. Tech. Notes 611(1931)

Chapter 6

6.1 Claude, G.: Air liquide, oxygène, azote, gas rares. 2. ed. Paris: Dunod 1926
6.2 Lucas, K.: Die laminare Filmkondensation binärer Dampfgemische. Habil.-Schrift, Ruhr-Univ. Bochum 1974
6.3 Tamir, A.; Taitel, Y.; Schlünder, E. U.: Direct contact condensation of binary mixtures. Int. J. Heat Mass Transfer 17(1974) 1253–1260
6.4 Ford, J. D.; Missen, R. W.: On the conditions for stability of falling films subject to surface tension disturbances; the condensation of binary vapors. Can. J. Chem. Eng. 46(1968) 309–312
6.5 Ford, J. D.: Ph.D. Thesis, Univ. of Toronto 1967
6.6 Tamir, A.: "Mixed" pattern condensation of multicomponent mixtures. Chem. Eng. J, 17(1979) 141–156
6.7 Ackermann, G.: Wärmeübergang und molekulare Stoffübertragung im gleichen Feld bei großen Temperatur- und Partialdruckdifferenzen. VDI-Forschungsh. 382(1937) 1–16
6.8 Silver, L.: Gas cooling with aqueous condensation. Trans. Inst. Chem. Eng. 25(1947) 30–42
6.9 Bell, K. J.; Ghaly, M. A.: An approximate generalized design method for multicomponent/partial condensers. Am. Inst. Chem. Eng. Symp. Ser. 69(1972) 72–79
6.10 Ward, D. J.: How to design a multicomponent partial condenser. Petro-Chem. Eng. 32(1960) C42–C48
6.11 Butterworth, D.: A calculation method for shellside heat exchangers in which the overall coefficient varies along the length. Nat. Eng. Lab. (U.K.) Rep. 510 (1975) 56–71
6.12 Emerson, W. H.: Effective tube-side temperature differences in multipass heat exchangers with nonuniform heat transfer coefficients and specific heats. Nat. Eng. Lab. (U.K.) Rep. 590(1975) 32–55
6.13 Roetzel, W.: Näherungsverfahren zur Berechnung von Kondensatoren für Dampfgemische. Wärme Stoffübertrag. 8(1975) 211–218
6.14 Stephan, K.; Laesecke, A.: The influence of suction on heat and mass transfer in condensation of mixed vapors. Wärme Stoffübertrag. 13(1980) 115–123
6.15 Sparrow, E. M.; Minkowycz, W. J.; Saddy, M.: Forced convection condensation in the presence of noncondensables and interfacial resistance. Int. J. Heat Mass Transfer 10(1967) 1829–1845
6.16 Bird, R. B.; Stewart, W. E.; Lightfoot, E. N.: Transport Phenomena. New York: Wiley 1962, p. 663
6.17 Colburn, A. P.; Hougen, O. A.: Design of cooler condenser for mixtures of vapors with noncondensing gases. Ind. Eng. Chem. 26(1934) 1178–1182
6.18 Krishna, R.; Standart, G. L.: A multicomponent film model incorporating a general matrix method of solution to the Maxwell–Stefan equation. Am. Inst. Chem. Eng. J. 22(1976) 383–389
6.19 Hirschfelder, J. O.; Curtiss, C. F.; Bird, R. B.: Molecular theory of gases and liquids. New York: Wiley 1967, p. 516

Chapter 7

7.1 Bernhardt, S. H.; Sheridan, J. J.; Westwater, J. W.: Condensation of immiscible mixtures. Am. Inst. Chem. Eng. Symp. Ser. 68(1972) 21–37
7.2 Akers, W. W.; Turner, M. M.: Condensation of vapors of immiscible liquids. Am. Inst. Chem. Eng. J. 8(1962) 587–589
7.3 Stephan, K.; Mayinger, F.: Thermodynamik. Bd. 2. 12. Aufl. Berlin: Springer 1987, S. 136
7.4 Tamir, A.; Taitel, Y.; Schlünder, E. U.: Direct contact condensation of binary mixtures. Int. J. Heat Mass Transfer 17(1974) 1253–1260
7.5 Tamir, A.; Rachmilev, I.: Direct contact condensation of an immiscible vapour on a thin film of water. Int. J. Heat Mass Transfer 17(1974) 1241–1251

Chapter 8

8.1 Renz, U.: Maßnahmen zur Verbesserung des Wärmeübergangs bei der Kondensation. In: Wärmeaustauscher, Neuere Entwicklungen und Berechnungsmethoden. Düsseldorf: VDI-Verlag 1983, S. 239–276
8.2 Berman, L. D.: Influence of vapour velocity on heat transfer with filmwise condensation on a horizontal tube. Therm. Eng. 26(1978) 16–20
8.3 Spencer, D. L.; Ibele, W. E.: Laminar film condensation of a saturated and superheated vapor on a surface with a controlled temperature distribution. Proc. IIIrd Int. Heat Transfer Conf., Chicago 1966, Vol. II, p. 337–347
8.4 Henrici, H.: Kondensation von R11, R12 und R22 an glatten und berippten Rohren. Kältetechnik 15(1963) 251–256
8.5 Katz, D. L.; Hope, R. E.; Zatsko, St. C.; Robinson, D. B.: Condensation of Freon 12 with finned tubes. Refrig. Eng. 53(1947) 211–215
8.6 Gregorig, R.: Hautkondensation an feingewellten Oberflächen bei Berücksichtigung der Oberflächenspannungen. Z. Angew. Math. Phys. 5(1954) 36–49
8.7 Webb, R. L.: Generalized procedure for the design and optimization of fluted Gregorig condensing surfaces. Trans. Am. Soc. Mech. Eng., Ser. C. J. Heat Transfer 101(1979) 335–339
8.8 Mori, Y.; Hijikata, K.; Hirasawa, S.; Nakayama, W.: Optimized performance of condensers with outside condensing surface, condensation heat transfer. Am. Soc. Mech. Eng. 18th Nat. Heat Transfer Conf., San Diego 1979, p.55–62
8.9 Panchal, C. B.; Bell, K. J.: Analysis of Nusselt-type condensation on a triangular fluted surface. Int. J. Heat. Mass Transfer 25(1982) 1909–1911
8.10 Lin Ji-fang, Hsu Tung-chi, Pei Jue-min: Heat transfer of condensation on a vertical V-type corrugated tube—A new physical model. 7th Int. Heat Transfer Conf., München 1982, Vol. V, p. 119–124
8.11 Barnes, C. G.; Rohsenow, W. M.: Vertical fluted tubes condenser performance prediction. 7th Int. Heat Transfer Conf., München 1982, Vol. V, p. 39–43
8.12 Adamek, Th.: Kondensation an Profilrohren. In: Wärmeaustauscher, Neuere Entwicklungen und Berechnungsmethoden. Düsseldorf: VDI-Verlag 1983, S. 277–299
8.13 Marto, P. J.; Reilly, D. J.; Fenner, J. A.: An experimental comparison of enhanced heat transfer condenser tubing. Advances in Enhanced Heat Transfer. Am. Soc. Mech. Eng. 18th Nat. Heat Transfer Conf., San Diego, 1979, p. 183–191
8.14 Azer, N. Z.; Said, S. A.: Augmentation of condensation heat transfer by internally finned tubes and twisted tape inserts. Advances in Enhanced Heat Transfer. Am. Soc. Mech. Eng. 18th Nat. Heat Transfer Conf., San Diego 1979, p. 33–38
8.15 Royal, J. H.; Bergles, A. E.: Augmentation of horizontal in-tube condensation by means of twisted tape inserts and internally finned tubes. Trans. Am. Soc. Mech. Eng., Ser. C. J. Heat Transfer 100(1978) 17–24
8.16 Akers, W. W.; Deans. H. A.; Crosser, O. K.: Condensing heat transfer within horizontal tubes. Chem. Eng. Progr. Symp. Ser. 55(1959) 19, 171–176
8.17 Khanpara, J. C.; Bergles, A. E.; Pate, M. B.: Augmentation of R-113 in-tube condensation with micro-fintubes. Trans. Am. Soc. Heating Refrig. Air Cond. Eng. (ASHRAE) 92, part 2B (1986) 506–524
8.18 Schlager, L. M.; Pate, M. B.; Bergles, A. E.: Evaporation and condensation of refrigerant-oil-mixtures in a smooth tube and a micro-fin tube. Trans. Am. Soc. Heating Refrig. Air Cond. Eng. (ASHRAE) 94, part 2 (1988) 149–166

8.19 Schlager, L. M.; Pate, M. B.; Bergles, A. E.: Evaporation and condensation of refrigerant-oil-mixtures in a low-fin tube. Trans. Am. Soc. Heating Refrig. Air Cond. Eng. (ASHRAE) 94, part 2 (1988) 1176–1194
8.20 Schlager, L. M.; Pate, M. B.; Bergles, A. E.: Heat transfer and pressure drop during evaporation and condensation of R22 in horizontal micro-fin tubes. Int. J. of Refrigeration 12(1989) 6–14
8.21 Chun, K. R.; Seban, R. A.: Heat transfer to evaporating liquid films. Trans. Am. Soc. Mech. Eng., Ser. C. J. Heat Transfer 93(1971) 391–396

Chapter 9

9.1 Prüger, W.: Die Verdampfungsgeschwindigkeit von Flüssigkeiten. Z. Phys. 115(1949) 202–244
9.2 Jakob, M.; Fritz, W.: Versuche über den Verdampfungsvorgang. Forsch. Ingenieurwes. 2(1931) 435–447
9.3 Jakob, M.: Kondensation und Verdampfung. VDI Z. 76(1931) 1161–1170
9.4 Jakob, M.; Linke, W.: Der Wärmeübergang von einer waagerechten Platte an siedendes Wasser, Forsch. Ingenieurwes. 4(1933) 75–81
9.5 Jakob, M.; Linke, W.: Der Wärmeübergang beim Verdampfen von Flüssigkeiten an senkrechten und waagerechten Flächen. Phys. Z. 36(1935) 267–280
9.6 Jakob, M.: Heat transfer in evaporation and condensation. Mech. Eng. 58(1936) 643–660 u. 729–739
9.7 Fritz, W.: Wärmeübergang an siedende Flüssigkeiten. VDI Z. Beihefte Verfahrenstech. 5(1937) 149–155
9.8 Fritz, W.: Verdampfen und Kondensieren. VDI Z. Beihefte Verfahrenstech. 1(1943) 1–14

Chapter 10

10.1 Mitrović, J.; Stephan, K.: Gleichgewichtsradien von Dampfblasen und Flüssigkeitstropfen. Wärme Stoffübertrag. 13(1980) 171–176
10.2 Bashfort, Fr.; Adams, J.: An attempt to test the theory of capillary action. Cambridge: Cambridge University Press 1883
10.3 Fritz, W.: Berechnung des Maximalvolumens von Dampfblasen. Phys. Z. 36(1935) 379–384
10.4 Fritz, W.; Ende, W.: Über den Verdampfungsvorgang nach kinematographischen Aufnahmen an Dampfblasen. Phys. Z. 37(1936) 391–401
10.5 Kabanow, W.; Frumkin, A.: Nachtrag zu der Arbeit "Über die Größe elektrisch entwickelter Gasblasen". Z. Phys. Chem. (A) 166(1933) 316–317
10.6 Stephan, K.: Beitrag zur Thermodynamik des Wärmeübergangs beim Sieden. Abh. Dtsch. Kältetech. Ver. Nr. 18. Karlsruhe: Müller 1964
10.7 König, A.: Der Einfluß der thermischen Heizwandeigenschaften auf den Wärmeübergang bei der Blasenverdampfung. Wärme Stoffübertrag. 6(1973) 38–44
10.8 von Ceumern, W. C.: Abreissdurchmesser und Frequenzen von Dampfblasen in Wasser und wässrigen NaCl-Lösungen beim Sieden an einer horizontalen Heizfläche. Diss. TU Braunschweig 1975
10.9 Mitrović, J.: Das Abreissen von Dampfblasen an festen Heizflächen. Int. J. Heat Mass Transfer 26(1983) 955–963
10.10 Stephan, K.: Bubble formation and heat transfer in natural convection boiling. In: Hahne, E.; Grigull, U.: Heat transfer in boiling. Washington: Hemisphere 1977, 3–20
10.11 Tokuda, N.: Dynamics of vapor bubbles in binary liquid mixtures with translatory motion. IVth Int. Heat Transfer Conf., Paris 1970, B 7.5
10.12 Kutateladze, J. J.; Gogonin, I. I.: Growth velocity and departure diameter of vapor bubbles in saturated liquids in free convective flow. Teplofiz. Vys. Temp. 17 (1979) 792–797
10.13 Mamontova, N. N.: Diss. Novosibirsk 1967
10.14 Golovin, V. S.; Kolcugin, B. A.; Sacharova, E. A.: Reports. ENIN 35(1976) 30
10.15 Jagov, V. V.; Gorodov, A. K.; Labunzov, D. A.: Collected papers 1968–1969. ENIN, 1969
10.16 Cole, R.: Bubble frequencies and departure volumes at subatmospheric pressures. Am. Inst. Chem. Eng. J. 13(1967) 779–783
10.17 Labunzov, D. A.; Kolcugin, B. A.; Golovin, V. S.; Sacharova, E. A.; Vladimirova, L. N.: Collected papers "Heat transfer in apparatus of energy engineering" Moscow: Nauka 1966

10.18 Grigorjev, V. A.; Pavlov, Ju. M.; Ametistov, E. V.; Klimenkov, V. V.; Klimenkov, A. V.: Reports MEI 198(1974) 3
10.19 Nishikawa, K.; Fujita, Y.; Nawata, J.; Nishijama, T.: Heat Transfer Jpn. Res. 5(1976) 66
10.20 Gorodov, A. K.; Kabankov, O. N.; Labunzov, D. A.; Jagov, V. V.: Reports MEI 198(1974) 48
10.21 Zamilova, G. N.: Diss. LTIHP, Leningrad 1968
10.22 Nordmann, D.: Temperatur, Druck und Wärmetransport in der Umgebung kondensierender Blasen. Diss. TU Hannover 1980
10.23 Jakob, M.; Linke, W.: Der Wärmeübergang von einer waagerechten Platte an siedendes Wasser. Forsch. Ingenieurwes. 4(1933) 75–81
10.24 Zuber, N.: Nucleate boiling. The region of isolated bubbles and the similarity with natural convection. Int. J. Heat Mass Transfer 6(1963) 53–78
10.25 McFadden, P.; Grassmann, P: The relation between bubble frequency and diameter during nucleate boiling. Int. J. Heat Mass Transfer 5(1962) 169–173
10.26 Ivey, H. J.: Relationships between bubble frequency, departure diameter and rise velocity in nucleate boiling. Int. J. Heat Mass Transfer 10(1967) 1023–1040
10.27 Malenkov, I. G.: The frequency of vapor-bubble separation as a function of bubble size. Fluid Mech. Sov. Res. 1(1972) 36–42

Chapter 11

11.1 Jakob, M.; Linke, W.: Der Wärmeübergang beim Verdampfen von Flüssigkeiten an senkrechten und waagerechten Flächen. Phys. Z. 36(1935) 267–280
11.2 Nukijama, S.: Maximum and minimum values of heat transmitted from metal to boiling water under atmospheric pressure. J. Soc. Mech. Eng. Jpn. 37(1934) 53–54 and 367–374; compare also Int. J. Heat Mass Transfer 9(1966) 1419–1433
11.3 Stephan, K.: Stabilität beim Sieden. Brennst. Wärme Kraft 17(1965) 571–578
11.4 Jakob, M.; Linke, W.: Der Wärmeübergang von einer waagerechten Platte an siedendes Wasser. Forsch. Ingenieurwes. 4(1933) 75–81
11.5 Rohsenow, W. M.: A method for correlation heat-transfer data for surface boiling of liquids. Trans. Am. Soc. Mech. Eng., Ser. C. J. Heat Transfer 74(1952) 969–976
11.6 Forster, H. K.; Zuber, N.: Dynamics of vapor bubbles and boiling heat transfer. Am. Inst. Chem. Eng. J. 1(1955) 531–535
11.7 Forster, H. K.; Zuber, N.: Growth of a vapor bubble in superheated liquid. J. Appl. Phys. 25(1954) 474–478
11.8 Han, C. Y.; Griffith, P.: The mechanism of heat transfer in nucleate pool boiling. Part I, II. Int. J. Heat Mass Transfer 8(1965) 887–904, 905–914
11.9 Beer, H.: Contribution to Heat Transfer in Boiling. Progr. Heat Mass Transfer 2(1969) 311–370
11.10 Moore, F. D.; Mesler, R. B.: The measurement of rapid surface temperature fluctuations during nucleate boiling. Am. Inst. Chem. Eng. J. 7(1961) 620–624
11.11 van Stralen, S. J. D.: The mechanism of nucleate boiling in pure liquids and in binary mixtures. Part I, II. Int. J. Heat Mass Transfer 9(1966) 995–1020, 1021–1046
11.12 Frost, W.; Kippenhan, C. J.: Bubble growth and heat transfer mechanism in the forced convection boiling of water containing a surface active agent. Int. J. Heat Mass Transfer 10(1967) 931–949
11.13 Styrikovich, M. A.; Nevstrueva, E. J., et al.: Interconnection between mass and heat transfer in boiling. IVth Int. Heat Transfer Conf., Paris 1970, B 7.4
11.14 Stephan, K.: Beitrag zur Thermodynamik des Wärmeübergangs beim Sieden. Karlsruhe: Müller 1963
11.15 Müller, F.: Wärmeübergang bei der Verdampfung unter hohen Drücken. VDI-Forschungsh. 522(1967)
11.16 Fedders, H.: Messung des Wärmeübergangs beim Blasensieden von Wasser an metallischen Rohren. Diss. TU Berlin 1970
11.17 Danilowa, G. N.; Belskij, W. K.: Investigation of heat transfer during boiling of freons 113 and 12 on tubes of varying roughness (in Russian). Cholod. Techn. 4(1965) 24–28
11.18 Danilowa, G. N.: Correlation of heat transfer during evaporation of freons. Cholod. Tech. 8(1969) 79–85
11.19 Nishikawa, K.; Fujita, Y.; Ohita, H.; Hidaka, S.: Effect of system pressure and surface roughness on nucleate boiling heat transfer. Mem. Fac. Eng. Kyushu Univ. 42(1982) 95–123

11.20 Stephan, K.; Abdelsalam, M.: Heat transfer correlations for natural convection boiling. Int. J. Heat Mass Transfer 23(1980) 73–87
11.21 Stephan, K.; Preusser, P.: Wärmeübergang und maximale Wärmestromdichte beim Behältersieden binärer und ternärer Flüssigkeitsgemische. Chem. Ing. Tech. 51(1979) 37(Synopse MS 649/79)
11.22 Fritz, W.: In VDI-Wärmeatlas. Düsseldorf: VDI-Verlag 1963, Chapter Hb2
11.23 Gorenflo, D.: In VDI-Wärmeatlas. Düsseldorf: VDI-Verlag 1984, Chapter Ha4–Ha7
11.24 Danilowa, G. N.: Influence of pressure and temperature on heat transfer to boiling freons (in Russian). Cholod. Techn. 4(1965) 36–42
11.25 Haffner, H.: Wärmeübergang an Kältemittel bei Blasenverdampfung, Filmverdampfung und überkritischem Zustand des Fluids. Bundesminist. Bild. Wiss. Forschungsber. Kernforsch. 70–24, 1970
11.26 Bier, K.; Engelhorn, H. R.; Gorenflo, D.: Wärmeübergang beim Blasensieden im Bereich niedriger Siededrücke. Chem. Ing. Tech. 49(1977) 671 (Synopse 514/77)
11.27 Jakob, M.; Fritz, W.: Versuche über den Verdampfungsvorgang. Forsch. Ingenieurwes. 2(1931) 435–447
11.28 Stephan, K.: Mechanismus und Modellgesetz des Wärmeübergangs bei der Blasenverdampfung. Chem. Ing. Tech. 35(1963) 775–784
11.29 Nishikawa, K.; Fujita, Y.; Ohita, H.; Hidaka, S.: Effect of the surface roughness on the nucleate boiling heat transfer over the wide range of pressure. Proc. VIIth Int. Heat Transfer Conf., München 1982, Vol. 4, p. 61–66
11.30 Gorenflo, D.: Zur Druckabhängigkeit des Wärmeübergangs an siedende Kältemittel bei freier Konvektion. Chem. Ing. Tech. 40(1968) 757–762 and Diss. TH Karlsruhe 1966
11.31 Bier, K.; Engelhorn, H. R.; Gorenflo, D.: Wärmeübergang an tiefsiedende Halogenkältemittel. Klima + Kältetech. 4(1976) 499–506. Compare also Gorenflo, D.: Abh. Dtsch. Kälte- und Klimatechn. Ver. 22(1977) 31. Karlsruhe: Müller
11.32 Slipčević, B.: Wärmeübergang bei der Blasenverdampfung von Kältemitteln an glatten und berippten Rohren. Klima + Kältetech. 3(1975) 279–286
11.33 Nikolaev, G. P.: Skripov, V. P.: Calculation of the critical heat flux based upon thermodynamic similarity (in Russian). Inzh. Fiz. Zh. 15(1968) 46–51
11.34 Hahne, E.; Feuerstein, G.: Heat transfer in pool boiling in the thermocritical region: Effect of pressure and geometry. In: Hahne, E.; Grigull, U.: Heat transfer in boiling. Washington: Hemisphere 1977, p. 159–200
11.35 Hesse, G.: Wärmeübergang bei Blasenverdampfung, bei maximaler Wärmestromdichte und im Übergangsbereich zur Filmverdampfung. Diss. TU Berlin 1972
11.36 Kutateladze, S. S.: Critical heat flux during a subcooled liquid flow(in Russian). Energetica 7(1959) 229–239 and Izv. Akad. Nauk. Otd. Tekh. Nauk 4(1951) 529
11.37 Zuber, N.: On the stability of boiling heat transfer. Trans. Am. Soc. Mech. Eng., Ser. C. J. Heat transfer 80(1958) 711
11.38 Bromley, L. A.: Heat transfer in stable film boiling. Chem. Eng. Progr. 46(1950)221–227
11.39 Roetzel, W.: Berechnung der Leitung und Strahlung bei der Filmverdampfung and der ebenen Platte. Wärme Stoffübertrag. 12(1979) 1–4

Chapter 12

12.1 Chun, K. R.; Seban, R. A.: Heat transfer to evaporating liquid films. Trans. Am. Soc. Mech. Eng., Ser. C. J. Heat Transfer 93(1971) 391–396
12.2 Zozulja, N. V., In: Kutateladze, S. S.: Fundamentals of heat transfer. Nauka, Nowosibirsk, 4th enlarged edition, (1970) 353 (in Russian)
12.3 Fujita, T.; Ueda, T.: Heat transfer to falling liquid films and film breakdown-II. Saturated films with nucleate boiling. Int. J. Heat Mass Transfer 21(1978) 108–118
12.4 Palen, J. W.: Falling film evaporation in vertical tubes, In: Heat exchanger technology (Chisholm, D. ed.), Elsevier Appl. Science, London, New York (1988) 208
12.5 Struve, H.: Der Wärmeübergang an einem verdampfenden Rieselfilm. VDI-Forschungh. 534(1969)
12.6 Thoma, R.: Messungen und Berechnung des Wärmeübergangs und des Druckverlustes in einem Fallfilmverdampfer. Diss. TH Karlsruhe 1966
12.7 Zaletnev, A. F.; Aksel'rod,A. F.; Tikhonov, A. V.: Prediction of heat transfer in surface boiling of water in tubes. Heat Transfer Sov. Res. 8(1976) 79–83

12.8 Slesarenko, V.: Hydrodynamics and heat transfer during seawater boiling in thin-film desalination plants. Desalination 21(1977) 275–283

Chapter 13

13.1 Collier, J. G.: Convective boiling and condensation. New York: McGraw-Hill 1972
13.2 Baker, O.: Simultaneous flow of oil and gas. Oil Gas J. 53(1954) 184–195
13.3 Hewitt, G. F.; Roberts, D. N.: Studies of two phase flow patterns by simultaneous X-ray and flash photography. UK At. Energy Agency Rep. No. AERE-M 2159, H.M.S.O., 1969
13.4 Taitel, Y.; Dukler, A. E.: A model for predicting flow regime transitions in horizontal and near horizontal gas-liquid flow. Am Inst. Chem. Eng. J. 22(1976) 47–55
13.5 Rouhani, S. Z.: Void measurements in the region of subcooled and low quality boiling. Atomenergy AE-239, part 2, 1966
13.6 Plummer, D. N.: Post critical heat transfer to flowing liquid in a vertical tube. Ph. D. thesis, Mass. Inst. Technol. 1974
13.7 Heat Exchanger Design Handbook. Editor E. U. Schlünder, Hemisphere Publ. Co., Washington, New York, London (1986), Vol. 2, Chapts. 2–3
13.8 Lockhart, R. W.; Martinelli, R. C.: Proposed correlation data for isothermal two-phase, two-component flow in pipes. Chem. Eng. Prog. 45(1949) 39–48
13.9 McAdams, W. H.; Wood, W. K.; Heroman, L. C.: Vaporization inside horizontal tubes-II-benzene-oil mixtures. Trans. Am. Soc. Mech. Eng., 64(1942) 193–200
13.10 Cicchitti, A.; Lombardi, C.; Silvestri, M.; Soldaini, G.; Zavatarlli, R.: Two-phase cooling experiments-pressure drop, heat transfer and burnout measurements. Energia Nucleare 7(1960) 407–425
13.11 Dukler, A. E.; Wicks, M.; Cleveland, R. G.: Pressure drop and hold-up in two-phase flow. Part A: A comparison of existing correlations, Part B: An approach through similarity analysis. Am. Inst. Chem. Eng. J. 10(1964) 38–51
13.12 Martinelli, R. C.; Nelson, D. B.: Prediction of pressure drop forced circulation boiling of water. Trans. Am. Soc. Mech. Eng., 70(1948) 695–702
13.13 Baroczy, C. J.: A systematic correlation for two-phase pressure drop. Chem. Eng. Prog. Symp. Ser. 62(1966) 232–249
13.14 Chisholm, D.: Pressure drop due to friction during the flow of evaporating two-phase mixtures in smooth tubes and channels. Int. J. Heat Mass Transfer 16(1973) 347–358
13.15 Theissing, P.: Eine allgemeingültige Methode zur Berechnung des Reibungsdruckverlustes der Mehrphasenströmung. Chem. Ing. Techn. 52(1980) 356–357
13.16 Chawla, J. M.: Wärmeübergang und Druckabfall in waagerechten Rohren bei der Verdampfung von Kältemitteln. Verein Deutscher Ing. Forsch. heft 532(1967), VDI-Verlag, Düsseldorf
13.17 VDI-Wärmeatlas, Berechnungsblätter für den Wärmeübergang. 5. Aufl. Düsseldorf: VDI-Verlag 1988, Chapter LG1 to LG3
13.18 Lombardi, C.; Pedrocchi, E.: A pressure drop correlation in two-phase flow. Energia Nucleare, 19(1972) 91–99
13.19 Friedel, L.: Druckabfall bei der Strömung von Gas/Dampf-Flüssigkeits-Gemischen in Rohren. Chem. Ing. Tech., 50(1978) 167–180
13.20 Teichel, H.: Druckabfall, Dampfgehalt und turbulenter Queraustausch in Wasser- und Wasserdampfströmungen. Brennst.-Wärme-Kraft, 30(1978) 334–340
13.21 Idsinga, W.; Todreas, N.; Bowring, R. W.: An assessment of two-phase pressure drop correlations for steam-water systems. Int. J. Multiphase Flow, 3(1977) 401–413
13.22 Hsu, Y. Y.: On the size of range of active nucleation cavities on a heating surface. Trans. Am. Soc. Mech. Eng., Ser. C. J. Heat Transfer 84(1962) 207–216
13.23 Davis, E. J.; Anderson, G. H.: The incipience of nucleate boiling in forced convection flow. Am. Inst. Chem. Eng., J. 12(1966) 774–780
13.24 Bergles, A. E.; Rohsenow, W. M.: The determination of forced-convection surface boiling heat transfer. Trans. Am. Soc. Mech. Eng., Ser. C. J. Heat Transfer 86(1964) 365–372
13.25 Bankoff, S. G.: Entrapment of gas in the spreading of liquid on a rough surface. Am. Inst. Chem. Eng. J. 4(1958) 24–26
13.26 Bowring, R. W.: Physical model based on bubble detachment and calculation of steam voidage in the subcooled region of a heated channel. OECD Halden Reactor Project Report HPR-10(1962)

13.27 Engelberg-Forster, K.; Grief, R.: Heat transfer to a boiling liquid—mechanism and correlations. Trans. Am. Soc. Mech. Eng., Ser. C. J. Heat Transfer 81(1959) 43–53
13.28 Bergles, A. E.; Rohsenow, W. M.: The determination of forced convection surface boiling heat transfer. Paper 63-HT-22. VIth Nat. Heat Transfer Conf. Am. Soc. Mech. Eng.—Amer. Soc. Chem. Eng., Boston 1963 see also [13.24]
13.29 Jens, W. H.; Lottes, P. A.: Analysis of heat transfer burnout, pressure drop and density data for high pressure water. Rep. ANL-4627, 1951
13.30 Thom, J. R. S.; Walker, W. M.; Fallon , T. A.; Reising G. F. S.: Boiling in subcooled water during flow up heated tubes or annuli. Symp. Inst. Mech. Eng. Paper 6. London: 1965
13.31 Dengler, C. E.; Addams, J. N.: Heat transfer mechanism for vaporization of water in a vertical tube. Chem. Eng. Progr. Symp. Ser. 52(1956) 95–103
13.32 Chawla, J. M.: Wärmeübergang und Druckabfall in waagerechten Rohren bei der Strömung von verdampfenden Kältemitteln. VDI-Forschungsh. 523(1967)
13.33 Stephan, K.; Auracher, H.: Correlations for nucleate boiling heat transfer in forced convection. Int. J. Heat Mass Transfer 24(1981) 99–107
13.34 Steiner, D.: Wärmeübergang beim Sieden gesättigter Flüssigkeiten. In: VDI-Wärmeatlas, 4. Aufl. Düsseldorf: VDI-Verlag 1984, Chapter Hbb
13.35 Pujol, L.: Boiling heat transfer in vertical upflow and downflow tubes. Ph. D. thesis, Lehigh Univ. 1968
13.36 Slipčević, B.: Wärmeübergang beim Sieden von R-Kältemitteln in horizontalen Rohren. Kältetech. Klim. 24(1972) 345–351
13.37 Chen, J. C.: Correlation for boiling heat transfer to saturated liquids in convective flow. Ind. Eng. Chem. Proc. Des. Dev. 5(1966) 322–329
13.38 Forster, H. K.; Zuber, N.: Dynamics of vapor bubbles and boiling heat transfer. Am. Inst. Chem. Eng. J. 4(1955) 531–535
13.39 Jallouk, P. A.: Two-phase pressure drop and heat transfer characteristics of refrigerants in vertical tubes. Ph. D. thesis, Univ. Tennessee, U. Microfilms 75–11–171, 1974
13.40 Gungor, K. E.; Winterton, R. H. S.: A general correlation for flow boiling in tubes and in annuli. Int. J. Heat Mass Transfer 29(1986) 351–358
13.41 Cooper, M. G.: Saturation nucleate pool boiling. A simple correlation. 1st UK Nat. Conf. Heat Transfer 2(1984) 785–793
13.42 Guerrieri, S. A.; Talty, R. D.: A study of heat transfer to organic liquids in single-tube natural circulation vertical-tube boilers. Chem. Eng. Progr. Symp. Ser. 52(1956) 69–77
13.43 Schrock, V. E.; Grossman, L. M.: Forced convection boiling studies. Univ. California. Inst. Eng. Res. Berkeley. Final Rep. Ser. No. 73308-UCX 2182, TID-14632, 1959
13.44 Bennett, J. A. R.; Collier, J. G.; Pratt, H. R. C.; Thornton, J. D.: Heat transfer to two-phase gas-liquid systems. Part I: Steam-water mixtures in the liquid-dispersed region in an annulus. Trans. Inst. Chem. Eng. 39(1961) 113–126
13.45 Wright, R. M.; Somerville, G. F.; Sani, R. L.; Bromley, L. A.: Downflow boiling of water and n-butanol in uniformly heated tubes. Chem. Eng. Progr. Symp. Ser. 61(1965) 220–229
13.46 Somerville, G. F.: Downflow boiling of n-butanol in an uniformly heated tube. Univ. California, Lawrence Radiation Lab. 10527, Oct. 1962
13.47 Collier, J. G.; Lacey, P. M. C.; Pulling, D. J.: Heat transfer to two-phase gas-liquid systems. Part II: Further data on steam-water mixtures in the liquid dispersed region in an annulus. Trans. Inst. Chem. Eng. 42(1964) T127–139
13.48 Pujol, L.; Stenning, A. H.: Effect of flow direction on the boiling heat transfer coefficient in vertical tubes. Proc. Int. Symp. Cocurrent Gas-Liquid Flow. Univ. Waterloo, Canada, Sept. 1968, p. 401–453
13.49 Sani, R. L.: Downflow boiling and non-boiling heat transfer in a uniformly heated tube. Univ. California, Lawrence Radiation Lab. 9023, Jan. 1960
13.50 Chaddock, J. B.: Forced convection evaporation in tubes. Am. Soc. Heating Refrig. Air Cond. Eng. (ASHRAE). Handbook of Fundamentals. Chapt. 3, New York 1972
13.51 Borishanskij, B. M.; Andeevskij, A. A.; Fromzel, V. N.; Fokin, B. S.; Cistgakov, V. A.; Danilowa, G. N.; Bikov, G. S.: Heat transfer during two-phase flows (in Russian). Teploenergetika 11(1971) 68–69
13.52 Shah, M. M.: A new correlation for heat transfer during boiling flow through pipes. Trans. Am. Soc. Heating Refrig. Air Cond. Eng. (ASHRAE) 82(1976) 66–86
13.53 Shah, M. M.: Chart correlation for saturated boiling heat transfer: Equations and further study. Trans. Am. Soc. Heating Refrig. Air Cond. Eng. (ASHRAE). Preprint no. 2673, 1982
13.54 Weatherhead, R. J.: Nucleate boiling characteristics and the critical heat flux occurrence in subcooled axial flow water systems. ANL 6675, 1963

13.55 Silvestri, M.: Two-phase (steam and water) flow and heat transfer. Part II. IInd Int. Heat Transfer Conf., Boulder 1961
13.56 Lee, D. H.; Obertelli, J. D.: An experimental investigation of forced convection boiling in high pressure water. Part I. AEEW—Rep. 213, 1963
13.57 Lee, D. H.: An experimental investigation of forced convective boiling in high pressure water. Part III. AEEW—Rep. 355, 1965
13.58 Matzner, B.: Basic experimental studies of boiling fluid flow and heat transfer at elevated pressures. TID 18978, 1963
13.59 Doroshchuk, V. E.; Levitan, L. L.; Lantsmann, F. P.: Recommendations for calculating burnout in a round tube with uniform heat release (in Russian). Teploenergetika 22(1975) 66–70; compare also: Academy of Science, USSR: Tabular data for calculating burnout when boiling water in uniformly heated round tubes. Therm. Eng. 23(1977) 77–79
13.60 Macbeth, R. V.: Burn-out analysis. Part 4. Application of a local condition hypothesis to world data for uniformly heated round tubes and rectangular channels. AEEW—Rep. 267, 1963
13.61 Thomson, B.; Macbeth, R. V.: Boiling water heat transfer—burnout in uniformly heated round tubes: A compilation of world data with accurate correlations. AEEW—Rep. 356, 1964
13.62 Bowring, R. W.: A simple but accurate rôund tube uniform heat flux, dryout correlation over the pressure range 0,7-17 MN/m^2 (100–2500 psia). AEEW—Rep. 789, 1972
13.63 Drescher, G.; Köhler, W.: Die Ermittlung kritischer Siedezustände im gesamten Dampfgehaltsbereich für innendurchströmte Rohre. Brennst. Wärme Kraft 33(1981) 416–422
13.64 Kon'kov, H. S.: Experimental studies of the conditions under which heat exchange deteriorates when a steam-water mixture flows in heated tubes. Teploenergetika 12(1965) 77
13.65 Alad'yev, I. G.; Goslov, L. D.; Dodonov, L. D.; Fedynskij, O. S.: Heat transfer to boiling potassium in uniformly heated tubes. Heat Transfer Sov. Res. 1(1969) 14–26
13.66 Wallis, G. B.: One dimensional two-phase flow. New York: McGraw-Hill 1969
13.67 Watson, G. B. R.; Lee, A.; Wiener, M.: Critical heat flux in inclined and vertical smooth and ribbed tubes. Vth Int. Heat Transfer Conf. Tokyo 1974, Vol.IV, p. 275–280
13.68 Drescher, G.; Hein, D.; Katsaounis, A.; Köhler, W.; Ulrych, G.: Kritische Siedezustände strömender Flüssigkeiten. In: VDI-Wärmeatlas. 4.Aufl. Düsseldorf: VDI-Verlag 1984, Chapter Hbc
13.69 Lee, D. H.: Burnout in a channel with nonuniform circumferential heat flux. AEEW—Rep. 477, 1966
13.70 Alekseev, G. V.: Burnout heat fluxes under forced flow. IIIrd Int. Conf. on peaceful uses of atomic energy, Geneva, A/Conf. P. 28/P/327 a, 1964
13.71 Collier, J. G.: Convective boiling and condensation. New York: McGraw-Hill 1972, p. 273
13.72 Köhler, W.: Einfluß des Benetzungszustandes der Heizfläche auf Wärmeübergang und Druckverlust in einem Verdampferrohr. Diss. TU München 1984
13.73 Iloeje, O. C; Plummer, D. N.; Rohsenow, W. M.; Griffith, P.: A study of wall rewet and heat transfer in dispersed vertical flow. M I T Dept. Mech. Eng., Rep. 727(1974) 18–92
13.74 Collier, J. G.: Post-dryout heat transfer—A review of current position. Proc. Nato-Advanced Study Inst. on Two-Phase Flow Heat Transfer. New York: Hemisphere 1976
13.75 Mayinger, F.; Langner, M.: Post-dryout heat transfer. Proc. VIth Int. Heat Transfer Conf., Toronto 1978, Vol.6, p. 181–198
13.76 Bennett, A. W.; Hewitt, G. F.; et al.: Heat transfer to steam-water mixtures flowing in uniformly heated tubes in which the critical heat flux has been exceeded. AERE—Rep. 5373, 1967
13.77 Gnielinski, V.: Neue Gleichungen für den Wärme-und den Stoffübergang in turbulent durchströmten Rohren und Kanälen. Forsch Ingenieurwes. 41(1975) 8–16
13.78 Groeneveld, D. C.: An investigation of heat transfer in the liquid deficient regime. At. Energy Can. Ltd. AECL Rep. 3281, 1969
13.79 Müller, H. J.: Beitrag zur Untersuchung des Wärmeübergangs an einer simulierten Sekundärkühlzone beim Stranggießverfahren. Diss. TU Clausthal 1972
13.80 Bolle, L.; Moureau, J. C.: Spray cooling of surfaces. In: Multiphase science and technology. Vol. 1. New York: McGraw-Hill 1982, p. 1–97
13.81 Wachters, L. H.; Westerling, N. A.: The heat transfer from a hot wall to impinging water drops in the spheroidal state. Chem. Eng. Sci. 21(1966) 1047–1056
13.82 Wachters, L. H.: De warmteoverdracht van een hete wand naar druppels in de sferoidale toestand. Thesis, TH Delft 1965
13.83 Moureau, J. C.: Le refroidissement des parois métalliques très chaudes par pulvérisation d'eau. Thèse. Univ. Catholique de Louvain 1978

13.84 Savic, P.: The cooling of a hot surface by drops boiling in contact with it. Nat. Res. Coun. Can. Rep. MT-37, 1958
13.85 Grigull, U.; Sandner, H.: Wärmeleitung. Berlin: Springer 1979, p. 66

Chapter 14

14.1 Bonilla, C. F.; Perry, C. W.: Heat transmission to boiling binary mixtures. Am. Inst. Chem. Eng. J. 37(1941) 685–705
14.2 Preusser, P.: Wärmeübergang beim Verdampfen binärer und ternärer Flüssigkeitsgemische. Diss. Ruhr-Univ. Bochum 1978
14.3 Stephan, K.; Mayinger, F.: Thermodynamik. Bd. 2. 12. Aufl. Berlin: Springer 1987, p. 200
14.4 Plesset, M. S.; Zwick, S. A. J.: The growth of vapor bubbles in superheated liquids. J. Appl. Phys. 25(1954) 493–500
14.5 Scriven, L. E.: On the dynamics of phase growth. Chem. Eng. Sci. 10(1959) 1–13
14.6 Stephan, K.; Körner, M.: Berechnung des Wärmeübergangs verdampfender binärer Flüssigkeitsgemische. Chem. Ing. Tech. 41(1969) 409–417
14.7 Stephan, K.: Wärmeübergang beim Verdampfen von Gemischen in natürlicher Strömung. Verfahrenstechnik Int. 14(1980) 470–474
14.8 Happel, O.; Stephan, K.: Heat transfer from nucleate to the beginning of film boiling in binary mixtures. Proc. Vth Int. Heat Transfer Conf., Tokyo 1974, Vol. IV, p. 340–344
14.9 Stephan, K.: Heat transfer in boiling of mixtures. Proc. VIIth Int. Heat Transfer Conf., München 1982, Vol.1, p. 59–81
14.10 Stephan, K.: Einfluß des Öls auf den Wärmeübergang von siedendem Frigen 12 und Frigen 22. Kältetechnik 16(1964) 162–166
14.11 Ivanov, O. P.: Experimental investigation of heat transfer during boiling of freon-oil mixtures (in Russian). Cholod. Techn. 42(1965) 32–35
14.12 Henrici, H.; Hesse, G.: Untersuchungen über den Wärmeübergang beim Verdampfen von R114 und R114-Öl-Gemischen an einem horizontalen Glattrohr. Kältetech. Klim. 23(1971) 54–58
14.13 Preusser, P.: Wärmeübergang beim Verdampfen binärer und ternärer Flüssigkeitsgemische. Diss. Ruhr-Univ. Bochum 1978
14.14 Grigorjev, L. N.; Usmanov, A. G.: Heat transfer during boiling of binary mixtures (in Russian). J. Tech. Phys. 28(1958) 325–332
14.15 de Dood, J.: Nucleate pool boiling of pure liquids, liquid mixtures and polymer solutions at subatmospheric conditions. Diss. Univ. Amsterdam 1981
14.16 Bier, K.; Schmadl, J.; Gorenflo, D.: Pool boiling heat transfer to mixtures of SF_6 and CF_3Br at elevated saturation pressures. Proc. VIIth Int. Heat Transfer Conf., München 1982, Vol. 4, p. 35–40
14.17 Schlünder, E. U.: Über den Wärmeübergang bei der Blasenverdampfung von Gemischen. Verfahrenstechnik 9(1982) 692–698
14.18 Stephan, K.; Preusser, P.: Heat transfer and critical heat flux in pool boiling of binary and ternary mixtures. German Chem. Eng. 2(1979) 161–169
14.19 Kutateladze, S. S.: Fundamentals of heat transfer. London: Arnold 1962
14.20 Badger, W. L.; Monrad, C. C.; Ziamonoc, H. W.: Evaporation of caustic soda to high concentrations by means of diphenyl vapors. Ind. Eng. Chem. 22(1930) 700
14.21 Kirschbaum, E.: Wärmeübergang im senkrechten Verdampferrohr in dimensionsloser Darstellung. Chem. Ing. Tech. 27(1955) 248–257
14.22 Rant, Z.: Verdampfen in Theorie und Praxis. 2. Aufl. Frankfurt: Sauerländer 1977
14.23 Sagan, I. I.: Further clarification of parametric relationships for the boiling of liquids in tubes (in Russian). Communications of the higher Institute of Food Technol. 1(1961) 106–110
14.24 Feldkamp, K.: Wärmeübergang beim Sieden von wässrigen Lösungen. Preprints IVth Int. Heat Transfer Conf., Paris 1970, Vol. VI, B 7.2
14.25 van Stralen, S. J. D.: Heat transfer to boiling binary mixtures. Brit. Chem. Eng. 4(1959) Part I, 8–17; Part II, 78–82; Part III, 834–840
14.26 van Stralen, S. J. D.: The mechanism of nucleate boiling in pure liquids and in binary mixtures. Int. J. Heat Mass Transfer 9(1966) 995–1046
14.27 van Wijk, W. R.; van Stralen, S. J. D.: Maximale Wärmestromdichte und Wachstumsgeschwindigkeit von Dampfblasen in siedenden Zweistoffgemischen. Chem. Ing. Tech. 37(1965) 509–517

14.28 Afgan, N. H.: Boiling heat transfer and burnout heat flux of ethylalcohol-benzene mixtures. Proc. IIIrd Int. Heat Transfer Conf., Paris 1966, Vol.3, p. 175–185
14.29 Huber, D. A.; Hoehne, J. C.: Pool boiling of benzene, diphenyl and benzene-diphenyl-mixtures. Trans. Am. Soc. Mech. Eng., Ser. C. J. Heat Transfer 85(1963) 215–220
14.30 Krushilin, G. N.; Subbotin, V. I.: Proc. IInd Int. Conf. for peaceful use of atomic energy, Geneva, paper no. 2144
14.31 Wright, R. D.; Colver, C. Ph.: Saturated pool boiling burnout of ethane–ethylene mixtures. Chem. Eng. Progr. Symp. Ser. 65(1969) 204–210
14.32 Noyes, R. C.: An experimental study of sodium pool boiling heat transfer. Trans. Am. Soc. Mech. Eng., Ser. C. J. Heat Transfer 85(1963) 125–131
14.33 Farrar, L. C.; Marschall, E.: Film boiling in a scaling liquid. Trans. Am. Soc. Mech. Eng., Ser. C. J. Heat Transfer 98(1976) 173–177
14.34 Bragg, J. R.; Westwater, J. W.: Film boiling of immiscible liquid mixtures on a horizontal plate. Preprints IVth Int. Heat Transfer. Conf., Paris 1970, Vol. VI, B7.1
14.35 Hamill, T. D.; Baumeister, U. J.: Effect of subcooling and radiation on film boiling heat transfer. NASA, Tech. Note D-3925, 1967
14.36 VDI-Wärmeatlas, Berechnungsblätter für den Wärmeübergang. 4. Aufl. Düsseldorf: VDI-Verlag 1983, Chapter Fa

Chapter 15

15.1 Shock, R. A. W.: Evaporation of binary mixtures in upward annular flow. Int. J. Multiphase Flow 2(1976) 411–433
15.2 Bennett, D. L.; Chen, J. C.: Forced convective boiling in vertical tubes for saturated pure components and binary mixtures. Am. Inst. Chem. Eng. J. 26(1980) 454–462
15.3 Butterworth, D.: Unresolved problems in heat exchanger design. In: Heat exchangers, thermal-hydraulic fundamentals and design. Kakaç, S., Bergles, A. E.; Mayinger, F. (Eds.) New York: Hemisphere 1981, p. 1095
15.4 Varma, H. K.; Sharma, C. P.: Heat transfer coefficients during forced convective evaporation of R12 and R22 mixtures in annular flow regime. Int. Congr. Refrig., Venice 1979, B1-46

Chapter 16

16.1 Jakob, M.; Fritz, W.: Versuche über den Verdampfungsvorgang. Forsch. Ingenieurwes. 2(1931) 435–447
16.2 Fedders, H.: Messungen des Wärmeübergangs beim Blasensieden von Wasser an metallischen Rohren. Diss. D 83, TU Berlin 1971
16.3 Danilowa, G. N.; Belskij, W. K.: Investigation of heat transfer during boiling of freons 113 and 12 in tubes of varying roughness (in Russian). Cholod. Techn. 4 (1965) 24–31
16.4 Danilowa, G. N.: Influence of pressure and temperature on heat transfer to boiling freons (in Russian). Cholod. Techn. 4(1965) 36–42
16.5 Berenson, P. J.: Experiments on pool-boiling heat transfer. Int. J. Heat Mass Transfer 5(1962) 985–999
16.6 Schimmelpfennig, K.: Über den Einfluß künstlicher Dampfblasenkeimstellen auf den Siedevorgang. Diss. D 83, TU Berlin 1973
16.7 Nix, G. H.; Vachon, R. I.; Hall, D. M.: A scanning and transmission electron microscopy study of pool boiling surfaces. Preprints IVth Int. Heat Transfer Conf., Paris 1970. Vol. V, B 1.6
16.8 Almgren, D. W.; Smith, J. L.: The inception of nucleate boiling with liquid nitrogen. Trans. Am. Soc. Mech. Eng., Ser. B. J. Eng. Ind. 91(1969) 1210–1216
16.9 Vachon, R. I.; Nix, G. H.; Tanger, G. E.; Cobb, R. O.: Pool boiling heat transfer from Teflon-coated stainless steel. Trans. Am. Soc. Mech. Eng., Ser. C. J. Heat Transfer 91(1969) 364–370
16.10 Young, R. K.; Hummel, R. L.: Improved nucleate boiling heat transfer. Chem. Eng. Progr. Symp. Ser. 61(1965) 264–270
16.11 Huelle, Z. R.: Leistungssteigerung eines Luftkühlers durch Beeinflussung der Kältemittelströmung. Kältetech. Klim. 23(1971) 198–202
16.12 Offenlegungsschrift Deutsches Patentamt 1919556, Patent application of Union Carbide, USA

16.13 Vachon, R. I.; Tanger, G. E.; Davis, D. L.; Nix, G. H.: Pool boiling on polished and chemically etched stainless-steel surfaces. Trans. Am. Soc. Mech. Eng., Ser. C. J. Heat Transfer 90(1968) 321–328
16.14 Union Carbide High Flux Tubing. Manufacturer's literature from Union Carbide, USA
16.15 German Patent DBP 155142
16.16 Stephan, K.; Mitrović, J.: Heat transfer in natural convective boiling of refrigerants and refrigerant-oil-mixtures of T-shaped finned tubes. Adv. Enhanced Heat Transfer, Am. Soc. Mech. Eng. HTD-vol. 18(1981) 131–146
16.17 Kirschbaum, E.: Neues zum Wärmeübergang mit und ohne Änderung des Aggregatzustandes. Chem. Ing. Tech. 24(1952) 393–400
16.18 Brooks, C. H.; Badger, W. L.: Heat transfer coefficients in the boiling section of a long-tube natural circulation evaporator. Trans. Am. Soc. Chem. Eng. 33(1937) 392–416
16.19 Elrod, W. C.; Clark, J. A.; Lady, E. R.; Merte, H.: Boiling heat transfer data at low heat flux. Trans. Am. Soc. Mech. Eng., Ser. C. J. Heat Transfer 88(1967) 235–243
16.20 de Dood, J.; Nucleate pool boiling of pure liquids, liquid mixtures and polymer solutions at subatmospheric conditions. Diss. Univ. Amsterdam 1981
16.21 Papaioannon, A. T.; Konmolitsoi, N. G.: The effect of polymer additives on nucleate boiling. Proc. VIIth Int. Heat Transfer Conf., München 1982, Vol.4, p. 67–72
16.22 Hampe, M. J.: Zur Thermodynamik der Transportprozesse in Grenzflächensystemen. Diss. TU München 1980

Index of Names

Subject Index

GPSR Compliance

The European Union's (EU) General Product Safety Regulation (GPSR) is a set of rules that requires consumer products to be safe and our obligations to ensure this.

If you have any concerns about our products, you can contact us on ProductSafety@springernature.com

In case Publisher is established outside the EU, the EU authorized representative is:

Springer Nature Customer Service Center GmbH
Europaplatz 3
69115 Heidelberg, Germany

Batch number: 10620143

Printed by Printforce, the Netherlands